Lecture Notes in Bioinformatics 9308

Subseries of Lecture Notes in Computer Science

More information about this series at http://www.springer.com/series/5381

Olivier Roux · Jérémie Bourdon (Eds.)

Computational Methods in Systems Biology

13th International Conference, CMSB 2015
Nantes, France, September 16–18, 2015
Proceedings

 Springer

Editors
Olivier Roux
École Centrale de Nantes
Nantes
France

Jérémie Bourdon
Université de Nantes
Nantes
France

ISSN 0302-9743   ISSN 1611-3349   (electronic)
Lecture Notes in Bioinformatics
ISBN 978-3-319-23400-7   ISBN 978-3-319-23401-4   (eBook)
DOI 10.1007/978-3-319-23401-4

Library of Congress Control Number: 2015947797

LNCS Sublibrary: SL8 – Bioinformatics

Springer Cham Heidelberg New York Dordrecht London
© Springer International Publishing Switzerland 2015

Printed on acid-free paper

Springer International Publishing AG Switzerland is part of Springer Science+Business Media
(www.springer.com)

Preface

This volume contains the articles presented at CMSB 2015. The 13th International Conference on Computational Methods in Systems Biology (CMSB) was held during September 16–18, 2015, at La Cité - Nantes Event Center in Nantes, France.

The CMSB annual conference series, created in 2003, provides a unique forum of discussion for computer scientists, biologists, mathematicians, engineers, and physicists interested in a system-level understanding of biological processes.

Topics of interest include formalisms for modeling biological processes; models and their biological applications; frameworks for model verification, validation, analysis, and simulation of biological systems; high-performance computational systems biology and parallel implementations; model inference from experimental data; model integration from biological databases; multi-scale modeling and analysis methods; and computational approaches for synthetic biology. Case studies in systems and synthetic biology were especially encouraged.

The 2015 conference was an opportunity to hear about research on the analysis of biological systems, networks, and data ranging from intercellular to multi-scale. This year, keyword lists often contain such words as: model checking, stochastic analysis, hybrid systems, circadian clock, time series data, logic programming, constraints solving. . . emphazing the wide spectrum of interests of the CMSB community.

The 2015 edition of CMSB received 48 regular submissions. Amongst them, 5 were withdrawn by the authors for various reasons, so 43 complete submissions were reviewed by the Program Committee, each submission being reviewed by 3 Program Committee members. In the end, 20 articles were selected for presentation at the conference and publication in the proceedings (acceptance rate: 46 %).

This volume contains in addition the abstracts of four invited speakers: Marta Kwiatkowska, David Harel, Gilles Bernot, and David Fell. Moreover, there are 2 short papers selected out of 4 submitted, and several posters were presented at the conference.

As program co-chairs, we have many people to thank. We would first like to thank the Program Committee members and the external reviewers for their peer reviews and the valuable feedback they provided to the authors. Our special thanks goes to François Fages and Pedro Mendes for their useful advice on matters related to the organization of the conference. We acknowledge the support of the EasyChair conference system during the reviewing process and the production of these proceedings (http://www. easychair.org). We also thank Kaushik Chowdhury and the IEEE Computer Society Technical Committee on Simulation for supporting the best student paper award. Our gratitude goes to all members of the Organizing Committee for their help, support, and spirited participation before, during, and after the conference. It is our pleasant duty to acknowledge the financial support from our sponsors. Finally, we would like to thank all the participants of the conference. It was the quality of their presentations and their contributions to the discussions that made the meeting a scientific success.

July 2015   Olivier Roux
                                                                       Jérémie Bourdon

Organization

Program Committee

Tatsuya Akutsu	Kyoto University, Japan
Julio Banga	IIM-CSIC, Spain
Gregory Batt	INRIA Paris-Rocquencourt, France
Gilles Bernot	University of Nice Sophia Antipolis, France
Alexander Bockmayr	Freie Universität Berlin, Germany
Jérémie Bourdon	LINA, France
Luca Cardelli	Microsoft Research, UK
Hidde De Jong	Inria, France
Diego Di Bernardo	Telethon Institute of Genetics and Medicine, Italy
Finn Drablos	NTNU, Norway
François Fages	Inria Paris-Rocquencourt, France
David Fell	Oxford Brookes University, UK
Radu Grosu	Stony Brook University, USA
Calin Guet	IST, Austria
Ashutosh Gupta	TIFR, India
Monika Heiner	Brandenburg University at Cottbus, Germany
Heinz Koeppl	TU Darmstadt, Germany
Marta Kwiatkowska	University of Oxford, UK
Pietro Lio'	University of Cambridge, UK
Wolfgang Marwan	Otto-von-Guericke-Universität Magdeburg, Germany
Hiroshi Matsuno	Yamaguchi University, Japan
Tommaso Mazza	IRCCS Casa Sollievo della Sofferenza - Mendel, Italy
Pedro Mendes	The University of Manchester, UK and University of Connecticut Health Center, USA
Stéphane Minvielle	CRCNA (UMR 892 INSERM/6299 CNRS), France
Satoru Miyano	University of Tokyo, Japan
Zoran Nikoloski	Max-Planck Institute of Molecular Plant Physiology, Germany
Loïc Paulevé	CNRS/LRI, France
Alberto Policriti	University of Udine, Italy
Ovidiu Radulescu	University of Montpellier 2, France
Olivier Roux	IRCCyN (UMR 6597 Ecole Centrale de Nantes, CNRS), France
Heike Siebert	Freie Universität Berlin, Germany
Anne Siegel	IRISA – CNRS, France
Evangelos Simeonidis	Luxembourg Centre for Systems Biomedicine, Luxembourg
Joerg Stelling	ETH Zurich, Switzerland

Carolyn Talcott	SRI International, USA
P.S. Thiagarajan	National University of Singapore, Singapore
Denis Thieffry	IBEns (UMR CNRS 8197 - INSERM 1024), France
Adelinde Uhrmacher	Universität Rostock, Germany
Jose Vilar	Biophysics Unit (CSIC-UPV/EHU), Spain
Verena Wolf	Saarland University, Germany

Steering Committee

Finn Drablos	NTNU, Norway
François Fages	Inria Paris-Rocquencourt, France
David Harel	Weizmann Institute of Science, Israel
Monika Heiner	Brandenburg University at Cottbus, Germany
Thomas Henzinger	IST Austria (Institute of Science and Technology Austria)
Tommaso Mazza	IRCCS Casa Sollievo della Sofferenza - Mendel, Italy
Pedro Mendes	The University of Manchester, UK and University of Connecticut Health Center, USA
Satoru Miyano	University of Tokyo, Japan
Gordon Plotkin	University of Edinburgh, UK
Corrado Priami	University of Trento, Italy
Carolyn Talcott	SRI International, USA
Adelinde Uhrmacher	Universität Rostock, Germany

Organizing Committee

Michèle-Anne Audrain, IRCCyN
Audrey Bihouée, BIRD
Jérémie Bourdon, LINA
Virginie Dupont, IRCCyN
Damien Eveillard, LINA
Julien Gras, LINA & BIRD

Elodie Guidon, LINA
Carito Guziolowski, IRCCyN
Abdelhalim Larhlimi, LINA
Morgan Magnin, IRCCyN
Olivier Roux, IRCCyN

Additional Reviewers

Akshay, S.
Basset, Nicolas
Bing, Liu
Ceska, Milan
Cinquemani, Eugenio
Donzé, Alexandre
Fauré, Adrien
Gonzalez, Aitor

Gyori, Benjamin
Helms, Tobias
Klarner, Hannes
Koch, Ina
Köksal, Ali Sinan
Llamosi, Artemis
Lukina, Anna
Monteiro, Pedro T.

Naldi, Aurélien
Napolitano, Francesco
Ono, Hiromasa
Pagliarini, Roberto
Paoletti, Nicola
Parvu, Ovidiu
Peng, Danhua
Petrov, Tatjana
Remy, Elisabeth

Rodionova, Alena
Rohr, Christian
Sandmann, Werner
Stoll, Gautier
Streck, Adam
Tamura, Takeyuki
Traynard, Pauline
Videla, Santiago

Invited Talks

Estimation and Verification of Hybrid Heart Models for Personalised Medical and Wearable Devices

Benoît Barbot, Marta Kwiatkowska,
Alexandru Mereacre and Nicola Paoletti

Department of Computer Science, University of Oxford, Oxford, UK
{benoit.barbot,marta.kwiatkowska,alexandru.mereacre,
nicola.paoletti}@cs.ox.ac.uk

Abstract. We are witnessing a huge growth in popularity of wearable and implantable devices equipped with sensors that are capable of monitoring a range of physiological processes and communicating the data to smartphones or to medical monitoring devices. Applications include not only medical diagnosis and treatment, but also biometric identification and authentication systems. An important requirement is personalisation of the devices, namely, their ability to adapt to the physiology of the human wearer and to faithfully reproduce the characteristics in real-time for the purposes of authentication or optimisation of medical therapies. In view of the complexity of the embedded software that controls such devices, model-based frameworks have been advocated for their design, development, verification and testing. In this paper, we focus on applications that exploit the unique characteristics of the heart rhythm. We introduce a hybrid automata model of the electrical conduction system of a human heart, adapted from Lian et al. [8], and present a framework for the estimation of personalised parameters, including the generation of synthetic ECGs from the model. We demonstrate the usefulness of the framework on two applications, ensuring safety of a pacemaker against a personalised heart model and ECG-based user authentication.

This research is supported by ERC AdG VERIWARE and PoC VERIPACE.

More Thoughts on the Whole Organism Challenge

David Harel

The Weizmann Institute of Science, Rehovot, Israel

Abstract. In 2002 I proposed a long-term "grand challenge" for the comprehensive and realistic modeling of biological systems, where we try to understand and analyze an entire system in detail, utilizing in the modeling effort all that is known about it. The proposal was to produce an interactive, dynamic, computerized model of an entire multi-cellular organism. Specifically, I suggested the *C. elegans* nematode, which is extremely complex despite its small size, but well-defined in terms of anatomy and genetics. In this talk I will review this challenge, and discuss some insights about its feasibility, based on some recent modeling efforts we have carried out, including the organogenesis of the pancreas, rat neural whisking, cancer tumor formation, and various projects regarding the C. elegans nematode worm.

A Genetically Modified Hoare Logic that Identifies the Parameters of a Gene Network

G. Bernot[1], J.-P. Comet[1], O. Roux[2]

[1] I3S laboratory, University Nice Sophia Antipolis, UMR CNRS 7271 CS 40121,
06903 Sophia Antipolis cedex, France
{bernot, comet}@unice.fr
[2] IRCCyN, UMR CNRS 6597, BP 92101, 1 rue de la Noë,
44321 Nantes Cedex 3, France
olivier.roux@irccyn.ec-nantes.fr

Abstract. The main difficulty when modelling gene networks is the identification of the parameters that govern the dynamics. Here we present a new approach based on Hoare logic and weakest preconditions (a la Dijkstra) that generates constraints on the parameter values: Once proper specifications are extracted from biological traces, they play a role similar to programs in the classical Hoare logic. We firstly remind the discrete modelling for genetic networks defined by René Thomas. Then, we define the Hoare/Dijkstra method extended to gene networks, that extracts the weakest precondition on parameter values.

Perspectives on Genome Scale Modelling of Metabolism

David A. Fell, Mark G. Poolman, and Hassan B. Hartman

Department of Biological and Medical Sciences, Oxford Brookes University,
Oxford OX3 0BP, UK
dfell@brookes.ac.uk
http://mudshark.brookes.ac.uk/

Abstract. Genome scale metabolic modelling is arguably the most successful current methodology at predicting a complex phenotype from genome sequences. Essentially it uses linear programming to predict optimal distributions of fluxes in a metabolic network reconstructed from a genome annotation, subject to constraints established from the requirement for mass balance in cellular metabolism in a dynamic steady state. The methodology, generally known as flux balance analysis (FBA) has increasing application in biotechnology and medicine. However, as the number of organisms and processes being modelled expands, new issues emerge that require innovation in model construction and analysis, and some methodologies that were developed for analysis of optimal microbial growth are less suitable for use in other contexts. In addition, there remain issues of data representation and interpretation in the bioinformatic and metabolic databases used in model construction that provide traps for the unwary and frustrate attempts at completely automated model building. Finally there are some persistent biochemical errors that are maintained by inheritance from older models.

Amongst the issues around modelling methodology is the choice of optimisation function. Whilst growth yield is often a justifiable choice for rapidly-growing microorganisms in a laboratory setting, it is not relevant to developed, differentiated tissue in multicellular organisms for example. Other modelling issues include the amount and nature of cellular maintenance metabolism, and the need to avoid the production of ATP form nothing (cellular perpetual motion).

As metabolic databases expand to become more comprehensive, there are more instances where the same overall reaction or metabolite is represented at different levels of detail. This can lead to entitites being incorporated into models more than once; in the case of reactions this can generate spurious cyclic routes, and in the case of metabolites to disruption of network connectivity. Indeed, both these issues affect other forms of metabolic network analysis as well.

The most common recurrent biochemical error is treating enzyme prosthetic groups as substrates and products of enzyme reactions, of which FAD and $FADH_2$ are the most frequent examples. This creates pool metabolites that could generate spurious redox interactions across the network that will not exist because these groups are contained and recycled entirely within a single enzyme reaction.

Contents

Short Papers

Invited Talks

Estimation and Verification of Hybrid Heart Models for Personalised Medical and Wearable Devices

Benoît Barbot, Marta Kwiatkowska[✉], Alexandru Mereacre,
and Nicola Paoletti

Department of Computer Science, University of Oxford, Oxford, UK
{benoit.barbot,marta.kwiatkowska,
alexandru.mereacre,nicola.paoletti}@cs.ox.ac.uk

Abstract. We are witnessing a huge growth in popularity of wearable and implantable devices equipped with sensors that are capable of monitoring a range of physiological processes and communicating the data to smartphones or to medical monitoring devices. Applications include not only medical diagnosis and treatment, but also biometric identification and authentication systems. An important requirement is personalisation of the devices, namely, their ability to adapt to the physiology of the human wearer and to faithfully reproduce the characteristics in real-time for the purposes of authentication or optimisation of medical therapies. In view of the complexity of the embedded software that controls such devices, model-based frameworks have been advocated for their design, development, verification and testing. In this paper, we focus on applications that exploit the unique characteristics of the heart rhythm. We introduce a hybrid automata model of the electrical conduction system of a human heart, adapted from Lian et al. [8], and present a framework for the estimation of personalised parameters, including the generation of synthetic ECGs from the model. We demonstrate the usefulness of the framework on two applications, ensuring safety of a pacemaker against a personalised heart model and ECG-based user authentication.

Recent technological advances have spurred a huge growth in apps and wearables for use in health monitoring. They employ a multiplicity of noninvasive sensors, e.g. accelerometers and miniature cameras, that can read physiological indicators, wirelessly send data to smartphones and analyse it not only to record trends (e.g. fitness bands), but also to support decision making for diagnosis and intervention. The success in miniaturisation of electronics has led to novel variants of traditional medical devices being introduced on the market, such as leadless cardiac pacemakers that can be implanted inside the human heart (e.g. Nanostim) and implantable glucose monitors that transmit data to a wristwatch to alert the wearer about any undesirable trends (e.g. Minimed). Applications are not limited to the medical field, and include also emerging technologies for biometric user identification and security, such as wristbands that periodically check

This research is supported by ERC AdG VERIWARE and PoC VERIPACE.

O. Roux and J. Bourdon (Eds.): CMSB 2015, LNBI 9308, pp. 3–7, 2015.
DOI: 10.1007/978-3-319-23401-4_1

the electrocardiogram (ECG) of the user to produce a template authentication signal (e.g. the Nymi band).

An important requirement for wearables is their personalisation, namely, the ability for the device to adapt to the physiology of the human wearer based on the person's individual characteristics. Personalisation is typically achieved via an appropriate parameterisation of a model of the physiological process, through parameter estimation and parameter synthesis techniques. Automation of personalised delivery of medical treatment is a major challenge; for example, rate-adaptive pacemakers are able to vary the rate of pacing depending on the activity and age of the patient [6], but insulin pumps still rely on human supervision. Another important role of personalised devices is in device safety assurance, where they can be used to faithfully reproduce the unique characteristics of the wearer in real-time for the purposes of testing.

Undoubtedly, personalised medical wearable and implantable devices are an important step towards achieving personalised healthcare. However, major advances are necessary to realise this vision, ranging from technological (miniaturisation, low-power circuits), software technologies (design automation, code generation, integration), to regulatory and legal frameworks (FDA approval, certification). This paper is concerned with model-based design and verification techniques for ensuring safety and effectiveness of personalised devices based on the bioelectrical activity of the heart.

We focus on the hybrid automata framework for closed-loop quantitative verification of cardiac pacemakers introduced in [4,7]. This was extended in [5] with techniques to automatically synthesise optimal timing delays to minimise energy consumption, and in [2] with a hardware-in-loop simulator to evaluate embedded pacemaker software on low-power hardware. However, personalisation was not supported.

In this paper, we extend the framework of [4,7] as follows. We introduce a new hybrid heart model encoded in Simulink/Stateflow and develop techniques to personalise the model through parameter estimation based on ECG data. We implement methods to produce synthetic ECGs that are characteristic for the given individual, and also to compare different ECG patterns. We consider two applications: verification of safety properties for a pacemaker against a personalised heart model, and biometric identification based on matching the wearer's signature with ECG data acquired for recognition. Further details on the methods and results are provided in the technical report [1].

1 Heart Model and Personalisation

We define a new heart model that includes the key components of the electrical conduction system of the human heart (Fig. 1) and is a hybrid automata translation of the model in [8].

The model can reproduce antegrade conduction (green arrows in the figure), arising when a stimulus is generated by the sinoatrial (SA) node and is propagated towards the ventricle passing through atrium and the atrio-ventricular

(AV) node. The impulse can also start from the ventricle (either intrinsically by component VRG or artificially by the pacemaker) and propagate in the opposite direction (retrograde conduction, red arrows). The transmission of cardiac waves between the atrium and ventricle is mediated by the AV node component (AVJ) and by intermediate conduction nodes (AVJOut, RAConductor and RVConductor). The model can reproduce, among others, ectopic beats (through components SANodeEctopic and VRGEctopic) and the collision of cardiac waves leading to fusion beats. The artificial pacemaker [3] is connected to the atrium and ventricle, and can both sense and stimulate them by delivering electrical impulses. An important feature of the model is the ability to generate *synthetic ECG signals*, which are used for parameter estimation and authentication. An ECG signal can be broken into five different waves, namely P, Q, R, S and T. Each wave is a simple bell-shaped curve which we reproduce in the synthetic signal by associating events in the heart model with Gaussian functions.

Estimation from ECG Data. To achieve a personalised model, we need to estimate parameter values so that the synthetic ECG is close to the input signal. The first steps are filtering and processing of the signal and detection of the ECG waves (Fig. 2a). Detected peak locations, widths

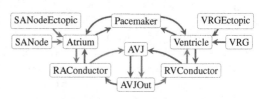

Fig. 1. Heart model (Color figure online).

and amplitudes can be directly mapped to some parameters of the model, e.g. the SA node frequency, overall AV conduction time and ventricular refractory time. Instead, some other parameters that cannot be inferred in this way are estimated using a Gaussian process optimisation (GPO) approach. Specifically, we seek to minimise the statistical distance between the input signal and the synthetic signal generated by the model with the parameters sampled in the GPO loop. In order to compute the distance, the signals are mapped into a single (statistical) ECG waveform centred around the R wave (the highest peak, see Fig. 2b).

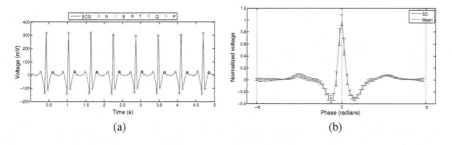

(a)   (b)

Fig. 2. Processed signal and detected peaks (a). Statistical ECG waveform (b).

Table 1. Results of pacemaker verification. *Aget* and *Vget* indicate the presence of an atrial and ventricular beat, respectively.

Property	Healthy	Arrhythmia	With pacemaker
$P_{=?}G^{<60000}(Vget \Rightarrow F^{<1100}Vget)$	0.99997 ± 0.0012	0.360607 ± 0.000015	$1 - 0.00003$
$P_{=?}G^{<60000}(Aget \Rightarrow F^{[100,200]}Vget)$	0.946454 ± 0.0005	$0.0 + 0.000005$	0.875494 ± 0.0008

2   Applications and Discussion

Pacemaker Verification. We study two properties related to two common heart conditions: bradycardia, i.e. slow heart rate, and AV block, i.e. conduction defect in the AV node. For the first property we query the probability that bradycardia episodes never occur, i.e. that the time between two consecutive ventricular events is always below some threshold. The second property requires correct conduction of the AV node, i.e. that the time between two consecutive atrial and ventricular events always lies in a given interval. Table 1 shows the results of the probabilistic verification for these properties on a healthy heart, a heart with arrhythmia (bradycardia for the first property and AV block for the second), and the same defective heart but with the pacemaker attached.

Note that the pacemaker can correct the two defective dynamics, since it ensures that the first property holds with probability 1 and the second with probability above 0.87.

Authentication. We show how the synthetic ECG generated by the personalised model can be used as a template for authentication purposes. This is based on computing its distance with the recognition ECG acquired for the identification. If the obtained score is small enough (e.g. not exceeding 50 % of the score obtained in the estimation phase), the authentication is successful. Figure 3a shows an example of successful identification when the ECGs for model estimation and authentication come from the same patient[1], while Fig. 3b shows how authentication failed with a signal from a different patient.

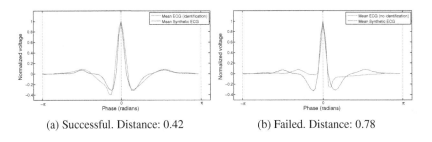

(a) Successful. Distance: 0.42   (b) Failed. Distance: 0.78

Fig. 3. Matching synthetic and recognition ECGs for authentication.

---

[1] MIT-BIH Normal Sinus Rhythm database, record $16265\,m^2$ record $17453\,m$.

Discussion. In this work, we presented methods to derive personalised heart models from data and showed their usefulness in the safety verification of pacemaker devices and in the ECG-based authentication. Besides enabling formal verification and synthesis [2,4], code generation and modularity, our formal model-based framework is sufficiently general to support, at the same time, other kinds of physiological systems and medical devices. This would enable improvement of the authentication performance by combining the ECG with other biometrics (e.g. fingerprints or iris) [9], and ultimately verification of the collective behaviour of multiple interconnected devices in a closed-loop with a highly-personalised model of the human physiological system.

References

1. Barbot, B., et al.: Estimation and verification of hybrid heart models for personalised medical and wearable devices. Technical report, Department of Computer Science, University of Oxford (2015)
2. Barker, C. et al.: Hardware-in-the-loop simulation and energy optimization of cardiac pacemakers. In: IEEE EMBC (2015) (to appear)
3. Boston Scientific: Pacemaker system specification (2007). http://sqrl.mcmaster.ca/_SQRLDocuments/PACEMAKER.pdf
4. Chen, T., et al.: Quantitative verification of implantable cardiac pacemakers over hybrid heart models. Inf. Comput. **236**, 87–101 (2014)
5. Diciolla, M., et al.: Synthesising optimal timing delays for timed I/O automata. In: EMSOFT 2014. ACM (2014)
6. Kwiatkowska, M., et al.: Formal modelling and validation of rate-adaptive pacemakers. In: ICHI 2014, pp. 23–32. IEEE (2014)
7. Kwiatkowska, M., Mereacre, A., Paoletti, N.: On quantitative software quality assurance methodologies for cardiac pacemakers. In: Margaria, T., Steffen, B. (eds.) ISoLA 2014, Part II. LNCS, vol. 8803, pp. 365–384. Springer, Heidelberg (2014)
8. Lian, J., et al.: Open source modeling of heart rhythm and cardiac pacing. Open Pacing Electrophysiol. Ther. J. **3**, 4 (2010)
9. Singh, Y.N., Singh, S.K.: Evaluation of electrocardiogram for biometric authentication. J. Inf. Secur. **3**, 39–48 (2012)

A Genetically Modified Hoare Logic that Identifies the Parameters of a Gene Network

Gilles Bernot[1][(⊠)], Jean-Paul Comet[1], and Olivier Roux[2]

[1] I3S Laboratory, University Nice Sophia Antipolis, UMR CNRS 7271 CS 40121,
06903 Sophia Antipolis Cedex, France
bernot,comet@unice.fr
[2] IRCCyN, UMR CNRS 6597, BP 92101, 1 rue de la Noë,
44321 Nantes Cedex 3, France
olivier.roux@irccyn.ec-nantes.fr

Abstract. The main difficulty when modelling gene networks is the identification of the parameters that govern the dynamics. Here we present a new approach based on Hoare logic and weakest preconditions (a la Dijkstra) that generates constraints on the parameter values: Once proper specifications are extracted from biological traces, they play a role similar to programs in the classical Hoare logic. We firstly remind the discrete modelling for genetic networks defined by René Thomas. Then, we define the Hoare/Dijkstra method extended to gene networks, that extracts the weakest precondition on parameter values.

1 Thomas' Gene Regulatory Networks with Multiplexes

Our formal framework [KCRB09] is based on the discrete approach of Thomas [TK01]: A gene network is a labelled directed graph (left part of Fig. 1) in which vertices are either *variables* (within circles) or *multiplexes* (within rectangles). Variables abstract genes or their products, and multiplexes contain propositional formulas that encode situations in which a group of variables (inputs of multiplexes) influence the evolution of some variables (outputs of multiplexes). In the figure the multiplex μ_2 expresses that the variable x can help the activation of the variable y when it is at least equal to 1. In general multiplexes can represent combined biological phenomena, one of the simplest being the formation of complexes (in which case the formula would contain a conjunction). In the figure, μ_1 reflects an auto-activation of x at level 2 which is controled by μ_3. Because μ_3 contains a negation, μ_1 is *inhibited* by y.

As shown in the right part of Fig. 1, this gives rise to 6 qualitative regions in the phase space, which we call (discrete) states. A *state* is an assignment of integer values to the variables. Such an assignment allows a natural evaluation of any formula within a multiplex: By replacing variables by their values we get a propositional formula whose atoms are the results of the integer inequalities. Then, we say that a multiplex m, predecessor of a variable v in the graph, is a *resource* of v iff

O. Roux and J. Bourdon (Eds.): CMSB 2015, LNBI 9308, pp. 8–12, 2015.
DOI: 10.1007/978-3-319-23401-4_2

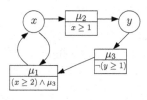 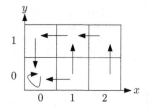

Fig. 1. (Left) Discrete gene network with variables x and y, multiplexes μ_1, μ_2 and μ_3 with associated formulas $\varphi_{\mu_1} \equiv ((x \geqslant 2) \wedge \mu_3)$, $\varphi_{\mu_2} \equiv (x \geqslant 1)$ and $\varphi_{\mu_3} \equiv \neg(y \geqslant 1)$. **(Right)** Its state graph obtained when choosing parameters $K_{x,\varnothing} = 0$, $K_{x,\{\mu_1\}} = 2$, $K_{y,\varnothing} = 0$, and $K_{y,\{\mu_2\}} = 1$.

its substituted formula is true: at the state $(x = 2, y = 1)$, μ_2 is the only resource of y whereas φ_{μ_1} is false and consequently, the set of resources of x is empty.

At a given state η, each variable v evolves in the direction of a specific level that only depends on the set of resources of v. This level is the integer value of a parameter $K_{v,\rho(\eta,v)}$, where $\rho(\eta, v)$ is the set of resources of v at η. Hence, at state η, v can increase if $\eta(v) < K_{v,\rho(\eta,v)}$, it can decrease if $\eta(v) > K_{v,\rho(\eta,v)}$, and it is stable if $\eta(v) = K_{v,\rho(\eta,v)}$. In the Thomas' method, the variables evolve asynchronously by unit steps toward their respective $K_{\dots}$. The dynamics of a gene network is then described by an asynchronous state graph (right part of Fig. 1).

2 Hoare Triples for Gene Networks

An *assertion* is a formula whose terms are sums or subtractions between integers, variables or parameters $K_{\dots}$ of the gene network, predicates are equalities or inequalities, and connectives are the usual ones of first order logic. A *Hoare triple* is an expression of the form "$\{P\}\, p\, \{Q\}$" where P and Q are assertions (pre- and post-conditions) and p is a *trace specification*.

Trace specifications are indeed the key concept to formalize the observations of a biologist during experiments. They are inductively defined as follows:

- For each variable v of the gene network, the expressions "$v+$", "$v-$" and "$v := n$" (where $n \in I\!N$) are trace specifications (increase, decrease or assignment of variable value).
- If e is an assertion then "$assert(e)$" is a trace specification.
- If p_1 and p_2 are trace specifications then so is $(p_1; p_2)$ (sequential composition).
- If p_1 and p_2 are trace specifications and if e is an assertion, then so is $(if\ e\ then\ p_1\ else\ p_2)$.
- If p is a trace specification and if e and I are assertions, then $(while\ e\ with\ I\ do\ p)$ is also a trace specification. The assertion I is called the invariant of the *while* loop.
- If p_1 and p_2 are trace specifications then so are $\forall(p_1, p_2)$ and $\exists(p_1, p_2)$ (quantifiers).

Conventionally, we call *the empty trace* $\varepsilon = assert(true)$.

For lack of space we do not define here the formal semantics of trace specifications [BCK+15]. Intuitively, "$v+$" (resp. "$v-$", "$v := n$") means that the expression level of variable v has been observed as increasing by one unit (resp. decreasing by one unit, or set to a particular value n by the experimental protocole). "$assert(e)$" expresses a property observed on the current state without change of state. The sequential composition concatenates two specifications whereas "if" chooses between two specifications according to the assertion e. The loop invariant I, as in classical Hoare logic, facilitates proofs through *while* loops. Finally, the quantifiers $\forall$ and $\exists$ group together several specifications.

3 A Hoare Logic for Gene Networks

We define our Hoare logic by giving the rule for each instruction of a trace specification. First, let us introduce a few conventional assertions.

If ω is a subset of the set of predecessors of a variable v in the network, the assertion Φ_v^ω characterizes the states such that ω is the set of resources of v:

$$\Phi_v^\omega \quad \equiv \quad (\bigwedge_{m \in \omega} \varphi_m) \wedge (\bigwedge_{m \notin \omega,\ m\ predecessor\ of\ v} \neg\varphi_m)$$

Then, the formula Φ_v^+ and Φ_v^- characterize the states such that v can increase/decrease:

$$\Phi_v^+ \equiv \bigwedge_{\omega \subset \{predecessors\ of\ v\}} (\Phi_v^\omega \Rightarrow K_{v,\omega} > v) \quad \Phi_v^- \equiv \bigwedge_{\omega \subset \{predecessors\ of\ v\}} (\Phi_v^\omega \Rightarrow K_{v,\omega} < v)$$

Our genetically modified Hoare logic is defined by the following inference rules, where v is a variable of the gene network.

Rules encoding Thomas' discrete dynamics:

$$\frac{}{\{\ \Phi_v^+ \wedge Q[v \leftarrow v+1]\ \}\ v+\ \{Q\}}\text{Incrementation}$$

$$\frac{}{\{\ \Phi_v^- \wedge Q[v \leftarrow v-1]\ \}\ v-\ \{Q\}}\text{Decrementation}$$

Rules for quantifiers:

$$\frac{\{P_1\}\ p_1\ \{Q\} \qquad \{P_2\}\ p_2\ \{Q\}}{\{P_1 \wedge P_2\}\ \forall(p_1,p_2)\ \{Q\}}\text{Universal}$$

$$\frac{\{P_1\}\ p_1\ \{Q\} \qquad \{P_2\}\ p_2\ \{Q\}}{\{P_1 \vee P_2\}\ \exists(p_1,p_2)\ \{Q\}}\text{Existential}$$

Other rules, directly inspired by Hoare Logic:

$$\frac{}{\{\ \Phi \wedge Q\ \}\ assert(\Phi)\ \{\ Q\ \}}\text{Assert} \qquad\qquad \frac{}{\{Q[v \leftarrow k]\}\ v := k\ \{Q\}}\text{Assignment}$$

$$\frac{\{P_1\}\ p_1\ \{P_2\} \qquad \{P_2\}\ p_2\ \{Q\}}{\{P_1\}\ p_1; p_2\ \{Q\}}\text{Sequential}$$

$$\frac{\{P_1\}\ p_1\ \{Q\} \qquad \{P_2\}\ p_2\ \{Q\}}{\{(e \wedge P_1) \vee (\neg e \wedge P_2)\}\ if\ e\ then\ p_1\ else\ p_2\ \{Q\}}\text{Conditional}$$

$$\frac{\{e \wedge I\}\ p\ \{I\} \qquad \neg e \wedge I \Rightarrow Q}{\{I\}\ while\ e\ with\ I\ do\ p\ \{Q\}}\text{Iteration} \qquad\qquad \frac{P \Rightarrow Q}{\{P\}\ \varepsilon\ \{Q\}}\text{Empty trace}$$

We have proved that this modified Hoare logic is correct, and complete assuming a proper choice of the loop invariants [BCK+15]. More precisely, the classical *backward strategy* of Dijkstra (where the Empty trace rule is never applied) computes the weakest precondition P_0 such that $\{P_0\}\ p\ \{Q\}$. Similarly to classical Hoare logic which reflects a partial correctness of imperative programs, the previous definition does not imply termination of *while* loops.

4   Example

In [Mt02] Uri Alon and co-workers have studied the most common *in vivo* patterns involving three genes. Among them, the *incoherent feedforward loop of type 1* is composed by a transcription factor a that activates a second transcription factor c, and a is an activator of b whereas c is an inhibitor of b (Fig. 2).

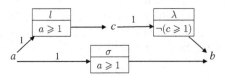

Fig. 2. Boolean variables: $\{a, b, c\}$. Multiplexes: $\{l, \lambda, \sigma\}$ with $\phi_l \equiv (a \geqslant 1)$, $\phi_\lambda \equiv (\neg(c \geqslant 1))$, $\phi_\sigma \equiv (a \geqslant 1)$. Unknown parameters: $K_{a,\emptyset}$, $K_{c,\varnothing}$, $K_{c,\{l\}}$, $K_{b,\varnothing}$, $K_{b,\{\sigma\}}$, $K_{b,\{\lambda\}}$ and $K_{b,\{\sigma,\lambda\}}$.

Uri Alon and many biologists consider that if a, b and c are equal to 0, the function of this feedforward loop is to ensure a *transitory activity* of b that signals when a has *switched* from 0 to 1: The idea is that a activates the productions of b and c, and then c stops the production of b. This is specified by the Hoare triple $\{P\}\ p\ \{Q_0\}$ where $P \equiv (a = 1\ \wedge\ b = 0\ \wedge\ c = 0)$, $p \equiv (b+; c+; b-)$ and $Q_0 \equiv (b = 0)$. The backward strategy using our genetically modified Hoare logic on this example gives the following successive conditions.

The weakest precondition through the last instruction "$b-$" is (Decrementation rule):

$$
\begin{cases}
\Phi_b^\varnothing \Rightarrow K_b < b \\
\Phi_b^\sigma \Rightarrow K_{b,\sigma} < b \\
\Phi_b^\lambda \Rightarrow K_{b,\lambda} < b \\
\Phi_b^{\sigma,\lambda} \Rightarrow K_{b,\sigma\lambda} < b \\
b - 1 = 0
\end{cases}
\equiv
\begin{cases}
(\neg\neg(c \geqslant 1) \wedge \neg(a \geqslant 1)) \Rightarrow K_b < b \\
(\neg\neg(c \geqslant 1) \wedge (a \geqslant 1)) \Rightarrow K_{b,\sigma} < b \\
(\neg(c \geqslant 1) \wedge \neg(a \geqslant 1)) \Rightarrow K_{b,\lambda} < b \\
(\neg(c \geqslant 1) \wedge (a \geqslant 1)) \Rightarrow K_{b,\sigma\lambda} < b \\
b - 1 = 0
\end{cases}
\Leftrightarrow
\begin{cases}
b = 1 \\
((c \geqslant 1) \wedge (a < 1)) \Rightarrow K_b = 0 \\
((c \geqslant 1) \wedge (a \geqslant 1)) \Rightarrow K_{b,\sigma} = 0 \\
((c < 1) \wedge (a < 1)) \Rightarrow K_{b,\lambda} = 0 \\
((c < 1) \wedge (a \geqslant 1)) \Rightarrow K_{b,\sigma\lambda} = 0
\end{cases}
$$

Then, the weakest precondition through "$c+$" is (Incrementation rule):

$$
\begin{cases}
\neg(a \geqslant 1) \Rightarrow K_c > c \\
a \geqslant 1 \Rightarrow K_{c,l} > c \\
b = 1 \\
((c + 1 \geqslant 1) \wedge (a < 1)) \Rightarrow K_b = 0 \\
((c + 1 \geqslant 1) \wedge (a \geqslant 1)) \Rightarrow K_{b,\sigma} = 0 \\
((c + 1 < 1) \wedge (a < 1)) \Rightarrow K_{b,\lambda} = 0 \\
((c + 1 < 1) \wedge (a \geqslant 1)) \Rightarrow K_{b,\sigma\lambda} = 0
\end{cases}
\Leftrightarrow
\begin{cases}
c = 0 \\
a < 1 \Rightarrow K_c = 1 \\
a \geqslant 1 \Rightarrow K_{c,l} = 1 \\
b = 1 \\
a < 1 \Rightarrow K_b = 0 \\
a \geqslant 1 \Rightarrow K_{b,\sigma} = 0
\end{cases}
$$

Lastly, through the first "$b+$" (Incrementation rule):

$$
\left\{
\begin{array}{l}
(\neg\neg(c \geqslant 1) \wedge \neg(a \geqslant 1)) \Rightarrow K_b > b \\
(\neg\neg(c \geqslant 1) \wedge (a \geqslant 1)) \Rightarrow K_{b,\sigma} > b \\
(\neg(c \geqslant 1) \wedge \neg(a \geqslant 1)) \Rightarrow K_{b,\lambda} > b \\
(\neg(c \geqslant 1) \wedge (a \geqslant 1)) \Rightarrow K_{b,\sigma\lambda} > b \\
c = 0 \\
a < 1 \Rightarrow K_c = 1 \\
a \geqslant 1 \Rightarrow K_{c,l} = 1 \\
b + 1 = 1 \\
a < 1 \Rightarrow K_b = 0 \\
a \geqslant 1 \Rightarrow K_{b,\sigma} = 0
\end{array}
\right.
\Leftrightarrow P_0 \equiv
\left\{
\begin{array}{l}
a < 1 \Rightarrow K_{b,\lambda} = 1 \\
a \geqslant 1 \Rightarrow K_{b,\sigma\lambda} = 1 \\
c = 0 \\
a < 1 \Rightarrow K_c = 1 \\
a \geqslant 1 \Rightarrow K_{c,l} = 1 \\
b = 0 \\
a < 1 \Rightarrow K_b = 0 \\
a \geqslant 1 \Rightarrow K_{b,\sigma} = 0
\end{array}
\right.
$$

Then, using the Empty trace rule to finish the correctness proof of the Hoare triple, we have to ensure $P \Rightarrow P_0$ and, after simplification, we get the correctness if and only if $K_{b,\sigma\lambda} = 1$ and $K_{c,l} = 1$ and $K_{b,\sigma} = 0$. So, under these three hypotheses and whatever the values of the other parameters, the system can exhibit a transitory production of b in response to a switch of a from 0 to 1.

References

[BCK+15] Bernot, G., Comet, J.-P., Khalis, Z., Richard, A., Roux, O.: A genetically modified hoare logic (2015). arXiv 1506.05887

[KCRB09] Khalis, Z., Comet, J.-P., Richard, A., Bernot, G.: The SMBioNet method for discovering models of gene regulatory networks. Genes Genomes Genomics **3**(Special issue 1), 15–22 (2009)

[Mt02] Milo, R., et al.: Network motifs: simple building blocks of complex networks. Science **298**, 824–827 (2002)

[TK01] Thomas, R., Kaufman, M.: Multistationarity, the basis of cell differentiation and memory. II. Logical analysis of regulatory networks in terms of feedback circuits. Chaos **11**, 180–195 (2001)

Regular Papers

SReach: A Probabilistic Bounded Delta-Reachability Analyzer for Stochastic Hybrid Systems

Qinsi Wang[1]([⊠]), Paolo Zuliani[2], Soonho Kong[1], Sicun Gao[3],
and Edmund M. Clarke[1]

[1] Computer Science Department, Carnegie Mellon University, Pittsburgh, USA
{qinsiw,soonhok,emc}@cs.cmu.edu
[2] School of Computing Science, Newcastle University, Newcastle upon Tyne, UK
paolo.zuliani@ncl.ac.uk
[3] CSAIL, Massachusetts Institute of Technology, Cambridge, USA
sicung@csail.mit.edu

Abstract. In this paper, we present a new tool *SReach*, which solves probabilistic bounded reachability problems for two classes of models of stochastic hybrid systems. The first one is (nonlinear) hybrid automata with parametric uncertainty. The second one is probabilistic hybrid automata with additional randomness for both transition probabilities and variable resets. Standard approaches to reachability problems for linear hybrid systems require numerical solutions for large optimization problems, and become infeasible for systems involving both nonlinear dynamics over the reals and stochasticity. *SReach* encodes stochastic information by using a set of introduced random variables, and combines δ-complete decision procedures and statistical tests to solve δ-reachability problems in a sound manner. Compared to standard simulation-based methods, it supports non-deterministic branching, increases the coverage of simulation, and avoids the zero-crossing problem. We demonstrate *SReach*'s applicability by discussing three representative biological models and additional benchmarks for nonlinear hybrid systems with multiple probabilistic system parameters.

1 Introduction

Stochastic hybrid systems (SHSs) are dynamical systems exhibiting discrete, continuous, and stochastic dynamics. Due to the generality, they have been widely used in various areas, including biological systems, financial decision problems, and cyber-physical systems [2,6]. One elementary question for the quantitative analysis of SHSs is the probabilistic reachability problem, considering that many verification problems can be reduced to reachability problems. It is to compute the probability of reaching a certain set of states. The set may represent certain unsafe states which should be avoided or visited only with some small probability, or dually, good states which should be visited frequently. This problem

---

This research was sponsored by the Air Force Office of Scientific Research (FA9550-12-1-0146) and the Office of Naval Research (N000141310090).

O. Roux and J. Bourdon (Eds.): CMSB 2015, LNBI 9308, pp. 15–27, 2015.
DOI: 10.1007/978-3-319-23401-4_3

is no longer a decision problem, as it generalizes that by asking what is the probability that the system reaches the target region. For SHSs with both stochastic and non-deterministic behavior, the problem results in general in a range of probabilities, thereby becoming an optimization problem.

To describe stochastic dynamics, uncertainties have been added to hybrid systems in various ways. One way expresses random initial values and stochastic dynamical coefficients using random variables, resulting in hybrid automata (HAs) [13] with parametric uncertainty. Another approach integrates deterministic flows with probabilistic jumps. When state changes forced by continuous dynamics involve discrete random events, we refer to such systems as probabilistic hybrid automata (PHAs) [20]. When continuous probabilistic events are also involved, we call them stochastic hybrid automata (SHAs) [9]. Other models substitute deterministic flows with stochastic ones, such as stochastic differential equations (SDEs) [1], where the random perturbation affects the dynamics continuously. When all such modifications have been applied, the resulting models are called general stochastic hybrid systems (GSHSs) [15]. Among these different models, of particular interest for this paper are HAs with parametric uncertainty and PHAs with additional randomness for both transition probabilities and variable resets. Note that, in the following, we use notations - HA_p and PHA_r - for these two model classes respectively.

When modeling real-world systems, such as biological systems and cyber-physical systems, using hybrid models, parametric uncertainty arises naturally. Although its cause is multifaceted, two factors are critical. First, probabilistic parameters are needed when the physics controlling the system is known, but some parameters are either not known precisely, are expected to vary because of individual differences, or may change by the end of the system's operational lifetime. Second, system uncertainty may occur when the model is constructed directly from experimental data. Due to imprecise experimental measurements, the values of system parameters may have ranges of variation with some associated likelihood of occurrence. Clearly, the HA_ps are suitable models considering these major causes. Note that, in both cases, we assume that the probability distributions of probabilistic system parameters are known and remain unchanged throughout the systems evolution.

As another interesting and more expressive class of models, PHAs extend HAs with discrete probability distributions. More precisely, for discrete transitions in a model, instead of making a purely (non)deterministic choice over the set of currently enabled jumps, a PHA (non)deterministically chooses among the set of recently enabled discrete probability distributions, each of which is defined over a set of transitions. Although randomness only influences the discrete dynamics of the model, PHAs are still very useful and have interesting practical applications [21]. In this paper, we consider a variation of PHAs, where additional randomness for both transition probabilities and resets of system variables are allowed. In other words, in terms of the additional randomness for jump probabilities, we mean that the probabilities attached to probabilistic jumps from one mode, instead of having a discrete distribution with predefined constant probabilities, can be expressed by equations involving random variables whose distributions can be either discrete or continuous. This extension is motivated by the fact that some transition probabilities can vary due to factors such as individual and environmental differences in real-world systems. When it comes to the randomness of variable resets, we allow that a system variable can be reset to a value obtained according to a known discrete or continuous distribution, instead of being assigned a fixed value.

In this paper, we describe our tool *SReach* which supports probabilistic bounded δ-reachability analysis for the above two model classes. It combines the recently proposed δ-complete bounded reachability analysis technique [11] with statistical testing techniques. *SReach* saves the virtues of the Satisfiability Modulo Theories (SMT) based Bounded Model Checking (BMC) for HAs [7,23], namely the fully symbolic treatment of hybrid state spaces, while advancing the reasoning power to probabilistic models. Furthermore, by utilizing the δ-complete analysis method, the full non-determinism of models will be considered. The coverage of simulation will be increased, as the δ-complete analysis method results in an over-approximation of the reachable set, whereas simulation is only an under-approximation of it. The zero-crossing problem can be avoided as, if a zero-crossing point exists, it will always return an interval containing it. By using statistical tests, *SReach* can place controllable error bounds on the estimated probabilities. We discuss three biological models - an atrial fibrillation model, a prostate cancer treatment model, and our synthesized Killerred biological model - to show that *SReach* can answer questions including model validation/falsification, parameter synthesis, and sensitivity analysis. To further demonstrate its applicability, we also apply it to additional real-world hybrid systems with parametric uncertainty.

Related Work. Hahn et al. promoted an abstraction-based method where the given PHA is abstracted into an n-player stochastic game [12], albeit being limited to linear dynamics. Fränzle et al. proposed a Stochastic SMT-based procedure [10]. But their tool SiSAT supports only discrete random variables. Ellen et al. [8] proposed a statistical model checking technique for verifying hybrid systems with continuous non-determinism, thereby expanding the class of systems analyzable, yet confined dynamics to (non-linear) pre-post conditions rather than ODEs. *SReach* supports both discrete and continuous random variables, and ODEs. ProbReach [19] also uses the δ-complete procedures and offers verified estimated probability interval containing the real probability, yet can only deal with hybrid systems with initial random variables. While *SReach* is able to handle probabilistic transitions as well.

The paper proceeds by introducing two model classes of SHSs under consideration in Sect. 2. Section 3 formally states probabilistic bounded δ-reachability problems and explains how *SReach* solves these problems by combining δ-complete decision procedures with statistical tests. Case studies and additional experiments are discussed in Sect. 4. Section 5 concludes the paper.

2   Stochastic Hybrid Models

Before introducing the algorithm implemented by *SReach* and the problems that it can handle, we first define two model classes that *SReach* considers formally. For HA$_p$s, we follow the definition of HAs in [13], and extend it to consider probabilistic parameters in the following way.

Definition 1 (HA$_p$). *A hybrid automaton with parametric uncertainty is a tuple $H_p = \langle (Q, E), V, RV, \mathsf{Init}, \mathsf{Flow}, \mathsf{Inv}, \mathsf{Jump}, \Sigma \rangle$, where*

- *The vertices $Q = \{q_1, \cdots, q_m\}$ is a finite set of discrete modes, and edges in E are control switches.*
- *$V = \{v_1, \cdots, v_n\}$ denotes a finite set of real-valued system variables. We write $\dot{V}$ to represent the first derivatives of variables during the continuous change, and write V' to denote values of variables at the conclusion of the discrete change.*
- *$RV = \{w_1, \cdots, w_k\}$ is a finite set of independent random variables, where the distribution of w_i is denoted by P_i.*
- Init, Flow, *and* Inv *are labeling functions over Q. For each mode $q \in Q$, the initial condition $Init(q)$ and invariant condition $Inv(q)$ are predicates whose free variables are from $V \cup RV$, and the flow condition $Flow(q)$ is a predicate whose free variables are from $V \cup \dot{V} \cup RV$.*
- Jump *is a transition labeling function that assigns to each transition $e \in E$ a predicate whose free variables are from $V \cup V' \cup RV$.*
- *Σ is a finite set of events, and an edge labeling function event $: E \to \Sigma$ assigns to each control switch an event.*

Another class is PHA$_r$s, which extend HAs with discrete probability transitions and additional randomness for transition probabilities and variable resets.

Definition 2 (PHA$_r$). *A probabilistic hybrid automaton with additional randomness H_r consists of $Q, E, V, RV, \mathsf{Init}, \mathsf{Flow}, \mathsf{Inv}, \Sigma$ as in Definition 1, and Cmds, which is a finite set of probabilistic guarded commands of the form:*

$$g \to p_1 : u_1 + \cdots + p_m : u_m,$$

where g is a predicate representing a transition guard with free variables from V, p_i is the transition probability for the ith probabilistic choice which can be expressed by an equation involving random variable(s) in RV and the p_i's satisfy $\sum_{i=1}^{m} p_i = 1$, and u_i is the corresponding transition updating function for the ith probabilistic choice, whose free variables are from $V \cup V' \cup RV$.

To illustrate the additional randomness allowed for transition probabilities and variable resets, an example probabilistic guarded command is $x \geq 5 \to p_1 : (x' = sin(x)) + (1 - p_1) : (x' = p_x)$, where x is a system variable, p_1 has a Uniform distribution $U(0.2, 0.9)$, and p_x has a Bernoulli distribution $B(0.85)$. This means that, the probability to choose the first transition is not a fixed value, but a random one having a Uniform distribution. Also, after taking the second transition, x can be assigned to either 1 with probability 0.85, or 0 with 0.15. In general, for an individual probabilistic guarded command, the transition probabilities can be expressed by equations of one or more new random variables, as long as values of all transition probabilities are within $[0, 1]$, and their sum is 1. Currently, all four primary arithmetic operations are supported. Note that, to preserve the Markov property, only unused random variables can be used, so that no dependence between the current probabilistic jump and previous transitions will be introduced.

3   SReach Algorithm

A recently proposed δ-complete decision procedure [11] relaxes the reachability problem for HAs in a sound manner: it verifies a conservative approximation of the system behavior, so that bugs will always be detected. The over-approximation can be tight (tunable by an arbitrarily small rational parameter δ), and a false alarm with a small δ may indicate that the system is fragile, thereby providing valuable information to the system designer (see [11] for details). We now define the probabilistic bounded δ-reachability problem based on the bounded δ-reachability problem defined in [11].

Definition 3. *The probabilistic bounded k step δ-reachability for a HA$_p$ H_p is to compute the probability that H_p reaches the target region T in k steps. Given the set of independent random variables $\mathbf{r}$, $Pr(\mathbf{r})$ a probability measure over $\mathbf{r}$, and Ω the sample space of $\mathbf{r}$, the reachability probability is $\int_\Omega I_T(\mathbf{r})dPr(\mathbf{r})$, where $I_T(\mathbf{r})$ is the indicator function which is 1 if H_p with $\mathbf{r}$ reaches T in k steps.*

Definition 4. *For a PHA$_r$ H_r, the probabilistic bounded k step δ-reachability estimated by SReach is the maximal probability that H_r reaches the target region T in k steps: $max_{\sigma \in E}Pr^k_{H_r,\sigma,T}(i)$, where E is the set of possible executions of H starting from the initial state i, and σ is an execution in the set E.*

After encoding uncertainties using random variables, *SReach* samples them according to the given distributions. For each sample, a corresponding intermediate HA is generated by replacing random variables with their assigned values. Then, the δ-complete analyzer *dReach* is utilized to analyze each intermediate HA M_i, together with the desired precision δ and unfolding depth k. The analyzer returns either unsat or δ-sat for M_i. This information is then used by a chosen statistical testing procedure to decide whether to stop or to repeat the procedure, and to return the estimated probability. The full procedure is illustrated in Algorithm 1, where *MP* is a given stochastic model, and *ST* indicates which statistical testing method will be used (See the tool website for various statistical tests that supported by *SReach* and the way to control the induced statistical error bounds). *Succ* and N are used to record the number of δ-sat instances and total samples generated so far respectively, and are then the inputs of *ST*. Note that, for a PHA$_r$, sampling and fixing the choices of all the probabilistic transitions in advance results in an over-approximation of the original PHA$_r$, where safety properties are preserved. To promise a tight

---

Algorithm 1. SReach

1: **function** SReach(MP, ST, δ, k)
2: **if** MP is a HA$_p$ **then**
3:      $MP \leftarrow EncRM_1(MP)$ ▷ encode uncertain system parameters
4: **else**   ▷ otherwise a PHA$_r$
5:      $MP \leftarrow EncRM_2(MP)$ ▷ encode probabilistic jumps and extra randomness
6: **end if**
7: $Succ, N \leftarrow 0$ ▷ number of δ-sat samples and total samples
8: $Assgn \leftarrow \emptyset$ ▷ record unique sampling assignments and dReach results
9: $RV \leftarrow \text{ExtractRV}(MP)$ ▷ get the RVs from the probabilistic model
10: **repeat** in parallel
11:     $S_i \leftarrow \text{Sim}(RV)$ ▷ sample the parameters
12:     **if** $S_i \in Assgn.sample$ **then**
13:         $Res \leftarrow Assgn(S_i).res$ ▷ no need to call dReach
14:     **else**
15:         $M_i \leftarrow \text{Gen}(MP, S_i)$ ▷ generate a dReach model
16:         $Res \leftarrow \text{dReach}(M_i, \delta, k)$ ▷ call dReach to solve k-step δ-reachability
17:     **end if**
18:     **if** $Res = \delta$-sat **then** $Succ \leftarrow Succ + 1$
19:     **end if**
20:     $N \leftarrow N + 1$
21: **until** $ST.done(Succ, N)$ ▷ perform statistical test
22: **return** $ST.output$
23: **end function**

---

over-approximation and correctness of estimated probabilities, *SReach* supports PHA$_r$s with no or subtle non-determinism. That is, in order to offer a reasonable estimation, for PHA$_r$s, *SReach* is supposed to be used on models with no or few non-deterministic transitions, or where dynamic interleaving between non-deterministic and probabilistic choices are not important, such as our KillerRed biological model. To improve the performance of *SReach*, each sampled assignment and its corresponding *dReach* result are recorded for avoiding redundant calls to *dReach*. This significantly reduces the total calls for PHA$_r$s, as the size of the sample space involving random variables describing probabilistic jumps is comparatively small. For the example PHA (as shown in Fig. 1), with this heuristic, the total checking time has been decreased from 11291.31 s for 658 samples (17.16 s per sample) to 3295.82 s (5.01 s per sample). Furthermore, a parallel version of *SReach* has been implemented using OpenMP, where multiple samples and corresponding HAs are generated, and passed to *dReach* simultaneously. Using this parallel *SReach* on a 4-core machine, the running time for the example PHA has been further decreased to 2119.55 s for 660 samples (3.33 s per sample).

Currently, *SReach* supports a number of hypothesis testing and statistical estimation techniques including: Lai's test [17], Bayes factor test [16], Bayes factor test with indifference region [25], Sequential probability ratio test (SPRT) [24], Chernoff-Hoeffding bound [14], Bayesian Interval Estimation with Beta prior [26],

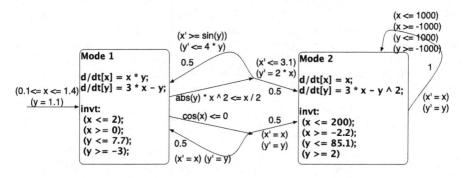

Fig. 1. An example probabilistic hybrid automaton

and Direct Sampling. All methods produce answers that are correct up to a precision that can be set arbitrarily by the user. See the tool website for more details about these statistical testing techniques. With these hypothesis testing methods, *SReach* can answer qualitative questions, such as "Does the model satisfy a given reachability property in k steps with probability greater than a certain threshold?" With the above statistical estimation techniques, *SReach* can offer answers to quantitative problems. For instance, "What is the probability that the model satisfies a given reachability property in k steps?" *SReach* can also handle additional types of interesting problems by encoding them as probabilistic bounded reachability problems. The model validation/falsification problem with prior knowledge can be encoded as a probabilistic bounded reachability question. After expressing prior knowledge about the given model as reachability properties, is there any number of steps k in which the model satisfies a given property with a desirable probability? If none exists, the model is incorrect regarding the given prior knowledge. The parameter synthesis problem can also be encoded as a probabilistic k-step reachability problem. Does there exist a parameter combination for which the model reaches the given goal region in k steps with a desirable probability? If so, this parameter combination is potentially a good estimation for the system parameters. The goal here is to find a combination with which all the given goal regions can be reached in a bounded number of steps. Moreover, sensitivity analysis can be conducted by a set of probabilistic bounded reachability queries as well: Are the results of reachability analysis the same for different possible values of a certain system parameter? If so, the model is insensitive to this parameter with regard to the given prior knowledge.

4   Experiments

Both sequential and parallel versions of *SReach* are available on https://github. com/dreal/SReach (see the tool website for its usage). Experiments for the following three biological models were conducted on a server with 2* AMD Opteron(tm) Processor 6172 and 32 GB RAM (12 cores were used), running on Ubuntu 14.04.1 LTS. In our experiments we used 0.001 as the precision for the δ-decision problem,

Table 1. Results for the 4-mode atrial fibrillation model ($k = 3$). For each sample generated, *SReach* analyzed systems with 62 variables and 24 ODEs in the unfolded SMT formulae. #RVs = number of random variables in the model, #S_S = number of δ-sat samples, #T_S = total number of samples, Est_P = estimated probability of property, A_T(s) = average CPU time of each sample in seconds, and T_T(s) = total CPU time for all samples in seconds. Note that, we use the same notations in the remaining tables.

Model	#RVs	EPI_TO1	EPI_TO2	#S_S	#T_S	Est_P	A_T(s)	T_T(s)
Cd_to1_s	1	U(6.1e-3, 7e-3)	6	240	240	0.996	0.270	64.80
Cd_to1_uns	1	U(5.5e-3, 5.9e-3)	6	0	240	0.004	0.042	10.08
Cd_to2_s	1	400	U(0.131, 6)	240	240	0.996	0.231	55.36
Cd_to2_uns	1	400	U(0.1, 0.129)	0	240	0.004	0.038	9.15
Cd_to12_s	2	N(400, 1e-4)	N(6, 1e-4)	240	240	0.996	0.091	21.87
Cd_to12_uns	2	N(5.5e-3, 10e-6)	N(0.11, 10e-5)	0	240	0.004	0.037	8.90

and Bayesian sequential estimation with 0.01 as the estimation error bound, coverage probability 0.99, and a uniform prior ($\alpha = \beta = 1$). All the details (including discrete modes, continuous dynamics that described by ODEs, non-determinism, and stochasticity) of models in the following case studies and additional benchmarks can be found on the tool website.

Atrial Fibrillation. The minimum resistor model reproduces experimentally measured characteristics of human ventricular cell dynamics [5]. It reduces the complexity of existing models by representing channel gates of different ions with one fast channel and two slow gates. However, due to this reduction, for most model parameters, it becomes impossible to obtain their values through measurements. After adding parametric uncertainty into the original hybrid model, we show that *SReach* can be adapted to synthesize parameters for this stochastic model, i.e., identifying appropriate ranges and distributions for model parameters. We chose two system parameters - *EPI_TO1* and *EPI_TO2*, and varied their distributions to see which ones allow the model to present the desired patterns. As in Table 1, when *EPI_TO1* is either close to 400, or between 0.0061 and 0.007, and *EPI_TO2* is close to 6, the model can satisfy the given bounded reachability property with a probability very close to 1.

Prostate Cancer Treatment. This model is a nonlinear hybrid automaton with parametric uncertainty. We modified the model of the intermittent androgen suppression (IAS) therapy in [22] by adding parametric uncertainty. The IAS therapy switches between treatment-on, and treatment-off with respect to the serum level thresholds of prostate-specific antigen (PSA), namely r_0 and r_1. As suggested by the clinical trials [4], an effective IAS therapy highly depends on the individual patient. Thus, we modified the model by taking parametric variation caused by personalized differences into account. In detail, according to clinical data from hundreds of patients [3], we replaced six system parameters with random variables having appropriate (continuous) distributions, including α_x (the proliferation rate of androgen-dependent (AD) cells), α_y (the proliferation rate

Table 2. Results for the 2-mode prostate cancer treatment model ($k = 2$). For each sample generated, *SReach* analyzed systems with 41 variables and 10 ODEs in the unfolded SMT formulae.

Model	#RVs	r_0	r_1	Est_P	#S_S	#T_S	A_T(s)	T_T(s)
PCT1	6	5.0	10.0	0.496	8226	16584	0.596	9892
PCT2	6	7.0	11.0	0.994	335	336	54.307	18247
PCT3	6	10.0	15.0	0.996	240	240	506.5	121560

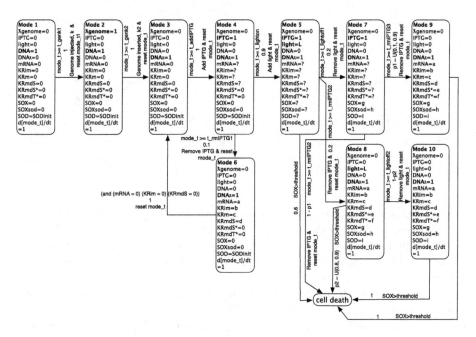

Fig. 2. A probabilistic hybrid automaton for synthesized phage-based therapy model

of androgen-independent (AI) cells), β_x (the apoptosis rate of AD cells), β_y (the apoptosis rate of AI cells), m_1 (the mutation rate from AD to AI cells), and z_0 (the normal androgen level). To describe the variations due to individual differences, we assigned α_x to be $U(0.0193, 0.0214)$, α_y to be $U(0.0230, 0.0254)$, β_x to be $U(0.0072, 0.0079)$, β_y to be $U(0.0160, 0.0176)$, m_1 to be $U(0.0000475, 0.0000525)$, and z_0 to be $N(30.0, 0.001)$. We used *SReach* to estimate the probabilities of preventing the relapse of prostate cancer with three distinct pairs of treatment thresholds (*i.e.*, combinations of r_0 and r_1). As shown in Table 2, the model with thresholds $r_0 = 10$ and $r_1 = 15$ has a maximum posterior probability that approaches 1, indicating that these thresholds may be considered for the general treatment.

Synthesized KillerRed Model. Due to the widespread misuse and overuse of antibiotics, drug resistant bacteria now pose significant risks to health, agriculture and the environment. An alternative to conventional antibiotics is

Table 3. Results for the 11-mode killerred model.

k	Est_P	#S_S	#T_S	A_T(s)	T_T(s)	k	Est_P	#S_S	#T_S	A_T(s)	T_T(s)
5	0.544	8951	16452	0.074	1219.38	8	0.004	0	240	0.004	0.88
6	0.247	3045	12336	0.969	11957.12	9	0.004	0	240	0.012	2.97
7	0.096	559	5808	5.470	31770.36	10	0.004	0	240	0.013	3.18

Table 4. Formal analysis results for our KillerRed hybrid model

$t_{lightON}$ (t.u.)	1	2	3	4	5	6	7	8	9	10
t_{total} (t.u.)	16	17.2	18.5	20	21.3	22.7	23.5	24.1	25	30
$t_{lightOFF_1}$ (t.u.)	1	2	3	4	5	6	7	8	9	10
Killed bacteria cells	Failed	Failed	Failed	Succ	Succ	Succ	Succ	Succ	Succ	Succ
t_{rmIPTG_3} (t.u.)	1	2	3	4	5	6	7	8	9	10
Killed bacteria cells	Succ	Succ	Succ	Succ	Succ	Succ	Succ	Succ	Succ	Succ
SOX_{thres} (M)	1e-4	2e-4	3e-4	4e-4	5e-4	6e-4	7e-4	8e-4	9e-4	1e-3
t_{total} (t.u.)	5.1	5.2	5.4	17	19	48	61	71	36	42

phage-based therapy. One approach to antibiotic resistance is to engineer a temperate phage λ with light-activated production of superoxide (SOX). The incorporated Killerred protein is phototoxic and provides another level of controlled bacteria killing [18]. A PHA_r with subtle non-determinism for this synthesized Killerred model (as shown in Fig. 2) has been constructed. Considering individual differences of bacterial cells and distinct experimental environments, additional randomness on transition probabilities have been considered. *SReach* was used to validate this model by estimating the probabilities of killing bacterial cells with different ks (see Table 3). We noticed that the probabilities of paths going through mode 6 to mode 11 are close to 0. This remains even after increasing the probability of entering mode 6, indicating that it is impossible for this model to enter mode 6. *SReach* was also used to find out (a) the relation between the time to turn on the light after adding the molecular biology reagent IPTG and the total time to kill bacterial cells with probability larger than 0.5 (see the first two rows of Table 4), (b) that the lower bound for the duration of exposure to light is 3 for successful bacterial killing with probability larger than 0.5 (see row 3–4 of Table 4), (c) that the time to remove IPTG is insensitive considering whether bacterial cells will be killed with probability larger than 0.5 (see row 5–6 of Table 4), and (d) that the upper bound of the necessary concentration of SOX to kill bacterial cells, with probability larger than 0.5, is 0.6667 (see from row 7–8 of Table 4). All these findings have been reported to biologists for further checking.

Additional Benchmarks. To further demonstrate *SReach*'s applicability, we also applied it to additional benchmarks including HA_ps, PHAs, and PHA_rs

Table 5. #Ms = number of modes, K indicates the unfolding steps, #ODEs = number of ODEs in the unfolded formulae, #Vs = number of total variables in the unfolded formulae, #RVs = number of random variables in the model, δ = precision used in *dReach*.

Benchmark	#Ms	K	#ODEs	#Vs	#RVs	δ	Est_P	#S_S	#T_S	A_T(s)	T_T(s)
BBK1	1	1	2	14	3	0.001	0.754	5372	7126	0.086	612.836
BBK5	1	5	2	38	3	0.001	0.059	209	3628	0.253	917.884
BBwDv1	2	2	4	20	4	0.001	0.208	2206	10919	0.080	873.522
BBwDv2K2	2	2	4	20	3	0.001	0.845	7330	8669	0.209	1811.821
BBwDv2K8	2	8	4	56	3	0.001	0.207	2259	10901	0.858	9353.058
Tld	2	7	2	33	4	0.001	0.996	227	227	0.213	48.351
Ted	2	7	4	50	4	0.001	0.996	227	227	12.839	2914.448
DTldK3	2	3	4	26	2	0.001	0.996	227	227	0.382	86.714
DTldK5	2	5	4	38	2	0.001	0.161	1442	8961	0.280	2509.078
W4mv1	4	3	8	26	6	0.001	0.381	5953	15639	0.238	3722.082
W4mv2K3	4	3	8	26	6	0.001	0.996	227	227	0.673	152.771
W4mv2K7	4	7	8	50	6	0.001	0.004	0	227	0.120	27.240
DWK1	2	1	4	14	5	0.001	0.996	227	227	0.171	38.817
DWK3	2	3	4	26	5	0.001	0.996	227	227	0.215	48.806
DWK9	2	9	4	62	5	0.001	0.996	227	227	5.144	1167.688
Que	3	2	3	13	4	0.001	0.228	2662	11677	0.095	1109.315
3dOsc	3	2	18	48	2	0.001	0.996	227	227	8.273	1877.969
QuadC	1	0	14	44	6	0.001	0.996	227	227	825.641	187420.507
exPHA01	2	2	4	20	2	0.001	0.524	345	658	5.01	3295.82
exPHA02	2	3	2	17	1	0.001	0.900	5361	5953	0.0004	2.35
KRk5	6	5	84	194	2	0.001	0.544	8946	16457	0.122	2015.64
KRk6	8	6	112	224	6	0.001	0.246	2032	8263	1.385	11444.22
KRk7	10	7	150	271	6	0.001	0.096	558	5795	16.275	94311.18
KRk8	7	8	105	303	6	0.001	0.004	0	227	0.003	0.58
KRk9	9	9	135	335	6	0.001	0.004	0	227	0.015	3.43
KRk10	11	10	165	367	6	0.001	0.004	0	227	0.026	5.92

with subtle non-determinism. Table 5 shows the results of these experiments. These experiments were conducted with the sequential version of *SReach* on a machine with 2.9 GHz Intel Core i7 processor and 8 GB RAM, running OS X 10.9.2. In our experiments we used 0.001 as the precision for the δ-decision problem; and Bayesian sequential estimation with 0.01 half-interval width, coverage probability 0.99, and uniform prior ($\alpha = \beta = 1$). In the following table, BB refers to the bouncing ball models, Tld the thermostat model with linear temperature decrease, Ted the thermostat model with exponential decrease, DT the dual thermostat models, W the watertank models, DW the dual watertank models, Que the model for queuing system which has both nonlinear functions and nondeterministic jumps, 3dOsc the model for 3d oscillator, and QuadC the model for quadcopter stabilization control. Following these hybrid systems with parametric uncertainty, we also consider two example PHAs - exPHA01 and exPHA02, and PHA$_r$s with trivial non-determinism - KR (our killerred models). Moreover, the detailed description of some of additional benchmarks and above case studies can be found on the tool website. The full descriptions of all the models that mentioned in this paper can be found on the tool website.

5   Conclusions and Future Work

We have presented a tool that combines δ-decision procedures and statistical tests. It supports probabilistic bounded δ-reachability analysis for HA_ps and PHA_rs with no or subtle non-determinism. This tool has been used to analyze three representative examples - a prostate cancer treatment model, a cardiac model, and a synthesized Killerred model - and other benchmarks, which are currently out of the reach of other formal tools. In the near future, we plan to extend support for more general stochastic hybrid models that include probabilistic jumps with continuous distributions, and stochastic differential equations.

References

1. Arnold, L.: Stochastic Differential Equations: Theory and Applications. Wiley - Interscience, New York (1974)
2. Blom, H.A., Lygeros, J., Everdij, M., Loizou, S., Kyriakopoulos, K.: Stochastic Hybrid Systems: Theory and Safety Critical Applications. Springer, Heidelberg (2006)
3. Bruchovsky, N., Klotz, L., Crook, J., Goldenberg, L.: Locally advanced prostate cancer: biochemical results from a prospective phase II study of intermittent androgen suppression for men with evidence of prostate-specific antigen recurrence after radiotherapy. Cancer **109**(5), 858–867 (2007)
4. Bruchovsky, N., Klotz, L., et al.: Final results of the Canadian prospective phase II trial of intermittent androgen suppression for men in biochemical recurrence after radiotherapy for locally advanced prostate cancer. Cancer **107**(2), 389–395 (2006)
5. Bueno-Orovio, A., Cherry, E.M., Fenton, F.H.: Minimal model for human ventricular action potentials in tissue. J. Theor. Biol. **253**(3), 544–560 (2008)
6. Clarke, E.M., Zuliani, P.: Statistical model checking for cyber-physical systems. In: Bultan, T., Hsiung, P.-A. (eds.) ATVA 2011. LNCS, vol. 6996, pp. 1–12. Springer, Heidelberg (2011)
7. Cordeiro, L., Fischer, B., Marques-Silva, J.: SMT-based bounded model checking for embedded ansi-c software. IEEE Softw. Eng. **38**(4), 957–974 (2012)
8. Ellen, C., Gerwinn, S., Fränzle, M.: Statistical model checking for stochastic hybrid systems involving nondeterminism over continuous domains. Int. J. Softw. Tools Technol. Transf. **17**, 1–20 (2014)
9. Fränzle, M., Hahn, E.M., Hermanns, H., Wolovick, N., Zhang, L.: Measurability and safety verification for stochastic hybrid systems. In: HSCC, pp. 43–52, April 2011
10. Fränzle, M., Hermanns, H., Teige, T.: Stochastic satisfiability modulo theory: a novel technique for the analysis of probabilistic hybrid systems. In: Egerstedt, M., Mishra, B. (eds.) HSCC 2008. LNCS, vol. 4981, pp. 172–186. Springer, Heidelberg (2008)
11. Gao, S., Kong, S., Chen, W., Clarke, E.M.: δ-complete analysis for bounded reachability of hybrid systems (2014). CoRR, arXiv:1404.7171
12. Hahn, E.M., Norman, G., Parker, D., Wachter, B., Zhang, L.: Game-based abstraction and controller synthesis for probabilistic hybrid systems. In: QEST, pp. 69–78. IEEE (2011)

13. Henzinger, T.A.: The Theory of Hybrid Automata. Springer, Berlin (2000)
14. Hoeffding, W.: Probability inequalities for sums of bounded random variables. J. Am. Stat. Assoc. **58**(301), 13–30 (1963)
15. Hu, J., Lygeros, J., Sastry, S.S.: Towards a theory of stochastic hybrid systems. In: Lynch, N.A., Krogh, B.H. (eds.) HSCC 2000. LNCS, vol. 1790, pp. 160–173. Springer, Heidelberg (2000)
16. Kass, R.E., Raftery, A.E.: Bayes factors. JASA **90**(430), 773–795 (1995)
17. Lai, T.L.: Nearly optimal sequential tests of composite hypotheses. AOS **16**(2), 856–886 (1988)
18. Wang, Q., Miskov-Zivanov, N., Telmex, C., Clarke, E.M.: Formal analysis provides parameters for guiding hyperoxidation in bacteria using phototoxic proteins. GLSVLSI 2015 (2015)
19. Shmarov, F., Zuliani, P.: Probreach: verified probabilistic delta-reachability for stochastic hybrid systems. In: HSCC (2015)
20. Sproston, J.: Decidable model checking of probabilistic hybrid automata. In: Joseph, M. (ed.) FTRTFT 2000. LNCS, vol. 1926, pp. 31–45. Springer, Heidelberg (2000)
21. Sproston, J.: Model checking for probabilistic timed and hybrid systems. Ph.D. thesis. SCS, University of Birmingham (2001)
22. Tanaka, G., Hirata, Y., Goldenberg, L., Bruchovsky, N., Aihara, K.: Mathematical modelling of prostate cancer growth and its application to hormone therapy. Phil. Trans. Roy. Soc. A Math. Phys. Eng. Sci. **368**(1930), 5029–5044 (2010)
23. Tinelli, C.: SMT-based model checking. In: NASA FM, p. 1 (2012)
24. Wald, A.: Sequential tests of statistical hypotheses. Ann. Math. Stat. **16**(2), 117–186 (1945)
25. Younes, H.L.: Verification and planning for stochastic processes with asynchronous events. Technical report, DTIC Document (2005)
26. Zuliani, P., Platzer, A., Clarke, E.M.: Bayesian statistical model checking with application to stateflow/simulink verification. Formal Methods Syst. Des. **43**(2), 338–367 (2013)

Experimental Design for Inference over the *A. thaliana* Circadian Clock Network

Daniel Trejo-Banos[1]([✉]), Andrew J. Millar[2,3], and Guido Sanguinetti[1,3]

[1] School of Informatics, University of Edinburgh, Edinburgh, Scotland
danieltba@hotmail.com
[2] SynthSys - Systems and Synthetic Biology, University of Edinburgh, Edinburgh, Scotland
[3] School of Biological Sciences, University of Edinburgh, Edinburgh, Scotland

Abstract. Planning experiments is a crucial step in successful investigations, which can greatly benefit from computational modeling approaches. Here we consider the problem of designing informative experiments for elucidating the dynamics of biological networks. Our approach extends previously proposed methodologies to the important case where the structure of the network is also uncertain. We demonstrate our approach on a benchmark scenario in plant biology, the circadian clock network of *Arabidopsis thaliana*, and discuss the different value of three types of commonly used experiments in terms of aiding the reconstruction of the unknown network.

1 Introduction

The execution of experiments to test a hypothesis is the essence of the scientific method. In the field of systems biology we are interested in testing and validating our hypotheses and predictions biochemical processes in living organisms, and our hypotheses are usually encoded in mathematical models which can adopt a variety of formalism. Modern biochemical experiments can be very complex and are often costly in both researcher time and other resources. For this reason, it is important to minimize the number of experiments while maximizing their information content.

Experimental design is the branch of statistics and operations research which is concerned with maximizing the information content of novel experiments. From a statistical point of view, the utility criterion for evaluating an experiment is a function of the probabilistic model chosen to represent the data-generating process. Depending on the objective of the experiment, the selection criterion can be either *maximize the information content of an experiment in order to*

DTB is funded by a Microsoft Research Studentship. GS acknowledges support from the European Research Council under grant MLCS30699. SynthSys was founded as a Centre for Integrative Biology by BBSRC/EPSRC award D19621 to AJM and others.

O. Roux and J. Bourdon (Eds.): CMSB 2015, LNBI 9308, pp. 28–39, 2015.
DOI: 10.1007/978-3-319-23401-4_4

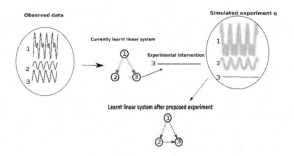

Fig. 1. Basic illustration of our experimental design approach. After a set of observations the distribution over the learnt system (blurred arrows) is used to draw samples of the experimental outcomes given an intervention (uncertainty over the outcomes is also represented by blurred functions). The aim is to choose the experiment that reduces the uncertainty over the learnt system (represented by the system with well defined arrows in the figure).

estimate a set of parameters, (*estimation criterion*) or improve the prediction qualities of a fitted model (*prediction criterion*).

In this paper we use a Bayesian approach to experimental design for dynamical models of biological systems. We restrict our attention to gene regulatory network (GRN) models, where the systems dynamics are generated by mutual interactions between genes which can modulate each others rate of expression; these models encompass a large fraction of the systems biology literature, and hence experimental design methods for this class of systems are of considerable interest. Dynamical systems such as ordinary differential equations (ODE) are widespread techniques for modeling GRNs. Previous work has considered experimental design and model selection techniques for non-linear ODE- based models of biological processes.

Liepe et al. [5] employ an approach based on mutual information which could be evaluated using Monte-Carlo simulations. This method is computationally intensive and crucially requires prior knowledge over the model components and their interactions: the structure and functional form of the equations defining the models is assumed known, and all the uncertainty is in the parametrisation. In reality, most models in systems biology are subject to considerable structural uncertainty, and clarifying the structure of interactions is the primary goal of systems biology experiments.

In this work we extend the Bayesian experimental design approach to models with structural uncertainty, formalized as hierarchical Bayesian models. We derive a *Bayesian experimental design score* for quantifying the information gain offered by different experiments. The abstract view of the method is shown on Fig. 1. We start by using some preliminary data (in the form of observed oscillatory expression levels) to learn a (posterior) probability distribution over a linear approximation of the system. Experimental interventions can be simulated by constraining some components of the model to fixed values (the specific details of how we model interventions are given later), obtaining predictions of the gene expression levels of all the other components given the experimental intervention (in the figure, the blurred lines represent uncertainty over the experimental outcomes). These enable us to quantify the information content of an intervention.

We illustrate our approach on a benchmark systems biology problem, the circadian clock of the *Arabidopsis thaliana* model plant [9]. We consider three classes of possible experiments: alterations to the light-dark input provided to the plant, direct measurements of regulatory links via chromatin immunoprecipitation (ChIP), and gene knock-outs. These commonly performed experiments are very different in terms of costs, and our preliminary results on their relative informativeness could be useful for practitioners.

2   Methods

Classical approaches to statistical experimental design have been primarily developed for linear regression models. Let an experiment q be given an experimental design Φ^q (usually a set of covariates and a model that accounts for the variables of the experiment) and parameters θ (which determine how each of these covariates determines the measured output of the experiment), and denote the experimental observations for experiment q as $\mathbf{y}^q$. The experimental outputs are assumed to be a linear combination of the covariates such that

$$\mathbf{y}^q = \Phi^q \theta + \epsilon \tag{1}$$

where ϵ is zero-mean Gaussian noise with variance σ^2. The probability of the observed outcomes given a set of parameters θ is known as *likelihood function* (it is a function of the parameters); we will denote it as

$$p\left(\mathbf{y}^q | \Phi^q, \theta\right) = \mathcal{N}\left(\mathbf{y}^q - \Phi^q \theta, \sigma^2\right) \tag{2}$$

The *Fisher information matrix* (FIM) quantifies how much a small change in the parameters θ is expected to affect the likelihood of the observations; mathematically, the FIM is defined as

$$\mathcal{I}_{i,j}(\theta) = E_{p(\mathbf{y}^q|\Phi^q,\theta)}\left[\frac{\partial p\left(\mathbf{y}^q|\Phi^q,\theta\right)}{\partial \theta_i}\frac{\partial p\left(\mathbf{y}^q|\Phi^q,\theta\right)}{\partial \theta_j}\right] \tag{3}$$

where E_q denotes expectation under the distribution q.

The FIM encodes interaction between the observed and the experimental covariates. The most common experimental design objective seeks to select a design Φ^q in order to attain the maximum FIM according to some ordering. For estimation purposes, the optimality criteria depends on the choice of matrix function from which to evaluate the information matrix. The most popular is the *D-optimal* criterion or maximize $\det\left(\mathcal{I}\left(\theta\right)/n\right)$. This criterion minimizes the volume of the confidence ellipsoid of the estimates [4]. A good review of D-optimal design and related criteria can be consulted in [10].

In order to accommodate further uncertainties about experimental covariates and model mis-specification, a different kind of statistical tools is needed. Bayesian methods employ a *prior distribution* over the parameters $p(\theta)$ to incorporate uncertainty in a principled way. This is incorporated with observations to compute the *posterior distribution* by applying Bayes rule which is

$$p\left(\theta|\mathbf{y}^q, \varPhi^q\right) = \frac{p\left(\mathbf{y}^q|\theta, \varPhi^q\right) p\left(\theta\right)}{p\left(\mathbf{y}^q\right)}. \tag{4}$$

The denominator in Eq. 4 is computed by integrating the likelihood over the prior distribution. *Bayesian experimental design* seeks to leverage prior information about the parameter distribution by averaging over the posterior distribution of the unobserved data samples [2]. For this, we employ the concept of *Mutual information*. In this context we can view the mutual information between θ and $\mathbf{y}^q$ as the reduction in uncertainty about θ that results from observing $\mathbf{y}^q$ [7]. Then, the Bayesian counterpart to D-optimal design maximizes the Mutual information between the parameters distribution and the experimental outcomes [2].

2.1 Bayesian Experimental Design

In his seminal work, Lindley [6] sets experimental design in a decision-theory framework. First he states that the previous knowledge over a system is encoded in the prior probability of its model parameters. The knowledge about parameters θ obtained after an experiment, given the observations y^q and experimental conditions ξ^q will be contained in the posterior distribution $p(\theta|y^q, \xi^q)$. Thus the information gained after an experiment can be expressed in terms of the expected *KL-divergence* between both distributions over the distribution of the observations

$$I\left(\theta; \mathbf{y}^q\right) = \int KL\left(p\left(\theta|\mathbf{y}^q\right) \| p\left(\theta\right)\right) p\left(\mathbf{y^q}\right) d\mathbf{y^q}.$$

Thus the *utility* of an experiment q with conditions ξ^q(which we will denote by $U\left(\theta; \mathbf{y}^q; \xi^q\right)$) is obtained by solving

$$U\left(\theta; \mathbf{y}^q; \xi^q\right) = \int \int \log \frac{p\left(\theta|\mathbf{y}^q, \xi^q\right)}{p\left(\theta\right)} p\left(\theta, \mathbf{y}^q|\xi^q\right) d\theta dy^q. \tag{5}$$

This utility function gives rise to what is known as *Bayesian D-optimal design* [2]. In order to choose the best experimental design, the objective is to maximize the value of the utility function $U\left(\theta, y^q, \xi^q\right)$ over the set of parameters and (unobserved) responses. Unlike classic optimal design, we aim at leveraging prior information encoded in the prior distribution of the parameters.

Whereas these ideas were introduced in the linear regression case, extending to different scenarios is conceptually trivial; however, the computational simplifications afforded by linear models are then lost, giving rise to an analytically intractable problem. Liepe et al. [5] employ the same utility criteria over a set of parameters for a nonlinear system of differential equations and then proceed to compute the utility function by Monte Carlo simulation. This requires at each step to simulate the experimental outcomes by solving the system, a procedure which may incur in severe computational overhead depending of the model size and parameters. Furthermore, the model structure is assumed fixed; introducing uncertainty in the model structure would add a further dimension to the already complex computational problem, ruling out all but the simplest problems.

In this work, we take the complementary approach of catering for structural uncertainty in the models, while simplifying the dynamics by assuming linearity and time invariance (LTI models). We approach the problem by adopting a probabilistic linear model of the frequency spectrum of the gene expression levels. In the case of oscillating networks, this linear model can offer a reasonable approximation to the system dynamics, and has been shown to be effective in capturing structural uncertainty in a network inference scenario [11]. The advantage of the LTI approximation is that sampling from the experimental outcomes "reduces" to sampling from a Multivariate Normal conditioned on a subset of variables, confining the need for Monte Carlo simulation to integrating out the structural uncertainty.

2.2   Frequency-Domain Model of Gene Expression Levels

We briefly review now the LTI approach to modelling GRN dynamics taken in [11]. We start by representing the LTI equations in frequency domain through the *Discrete Fourier Transform* (DFT). Under certain conditions the DFT is a discrete sample of the Fourier spectrum of the signal, see [8]. With this approximation we derive a matrix equation for the linearized network dynamics, this matrix equation is

$$\dot{\mathbf{X}}^q = \mathbf{X}^q \mathbf{A}^T + \mathbf{U}^q \mathbf{C}^T. \tag{6}$$

Here, matrix $\mathbf{X}^q$ is the matrix whose columns represent the DFT coefficients (spectrum) of the expression level samples of a set of N genes for an experiment q. Analogously, $\mathbf{U}^q$ will represent the DFT of the system inputs. We denote by $\dot{\mathbf{X}}^q$ the time derivative of the spectra, which can be computed by the matrix product $\mathbf{DX}$, being $\mathbf{D}$ a derivative operator. The DSS model presented in [11] proposes a Gaussian likelihood regression model for estimating coefficients $\mathbf{A}$ and $\mathbf{C}$ by the distribution of the residues $\mathbf{Q}^q = \dot{\mathbf{X}}^q - \mathbf{X}^q \mathbf{A}^T - \mathbf{U}^q \mathbf{C}^T$ such that

$$\mathrm{p}\left(\mathbf{Q}^q | \sigma_D\right) = \mathcal{N}\left(\dot{\mathbf{X}}^q - \begin{bmatrix} \mathbf{X}^q & \mathbf{U}^q \end{bmatrix} \begin{bmatrix} \mathbf{A}^T \\ \mathbf{C}^T \end{bmatrix}, \sigma_D^2\right).$$

In order to estimate the parameters $\{\mathbf{A}, \mathbf{C}\}$, a sparsity inducing prior is set over these parameters. This prior is a spike and slab distribution of the form presented in [3]: intuitively, this is a mixture distribution where parameters (LTI coefficients) can either be sampled from a distribution concentrated at zero (the spike) or a broad distribution (the slab). Thus, conditioning on data, spike and slab models carry out automatic feature selection by assigning the value zero to irrelevant features (in our case interaction coefficients between non-interacting genes).

This prior encodes the network topology through an adjacency matrix $\mathbf{H}$ within a Hierarchical Bayesian model. We call this model the DFT-Spike and slab (DSS) model of gene expression. The precise details of the model, as well as Bayesian algorithms for network inference within this framework, are provided in [11]. For the purposes of experimental design, it is sufficient to state that this framework provides us with a methodology to recast GRN dynamics in a (Bayesian) regression framework, where the (DFT projection) of the signal derivative is regressed upon the (DFT projection) of the signal. The Hierarchical Bayesian model then provides a structured prior distribution to capture the uncertainty over the underlying networks.

2.3 Experimental Design for Estimating Parameters of a DSS Model

Having specified the DSS family of models, we now discuss in detail the experimental design techniques for three classes of experiments. The starting point is a prior distribution over LTI coefficients, which in itself could be (and, generally, is) the posterior distribution from some previous experiments. The crucial problems are two, how can an experimental perturbation be encoded mathematically within the model? how can we compute the utility score for a perturbation?

The answer to these questions depends on the specific perturbation considered; here we focus on three commonly employed experiments. The first type are changes in the external input to the system, the U matrix in Eq. (6). We denote this class of experiments as *photo-period experiments*, since in the case study of *A. thaliana* the input matrix represents the light inputs to the circadian clock. The second type are mutagenic experiments, where a single gene is removed from the system (*knock-out*). The third type are observation experiments, where presence/absence of one or more edges is observed directly through experiments such as Chromatin Immunoprecipitation (ChIP) or any affinity-binding detection methods.

Notice that observation experiments are somewhat different from the other types, as they do not constitute a perturbation of the system; for this reason, in the following we describe experimental design methodologies for observation experiments separately.

Photo-Period Experiments and Knock-Out Experiments. In the DSS setting, we frame experimental design for photo-period and knock-out settings as choosing the best experiment q defined as *i*nterventions in matrix $[\mathbf{X}^q \mathbf{U}^q]$ that maximizes the *information gain* over the parameters $\mathbf{B} = [\mathbf{A}, \mathbf{C}]$ of the linear dynamical model of Eq. 6. An *i*ntervention consists of setting a column of $\mathbf{U}^q$ or $\mathbf{X}^q$ to a known value ξ^q (zero in case of knock-out experiments or the frequency spectrum for a light signal in the case of photo-period experiments). We will denote the intervened element as column(s) $\mathbf{X}_i^q$ and the rest of the columns as $\mathbf{X}_{\setminus i}^q$.

The utility function of Eq. 5 can be computed by calculating the KL-divergence between the current distribution of the LTI-coefficients (either prior distribution or posterior distribution of a previous experiment) and the posterior distribution over said parameters after performing the desired experiment. This implies that we have to be able to compute the expected value of the next experiment's observations, in order to compute the mutual information and thus the utility of the next experiment. Explicitly this utility function is

$$U\left(\mathbf{B}; \mathbf{X}^q; \xi^q\right) = \int \int p\left(\mathbf{X}^q_{\backslash i}, \mathbf{B} | \mathbf{X}^q_i = \xi^q\right) \log \frac{p\left(\mathbf{B} | \mathbf{X}^q_{\backslash i}, \mathbf{X}^q_i = \xi^q\right)}{p\left(\mathbf{B}\right)} d\mathbf{X}^q d\mathbf{B}$$

the prior (current knowledge) $p\left(\mathbf{B}\right)$ doesn't depend on the next, simulated experiment (we simulate using the current knowledge), as such, the selection criteria can be stated in terms of the numerator as the integral

$$\int \int \mathbf{p}\left(\mathbf{X}^q_{\backslash i}, \mathbf{B} | \mathbf{X}^q_i = \xi^q\right) \log \mathbf{p}\left(\mathbf{B} | \mathbf{X}^q_{\backslash i}, \mathbf{X}^q_i = \xi^q\right) d\mathbf{X}^q d\mathbf{B} \tag{7}$$

The conditional distribution $p\left(\mathbf{B} | \mathbf{X}^q_{\backslash i}, \mathbf{X}^q_i = \xi^q\right)$ as derived in [11] is a result of a Linear regression model with Gaussian likelihood. As such the conditional over the coefficients $\mathbf{B}$ can be obtained by factorizing, and is

$$\log p\left(\mathbf{B} | \mathbf{X}^q, \xi^q\right) \propto \log \left[\det \left(\sigma_D^{-2} \mathbf{\Sigma}^{-1}\right)^{-1/2}\right] - \frac{1}{2\sigma_D^2}\left(-2\bar{\eta}^T \bar{\mathbf{B}} + \bar{\mathbf{B}}^T \mathbf{\Sigma}^{-1} \bar{\mathbf{B}}\right) \tag{8}$$

with the terms

$$\bar{\eta} = vec\left(\sum_q \left[\mathbf{X}_q \mathbf{U}_q\right]^T \dot{\mathbf{X}}_q\right); \quad \mathbf{\Sigma}^{-1} = \mathbf{I} \otimes \left(\sum_q \left[\mathbf{X}_q \mathbf{U}_q\right]^T \left[\mathbf{X}_q \mathbf{U}_q\right]\right).$$

We evaluate Eq. (7) through Monte Carlo simulation by drawing a sample from the joint distribution

$$p\left(\mathbf{X}^q_{\backslash i}, \mathbf{B} | \mathbf{X}^q_i = \xi^q\right) = p\left(\mathbf{X}_{\backslash i}^q | \mathbf{B}, \mathbf{X}^q_i = \xi^q\right) p\left(\mathbf{B}\right) \tag{9}$$

The Monte Carlo algorithm will consist of integrating $U_{DSS}\left(\bar{\eta}, \mathbf{\Sigma}, \mathbf{B}\right)_{DSS}$ over both random variables

$$\frac{1}{S_1} \sum_{s_1=1}^{S_1} \left(\frac{1}{S_2} \sum_{s_2=1}^{S_2} \log p\left(\mathbf{B}^{(s_1)} | \mathbf{X}^{q(s_2)}_{\backslash i}, \mathbf{X}^q_i = \xi^q\right)\right) \tag{10}$$

we draw a sample $\mathbf{B}^{(s_1)}$ from $p\left(\mathbf{B}\right)$, then we evaluate Eq. 8 by drawing samples $\mathbf{X}^{q(s_2)}_{\backslash i}$ from the conditional distribution term of Eq. 9. We derive the conditional distribution $p\left(\mathbf{X}_{\backslash i}^q | \mathbf{B}, \mathbf{X}^q_i = \xi^q\right)$ from the Gaussian likelihood of the regression model in [11] by using the Kronecker product and the vectorization operator. We apply the technique of completing the square [1], so we can get the distribution over the frequency spectra, from which we can draw samples as it is a Gaussian of the form

$$p\left(\mathbf{X}^q|\mathbf{B},\sigma^2\right) \sim \mathcal{N}\left(\eta,\Lambda^{-1}\right) \tag{11}$$

with $\Lambda = \frac{1}{\sigma^2}\left(\mathbf{I}\otimes\mathbf{D} - \mathbf{A}^T\otimes\mathbf{I}\right)^T\left(\mathbf{I}\otimes\mathbf{D} - \mathbf{A}^T\otimes\mathbf{I}\right)$ and
$\eta = -\Lambda^{-1}\left(\mathbf{I}\otimes\mathbf{D} - \mathbf{A}^T\otimes\mathbf{I}\right)^T\bar{\mathbf{U}}\mathbf{C}.$

Experiments for Observing Interactions. As a complement to the previous scores, we wished to account for an additional source of information, direct observations over DNA-protein interactions. A result of this kind of experiment can be viewed as an observation over element h_{ij} of matrix $\mathbf{H}$.

Here the observed gene expression spectra are considered a fixed set $\mathbf{X}^q$. Having these observations, we aim at choosing which link h_{ij} possess the highest mutual information for learning parameters $\mathbf{B}$. This can be represented in terms of the conditional mutual information, which is a function of two conditional entropies such that $I\left(\mathbf{B}; h_{ij}|\mathbf{X}^q\right) = H\left(\mathbf{B}|\mathbf{X}^q\right) - H\left(\mathbf{B}|\mathbf{X}^q, h_{ij}\right)$.

The conditional entropy is not a function of the selected link, so its computation is not necessary for discriminating between links. Then we introduce the utility function U_h equal to the negative conditional entropy of variable $\mathbf{B}$ given the gene expressions $\mathbf{X}^q$ and the observed link h_{ij}

$$U_h\left(\mathbf{B},\mathbf{X}^q,h_{ij}\right) = \sum_{\gamma\in\{0,1\}} p\left(h_{ij}=\gamma\right)\int p\left(\mathbf{B}|\mathbf{X}^q,h_{ij}=\gamma\right)\log p\left(\mathbf{B}|\mathbf{X}^q,h_{ij}=\gamma\right)d\mathbf{B}$$

where $p\left(\mathbf{B}|\mathbf{X}^q, h_{ij}=\gamma\right)$ is the posterior distribution over $\mathbf{B}$ given a fixed value for link h_{ij} (either 0 or 1).

We evaluate the integral by drawing samples from the conditional posterior $p\left(\mathbf{B}|\mathbf{X}^q, h_{ij}=\gamma\right)$, for $\gamma\in\{0,1\}$, and evaluating $\log p\left(\mathbf{B}|\mathbf{X}^q, h_{ij}=\gamma\right)$. We integrate by Monte Carlo method, with samples s_3 and s_4 drawn from the posterior distribution $p\left(\mathbf{B}|\mathbf{X}^q, h_{ij}=\gamma\right)$. As such the utility criterion is

$$U_h\left(\mathbf{B};\mathbf{X}^q;h_{ij}\right) = \frac{\sum_{s_3=1}^{S_3}\log p\left(\mathbf{B}^{(s_3)}|\mathbf{X}^q,h_{ij}=0\right)}{2S_3} + \frac{\sum_{s_3=1}^{S_4}\log p\left(\mathbf{B}^{(s_4)}|\mathbf{X}^q,h_{ij}=1\right)}{2S_4} \tag{12}$$

2.4 *A. thaliana* Circadian Clock Model

In [9] we observe a state of the art model of the *A. thaliana* circadian clock network. It consists of the transcription factors LHY/CCA1 LHY (LATE ELONGATED HYPOCOTYL) and CCA1 (CIRCADIAN CLOCK ASSOCIATED 1), these execute an activating interaction with the transcriptional co-regulators PRR9, PRR7 and PRR5/NI (PSEUDO-RESPONSE Regulators 9, 7, 5/night inhibitor) which at the same time are interlocked in a negative feedback loop with LHY/CCA1. This feedback loop is thought to be the responsible for peak activity of day-time components.

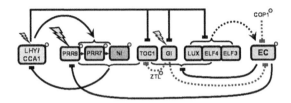

Fig. 2. Circadian clock model for *A. thaliana,* as shown in [9]. Transcriptional elements LHY, PRR579, GI, TOC1, LUX, ELF4 and ELF3 are assumed observed. While the expression levels of the Evening Complex (EC) is unobserved, along with other post-transcriptional interactions involving ZTL and COP1.

On the other hand we have the evening loop, thought to be driven by EC (Evening complex), composed by the binding of ELF3 (EARLY FLOWERING 3), ELF4 (EARLY FLOWERING 4) and the GARP transcription LUX (LUX ARRHYTHMO) which controls LHY expression by a double negative connection [9]. A graphical representation of the model is shown in Fig. 2.

3   Results

We simulate the *A. thaliana* circadian clock model, we selected and sub sampled the simulated data in order to get 12 samples over one light/dark cycle for a Wild Type population. We ran DSS and collected 10000 samples of the joint posterior over the model parameters. We executed DSS using standard parameters as in [11] and evaluated the mutual information criterion 10, we draw 1000 samples, thus setting parameter $S_1 = 1000$. We draw 100 samples for each gene expression level at each step, thus setting parameter $S_2 = 100$.

First, we chose photo-periods of 6/18, 8/16, 8/6 and 20/24, we computed the DFT of a $\{-1,1\}$ light input (ξ^q) and added it to the spectra matrix. Thus drawing samples from the conditional distribution $p\left(\mathbf{X}^q|\mathbf{B}, \sigma^2, \mathbf{U} = \xi^q\right)$.

Then we selected a set of knock out mutants commonly seen in experimental settings. In this way knock-out mutants ΔLHY, ΔLHY-GI, ΔLHY-TOC1 and ΔPRR7-PRR9 were simulated by conditioning the rest of the gene spectra given that the intervened genes have a constant spectrum of zero, that is $p\left(\mathbf{X}^q_{\backslash i}|\mathbf{B}, \sigma^2, \mathbf{X}^q_i = \mathbf{0}\right)$.

In Fig. 3 we present the results of evaluating Eq. 10 for these two set of experiments. The boxes go from the 25th to the 75th percentiles and the red bar indicates the median score. It shows photo-period experiments having a median score between 220 and 225, while the knock-out mutants show less median values ranging from 210 to 217. It is of interest that the lowest information gain looks to be accredited to the ΔLHY-TOC1 double mutant, being these two genes the main drivers of circadian oscillations. This may be due to the nature of the mutual information criterion, as it accounts for the reduction in uncertainty over the estimation of parameters. It seems plausible that the disruption of these two

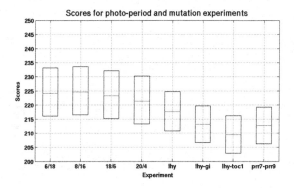

Fig. 3. Box plot for the evaluation of the DSS criterion, higher score means higher mutual information between experimental design and experimental outcomes. From left to right, photo periods of 6/18, 8/16, 18/6 and 20/4. Then knockout Mutants ΔLHY, ΔLHY-GI, ΔLHY-TOC1 and ΔPRR7-PRR9 (Color figure online).

components alters clock behavior enough that parameter inference is less reliable, as the score suggests that the uncertainty over the model behaviour grows. This may be in fact another source of information about the importance of these two clock components.

Complementary, we computed the conditional mutual information for Chip experiments according to Eq. 12. First we simulated Wild-type gene expression levels for 12 samples over a 24 hour period, using the same procedure as in the previous paragraph. Then, we selected a set of candidate links to observe, these include those known to be part of the true network, and those involving the EC components. Each one of these links was set to their possible values (one and zero), and the posterior distribution calculated for each case, this implies running DSS twice for each studied link with standard parameters as proposed in [11].

We show the resulting scores in Fig. 4. In this scatter plot, regulators are shown in the x axis, and the scores are presented through colored dots. Each dot is labeled according to the putative regulation tested (the regulators target is marked by a ->). Here we observe that the regulating interactions involving the elements of the EC complex (LUX, ELF4 and ELF3) as regulators show the lest information. This is not surprising as model assumptions are that the EC complex is the transcription factor involved in the evening regulation, and its effects even though essential, are not directly observable through its components. On the other hand we find that the most useful information seems to be related to the elucidation of the role of the light input over LHY and specially GI, with the highest score of 437, above of the mean value of 432.7. Another interesting interactions include that for LHY its most useful observation would be its regulation of TOC1, correspondingly, LHY would be the most informative interaction to observe for TOC1. As stated earlier, the interaction between these two components is the main driver of the morning oscillator.

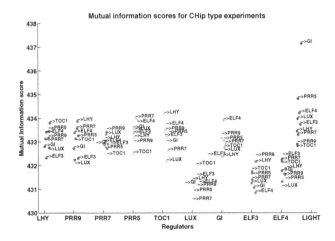

Fig. 4. Scatter plot of the conditional mutual information scores for observations over some edges. Each score is labeled with the represented interaction. The regulating interactions are symbolized by a "->" as "->targets", with the regulator being the label on the x axis tick. From left to right we have regulators LHY, PRR9, PRR7, PRR5, TOC1, LUX, GI, ELF3, ELF4 and photo-regulation in case of light inputs (Color figure online).

Taking in account these two complimentary criteria, some decisions about the utility of the experiments can be made. In these case, it seems to points towards light-related experiments, as the expected mutual information for all the photo-period experiments seems to be on par. This at the same time could be validated by the fact that light-input nodes of the network seem to be the most informative in first instances. Finally the LHY-TOC1 double mutant score suggest that the behaviour of the system under these circumstances is more uncertain, insight that may result useful for the researcher and thus an interesting experiment to execute.

4 Conclusions

We have presented a methodology for Bayesian experimental design in biological dynamical systems with structural uncertainty. Experimental design is a branch of classical computational statistics which is gaining increasing attention in systems biology, due to inherent complexity and uncertainty of biological systems. Adapting classical methods to modern systems biology is problematic, as sources of uncertainty are ubiquitous in systems biology data, leading to computationally intractable problems and/ or predictions with large associated uncertainty. In general, handling both parametric and structural uncertainty in nonlinear systems is highly problematic. Earlier work such as [5] chose to focus on non-linear systems without structural uncertainty. However, in many biological systems, such as oscillatory systems, it may be preferable to approximate the system dynamics to gain computational savings which will enable structural uncertainty to be considered in experimental design. Our results on the *A. thaliana* clock model show that this approach can be fruitful, highlighting potentially large differences in information content for different classes of experiments, and for different individual experiments in each class. These results are potentially precious for practitioners, whose prime preoccupation is often the prioritisation of experiments in the face of technical and resource limitations.

There are several directions along which the approach could be further developed. A simple, but potentially useful, extension would be to modify the utility function by explicitly accounting for the different costs of different experiments. It would also be of interest to develop strategies for planning multiple experiments, as the information gain is generally a non-linear function on the space of possible experiments. While the same apprxoach can be easily deployed for small sets of experiments, the general issue of multiple experimental design yields a very challenging discrete optimisation problem. We envisage that ideas from reinforcement learning could be effective in this scenario.

References

1. Bishop, C.M.: Bishop Pattern Recognition and Machine Learning. Springer, New York (2001)
2. Chaloner, K., Verdinelli, I.: Bayesian experimental design: a review. Statist. Sci. **3**, 273–304 (1995)
3. Ishwaran, H., Sunil Rao, J.: Spike and slab variable selection: frequentist and bayesian strategies. Ann. Stat. **33**(2), 730–773 (2005)
4. Kreutz, C., Timmer, J.: Systems biology: experimental design. FEBS J. **276**(4), 923–942 (2009)
5. Liepe, J., Filippi, S., Komorowski, M., Stumpf, M.P.H.: Maximizing the information content of experiments in systems biology. PLoS Comput. Biol. **9**(1), e1002888 (2013)
6. Lindley, D.V.: On a measure of the information provided by an experiment. Ann. Math. Statist. **4**, 986–1005 (1956)
7. MacKay, D.J.C.: Information Theory, Inference, and Learning Algorithms, vol. 7. Cambridge University Press, Cambridge (2003)
8. Pintelon, R., Schoukens, J.: System Identification: A Frequency Domain Approach, 2nd edn. Wiley-IEEE Press, New York (2012)
9. Pokhilko, A., Fernandez, A.P., Edwards, K.D., Southern, M.M., Halliday, K.J., Millar, A.J.: The clock gene circuit in Arabidopsis includes a repressilator with additional feedback loops. Mol. Syst. Biol. **8**, 574 (2012)
10. Pukelsheim, F.: Optimal Design of Experiments. Classics in Applied Mathematics. Society for Industrial and Applied Mathematics, Philadelphia (2006)
11. Trejo-Banos, D., Sanguinetti, G., Millar, A.J.: A Bayesian approach for structure learning in oscillating regulatory networks. arXiv:1504.06553 [stat.ML]

Efficient Stochastic Simulation of Systems with Multiple Time Scales via Statistical Abstraction

Luca Bortolussi[1,2,3]($\boxtimes$), Dimitrios Milios[4]($\boxtimes$), and Guido Sanguinetti[4,5]

[1] Modelling and Simulation Group, University of Saarland, Saarbrücken, Germany
[2] Department of Mathematics and Geosciences, University of Trieste, Trieste, Italy
[3] CNR/ISTI, Pisa, Italy
[4] School of Informatics, University of Edinburgh, Edinburgh, Scotland, UK
dmilios@inf.ed.ac.uk
[5] SynthSys, Centre for Synthetic and Systems Biology, University of Edinburgh, Edinburgh, Scotland, UK

Abstract. Stiffness in chemical reaction systems is a frequently encountered computational problem, arising when different reactions in the system take place at different time-scales. Computational savings can be obtained under time-scale separation. Assuming that the system can be partitioned into slow- and fast- equilibrating subsystems, it is then possible to efficiently simulate the slow subsystem only, provided that the corresponding kinetic laws have been modified so that they reflect their dependency on the fast system. We show that the rate expectation with respect to the fast subsystem's steady-state is a continuous function of the state of the slow system. We exploit this result to construct an analytic representation of the modified rate functions via statistical modelling, which can be used to simulate the slow system in isolation. The computational savings of our approach are demonstrated in a number of non-trivial examples of stiff systems.

1 Introduction

The presence of multiple scales, either temporal, spatial, or organisational, is one of the hallmarks of complexity of biological systems. Multi-scale systems present daunting challenges to their mathematical and computational treatment, as the cost of analysis and simulation is significantly increased. In order to tame such complexity, a common practice is to rely on abstraction techniques, simplifying some scales of the model, yet still capturing relevant features of the dynamics. Examples are the abstraction of the complex intra-cellular state as a finite state automaton, a typical approach to build cell population models, the abstraction

L. Bortolussi is partially supported by EU-FET project QUANTICOL (nr. 600708), by FRA-UniTS, and the German Research Council (DFG) as part of the Cluster of Excellence on Multimodal Computing and Interaction at Saarland University. D.M. and G.S. are supported by the ERC under grant MLCS 306999.

© Springer International Publishing Switzerland 2015
O. Roux and J. Bourdon (Eds.): CMSB 2015, LNBI 9308, pp. 40–51, 2015.
DOI: 10.1007/978-3-319-23401-4_5

of the local dynamics of epidemic spreading in country-level models [12], or the averaging of fast dynamics in enzyme kinetics [6,13]. The downside of such approaches is that the abstractions that are constructed are non-trivial and model-specific, and often require considerable efforts from the modellers.

In this paper, we explore the idea that model abstraction can be simplified by relying on statistical methodologies which can be learned automatically from (few) exploratory runs of the models. We focus on the specific sub-problem of multiple-time scales, related to stiffness, a well studied issue but still problematic, especially for stochastic systems. We build upon the two common theoretical frameworks of Quasi-Steady-State (QSSA) [11,13] and Quasi-Equilibrium (QE) [2] for stochastic models of chemical reaction networks. These approaches provide recipes to construct abstracted models, by decomposing a model in a fast and a slow subsystems (more time scales can be considered, but this generalisation is not considered here for simplicity). The fast subsystem is assumed to equilibrate at a time scale which is much faster than the characteristic time scale of the slow subsystem, hence it is abstracted by averaging out fast variables according to their equilibrium distribution, conditional on a fixed state of the slow subsystem. This averaging is performed on the kinetic rate functions of the slow subsystem. This theoretical recipe can produce accurate results, when the QSSA or QE assumptions are satisfied, yet it is very hard to obtain analytical expressions for the kinetic rates of the slow subsystem, which hinders its use in practice.

In this paper, we propose a method to circumvent the problem by exploiting ideas from machine learning, in particular Gaussian Processes [14], to learn the abstracted slow kinetic rates, as a function of slow variables. This approach allows us to construct statistical surrogates of the reduced rate functions in a fully automatic and computationally cheap way, without analytical efforts from the modeller side. It relies only on continuity properties of slow rates, which are also investigated in the paper. Such statistical abstraction of the slow model can then be used to perform simulation efficiently. In the paper we present the novel simulation algorithm, and assess its performance with respect to other slow scale simulation methods proposed in literature. Furthermore, our approach has another advantage: using the same learning strategy, and at a mild additional preprocessing cost, we can additionally learn slow rates as a function of some model parameters, enabling efficient parameter exploration in the stiffness regime.

The paper is organised as follows: in Sect. 2, we introduce the relevant background material and related work, as well as the QSSA and QE model reduction strategies. The continuity results and the statistical abstraction procedure, together with the resulting simulation algorithm, are presented in Sect. 4. Section 5 contains the experimental validation of the proposed approach, while final comments are discussed in Sect. 6. Throughout the paper we will use a simple enzyme-substrate model as a running example.

2   Background and Related Work

Chemical Reaction Networks. We will describe biochemical systems using the widespread formalism of (bio)Chemical Reaction Networks (CRN). The main entities involved are species and reactions.

- Each species represents a molecule described in the model; the vector $X(t) = (X_1(t), \ldots, X_n(t)) \in S \subseteq \mathbb{N}^n$ counts the number of molecules of each species in the system at time t.
- Reactions describe how the system state can change. Each reaction is of the form

$$r_1 X_1 + \ldots r_n X_n \xrightarrow{f(X)} s_1 X_1 + \ldots s_n X_n,$$

where the left hand side represent molecules that are consumed by the reaction, the right hand side describe which molecules are created, and $f(X)$ is the kinetic rate function, giving the speed of the reaction as a function of the system state. For each reaction R_j, we can define the vector r_j (respectively s_j), encoding how many agents are consumed (respectively produced) in the reaction, so that $v_j = s_j - r_j$ gives the net change of species.

We will consider the stochastic interpretation of biochemical reaction networks [9], in which the dynamics of the system is described by a Continuous-Time Markov Chain (CTMC), a Markovian (i.e. memoryless) stochastic process defined on a finite or countable state space S and evolving in continuous time [8]. In general, we can think of CTMCs as a collection of random variables $X(t)$ on the state space S, indexed by time $t \in [0, \infty)$.

Molecular systems described by CRN are located in a finite volume V, and one can reason on concentrations, rather than molecule numbers, by dividing variables by the volume V. We will indicate with capital letters X the molecular numbers and with small letters $x = X/V$ the concentrations. Rate functions can be expressed either in terms of molecular numbers or concentrations, modulo a rescaling of parameters [9]. We will denote with $f_j(X)$ and $f_j(x)$ the same rate function, expressed in molecular numbers or concentrations, respectively.

For most CRNs, it is impossible or prohibitively expensive to numerically solve the underlying CTMC directly, so it is a common practice to explore the system's behaviour via stochastic simulation. The standard simulation approach is known as the *Gillespie algorithm* [9], and it is exact in the sense that it simulates every single reaction event happening.

Running Example - Part I. We demonstrate the main concepts of the paper on a simple enzyme-substrate model [13]. The system state is represented as a vector $X = (X_E, X_S, X_{ES}, X_P)$ that denotes the populations for an enzyme E, a substrate S, the complex ES formed by the combination of the enzyme with the substrate, and a product P. The state can be changed by the reactions:

$$
\begin{aligned}
E + S &\xrightarrow{f_1(X)} ES, & f_1(X) &= c_1 X_E X_S \\
ES &\xrightarrow{f_2(X)} E + S, & f_2(X) &= c_2 X_{ES} \\
ES &\xrightarrow{f_3(X)} E + P, & f_3(X) &= c_3 X_{ES}
\end{aligned}
\tag{1}
$$

Related Work. The approach to model reduction exploiting time scale separation presented here falls within the scope of Quasi-Steady-State Approximation (QSSA) for stochastic models [3–6,10,13]. In these approaches, species are partitioned into fast and slow, and transitions are separated accordingly. Then, the fast system, conditional on the slow one, is averaged away assuming it is at steady state. The issue with all these approaches is that they require a-priori identification of fast and slow species, which is usually a choice left to intuition of the modeller. A similar approach, known as Quasi-Equilibrium [2], instead, starts by partitioning the transitions into fast and slow, and then separating species, possibly defining new species by taking a linear combination of the original ones. In both cases, the so obtained system satisfies the decomposition discussed in this section, hence our simulation algorithm can be applied.

A common characteristic of these earlier works on quasi-equilibrium reduction is that they rely on model-dependent expressions to calculate or approximate the rate expectations of the slow reactions, de facto limiting the applicability of the derived simulation algorithms [4–6]. In this work, we investigate the potential of automatically learning these expectations using a regression technique. Under the quasi-equilibrium assumption, our approach relies on no more assumptions regarding the form or the structure of the fast subsystem.

A generic approach to approximate the rate expectation for the slow reactions is prescribed in [15], where a *Nested Stochastic Simulation Algorithm* (Nested-SSA) is proposed to approximate the steady-state of the fast subsystem. We have implemented Nested-SSA following its description in the original paper, in order to produce some comparative results. The step parameter for Nested-SSA has been explored experimentally such that the efficiency of the two approximate simulation approaches has been roughly the same, in order to perform a fair comparison in terms of approximation quality. Another approach related to Nested-SSA has been recently proposed in [16].

3   Quasi-Equilibrium Reduction

Gillespie's exact simulation approach can have high computational costs in presence of stiffness, where a small number of reactions dominate computations. We will now introduce an approach to address such problems by partitioning the system in two separate subsystems with different time-scales. We will first discuss how to construct the reduced model, and then comment on how such fast and slow subsystems can be identified. We will make some strong assumptions on the structure, commenting later on how to relax them.

Partition of Species and Reactions. We assume that species $X = X_1, \ldots, X_n$ of the system can be partitioned in two disjoint subsets: fast species, denoted by $Y = Y_1, \ldots, Y_m$, and slow species, indicated with $Z = Z_1, \ldots, Z_s$, with $m + s = n$. Hence, the state space S is decomposed into the fast S_f and the slow S_s subspaces, so that $S = S_f \times S_s$. We will use this notation consistently in rates, writing $f_j(Y, Z)$ in place of $f_j(X)$.

Similarly, we assume that the set $\mathcal{R}$ of reactions is also partitioned into fast and slow subsets, denoted respectively $\mathcal{R}_{fast}$ and $\mathcal{R}_{slow}$. The idea is that fast reactions act only on fast variables (i.e. for each $R_j \in \mathcal{R}_{fast}$, v_j is zero in correspondence to slow variables), and quickly bring the fast subsystem to equilibrium. Hence, the evolution of slow variables will essentially sense the fast system only via its steady state distribution. Slow reactions, instead, can modify both fast and slow subsystems.

Reduced Model. Given a partition of species and reactions into fast and slow classes, we can construct the fast and the slow subsystems. The fast subsystem is defined conditionally on a fixed value of the slow variables $\mathbf{Z}$. It is a CRN with species $\mathbf{Y} = Y_1, \ldots, Y_m$ and reactions $\mathcal{R}_{fast}$. In particular, the rate functions of reactions in $\mathcal{R}_{fast}$ are computed by instantiating the slow variables with their fixed value. Here we assume that such kinetic rate functions depend on slow variables via their concentration, $f_j = f_j(\mathbf{Y}, \mathbf{z})$, hence the fast subsystem will be parameterised by the concentration $\mathbf{z}$ of slow species, which can take values in $\mathbb{R}_{\geq 0}^s$ or on a compact subset, if the state space S_s is finite. This dependency will be made explicit in the notation $\mathbf{Y}_{|\mathbf{z}}$.

At this stage, we need to make a crucial assumption for the method to work, namely that the conditional fast process $\mathbf{Y}_{|\mathbf{z}}(t)$ is an *irreducible* and *positive recurrent* CTMC on the fast subspace S_f, for any value of $\mathbf{z}$. This will guarantee *existence and uniqueness of the steady state distribution* $\mathbf{Y}_{|\mathbf{z}}(\infty)$ of $\mathbf{Y}_{|\mathbf{z}}(t)$. In the following, we will denote the conditional expectation of a function $f(\mathbf{Y}, \mathbf{z})$, with respect to the steady state distribution $\mathbf{Y}_{|\mathbf{z}}(\infty)$ of the conditional fast process by $\mathbb{E}_{|\mathbf{z}}[f(\mathbf{Y}, \mathbf{z})]$, to stress the fact that this will be a function of the concentration of slow species.

The slow subsystem, instead, is a CRN on the slow species $\mathbf{Z}$, with dynamics given by the slow reactions $\mathcal{R}_{slow}$ only. However, all reactions R_j in $\mathcal{R}_{slow}$ are modified by

1. removing fast species from the left and right hand side of the rule of R_j,[1]
2. replacing the rate function $f_j(\mathbf{Y}, \mathbf{z})$ by $\hat{f}_j(\mathbf{z}) = \mathbb{E}_{|\mathbf{z}}[f_j(\mathbf{Y}, \mathbf{z})]$, i.e. averaging out fast variables with respect to the steady state distribution of fast species, conditional on a given concentration of slow species.

Running Example: Part II. In the enzyme-substrate example, stiffness can easily arise if we assume that $c_1, c_2 \gg c_3$. In that case, the reactions in (1) can be partitioned into fast and slow subsets $\mathcal{R}_{fast} = \{R_1, R2\}$ and $\mathcal{R}_{slow} = \{R_3\}$ correspondingly. Consequently, we have fast species $\mathbf{Y} = (X_E, X_S, X_{ES})$ and slow species $\mathbf{Z} = (X_P)$. We therefore obtain the following fast subsystem:

$$
\begin{aligned}
E + S &\xrightarrow{f_1(\mathbf{Y}, \mathbf{z})} ES, & f_1(\mathbf{Y}, \mathbf{z}) &= c_1 X_E(N - X_{ES} - X_P) \\
ES &\xrightarrow{f_2(\mathbf{Y}, \mathbf{z})} E + S, & f_2(\mathbf{Y}, \mathbf{z}) &= c_2 X_{ES}
\end{aligned}
\tag{2}
$$

---

[1] This is a technically sound operation, as the fast subsystem has a unique steady state distribution, depending only on the state $\mathbf{z}$ of the slow subsystem, which is reached immediately after the firing of a slow reaction.

where N is a constant that denotes the total enzyme/substrate population in the system; in this way, the dependency on the slow system is reflected in the reaction rates. The slow subsystem is then described by the following reactions:

$$\emptyset \xrightarrow{\hat{f}_3(z)} P, \quad \hat{f}_3(z) = \mathbb{E}_{|z}[f_3(Y, z)] \tag{3}$$

4 Approximation of Rate Expectations

4.1 Continuity of Rates of the Slow System

We start by proving a crucial property for our method, namely that the rate functions of the reduced slow subsystem are continuous as a function of the concentration of slow species, taking values on the whole $\mathbb{R}^s_{\geq 0}$ (or on a compact connected subset). This property is a consequence of mild regularity properties of the original kinetic rate functions, and is captured by the following theorem, whose proof can be found in the appendix.

Theorem 1. *Let $f(Y, z)$ be a locally Lipschitz continuous function w.r.t. (normalised) slow variables. Assume that the fast process, conditional on a fixed concentration z of the slow variables, is irreducible and positive recurrent for each z. Then $\mathbb{E}_{|z}[f(Y, z)]$ is a continuous function of z.* □

Theorem 1 enables us to use powerful techniques based on statistical emulation, which will be discussed in the following subsection, and which are the key of our simulation algorithm.

4.2 Exploring Rate Expectation via Pre-simulation Runs

As discussed in Sect. 3, for many systems exhibiting time-scale separation, it is possible to obtain a good approximation of the system by introducing an auxiliary system where the time scales are separated. Hence, the slow variables are treated as *statistically independent* random variables from the fast variables, and the time-scale separation is equivalent to a mean-field approximation which replaces the true transition rates of the slow variables (which in general depend on the actual fast variables) with their averages with respect to the equilibrium distribution of the fast variables. While this approximation in principle offers huge computational savings, in practice for most systems the equilibrium distribution of the fast variables cannot be computed analytically, and its expectation can consequently be computed only from a set of simulations. Furthermore, in most cases the statistics of the equilibrium distribution of the fast variables will themselves depend on the slow variables. This feedback mechanism engenders stiffness which effectively negates the computational benefits of time-scale separation: for every simulation step in the slow variables, a whole (large) set of complete simulations for the fast variables must be executed to obtain reliable estimates of the equilibrium statistics of the fast variables.

A possible solution to this computational problem would be to explore the functional dependency of the equilibrium statistics of the fast variables on the state of the slow variables. This in principle would greatly facilitate computations, replacing the need for simulations of the fast variables with a lookup table for the statistics. However, in general the number of states visited by slow variables may be very high, resulting in a need for very long precomputing steps. To obviate this problem, we exploit the results of Sect. 4.1, which imply that the equilibrium statistics of the fast variables are a continuous function of the slow variables (rescaled to concentrations). This enables us to leverage powerful machine learning techniques to construct a statistical approximation to the equilibrium statistics from a potentially much smaller number of pre-simulation runs. We use Gaussian Processes (GP) regression, a flexible non-parametric Bayesian method for non-linear regression, although other methods are also possible in principle. GPs provide us with a fast analytical approximation to the unknown function from a set of precomputed values of the function; importantly, their flexibility guarantees that they can approximate arbitrarily well any continuous function [1]. We refer the reader to [14] for a comprehensive introduction to GP regression, which we do not provide for space reasons.

4.3   Stochastic Simulation via Statistical Abstraction

We propose a stochastic simulation algorithm via statistical abstraction (SA-SSA), which involves simulating the slow system only. The algorithm works in two phases. In an initialisation phase, we construct an analytical approximation of the rates of the slow subsystem. In the simulation phase, these approximate rates are used in place of the true slow kinetic rate functions to simulate the slow subsystem with standard Gillespie simulation [9]. As the simulation phase is standard, we shall focus on the first phase.

The construction of these analytic approximations during the *initialisation process* is broken down to two steps. The first step involves estimating the rate expectations $\hat{f}_j(\boldsymbol{z}), \forall R_j \in \mathcal{R}_{slow}$ for a grid of n population vectors, which correspond to n different states of the slow process. For each population vector, the fast subsystem is simulated until steady-state is reached, and the expectation of $f_j(\boldsymbol{Y}, \boldsymbol{z})$ is calculated as follows:

$$\hat{f}_j(\boldsymbol{z}) = 1/t_f \int_{t_0}^{t_0+t_f} f_j(\boldsymbol{Y}, \boldsymbol{z}) dt \tag{4}$$

where t_0 is the time required to reach equilibrium and t_f is sufficiently large to compute accurately the time average. This is estimated using a simple heuristic: the rate expectation is measured for regular subsequent time intervals, and steady-state is considered to have been reached if the change observed is less than 1%. Since we have assumed that the fast process is ergodic, there should be exactly one steady-state distribution, therefore the expectation can be calculated using a single trajectory for each of the n states. We stress that our approach is independent of the choice of the method to estimate the steady state, which can be safely replaced.

At the end of this pre-simulation process, we have a collection of n population vectors paired with n noisy observations of the rate expectation as a function of the state of the slow system. GP regression is a natural and fully automated choice to obtain estimations *for the the expectations for any point in the state-space*, since it transfers information across neighbouring points.

To comment on the cost of the initialisation process, we have to consider the cost of the pre-simulation runs and the regression step. One of the main assumptions of QE reduction is that steady-state is reached quickly for the fast subsystem, therefore pre-simulation avoids the excessive simulation of the fast system that occurs when stiffness is present. The cost of regression is dominated by the solution of a linear system, whose complexity is $O(n^2)$, where n is the number of training points[2]. This cost can be further reduced by employing sparse approximations to GPs, which is a subject well studied in the machine-learning community [14]. An important note on the initialisation cost is that it has to be paid only once, and then name trajectories can be efficiently sampled from the slow subsystem. If the rate expectations are learned as a function of the system parameters as well, then it is possible to approximate an entire family of stiff systems. The relationship between the initialisation cost and the computational savings achieved is demonstrated in the experiments of Sect. 5.

5   Experimental Evaluation

In order to demonstrate the computational savings and assess the approximation quality of our approach, we consider two stiff examples of bio-chemical reaction networks. We have generated samples from the distributions of the slow species, using both the standard Gillespie algorithm [9] and SA-SSA. The approximation quality is evaluated by in terms of the *histogram distance* between the samples from the exact and the approximate simulation process. To put the histogram distance in a context, this has to be compared with the corresponding self-distance. A distance value smaller than the self-distance implies that the two distributions are practically indistinguishable for a given number of samples. The self-distance is estimated using the following result of Cao &Petzold [7]: an upper bound for the average histogram self-distance is given by $\sqrt{(4K)/(\pi N)}$, for N samples and K intervals in the histogram. For the examples that follow, we consider $K = 50$.

5.1   Stiff Enzyme-Substrate Reaction

We perform numerical experiments on the enzyme-substrate example, given the partitioning described by Eqs. (2) and (3). We consider kinetic constants $c_1 = 0.01$, $c_2 = 1$ and $c_3 = 10^{-4}$, and initial state $\boldsymbol{X}_0 = (220, 3000, 0, 0)$. The rate expectation for R_3 in (3) has been approximated via GP regression. For the training set, we have sampled 1000 population values for the slow variable P between 0 and 3000.

---

[2] GP regression typically involves matrix inversion, but this can be avoided as we make no use of predictive variances.

Table 1. Enzyme-substrate model: histogram distances for 10^3 simulation runs (estimated self-distance: 0.252).

Time	P Nested-SSA	SA-SSA
5×10^4	0.290	0.246
10×10^4	0.250	0.204
18×10^4	1.016	0.160
20×10^4	0.940	0.142

The results of simulating the slow subsystem can be seen in Table 1, which summarises the histogram distances from the true distribution for the population of P, at four time-points. Most of the distances recorded are lower than the estimated upper bound for the average self-distance (i.e. 0.252). We also report the corresponding histogram distances for the Nested-SSA method of Weinan et al. [15], which was parametrised so that it has been as efficient as out method (see Table 2). For the given level of efficiency, our method resulted in lower values for the histogram distance in most cases. Most importantly, the simulation strategy that we propose has been significantly more efficient that exact Gillespie simulation, as can be seen in Table 2. We also report the time required for initialisation, which is broken down to pre-simulation runs, hyperparameter optimisation, and the training of the GP regression model.

Parameter Exploration. We demonstrate an example of learning the expected rates as a function of the slow state in combination with a parameter of the system. This practice allows us to pay the initialisation cost once and then simulate a range of stiff systems using our accelerated simulation approach. For the enzyme-substrate system we consider that c_1 varies in the range $[0.01, 1]$; note that for the values of c_1 considered, the system remains stiff, so the QE reduction is meaningful. We have randomly sampled a grid of 1000 values for $X_P \in [0, 3000]$ and $c_1 \in [0.01, 1]$, which was used as training set for a regression model. By fixing the parameter c_1 to a particular value, we were able to generate trajectories efficiently using SA-SSA. Table 3 summarises the relative mean error observed when approximating the mean value of X_P, for different

Table 2. Execution times in seconds for 10^3 simulation runs.

Method		Enzyme-substrate	Viral model
SA-SSA	Pre-simulation	0.291	26.11
	Hyperparam. opt.	1.484	1.68
	Training	0.080	0.05
	Total initialisation	1.855	27.84
	Simulation	153	316
Exact SSA		6947	2410
Nested-SSA		209	327

Table 3. Relative mean error values for approximating the mean value of X_P, for 10^3 simulation runs.

Time	P (RME)			
	$c_1 = 0.01$	$c_1 = 0.1$	$c_1 = 0.5$	$c_1 = 1$
5×10^4	1.83×10^{-3}	9.08×10^{-4}	2.35×10^{-3}	2.17×10^{-3}
10×10^4	1.20×10^{-3}	1.49×10^{-3}	1.94×10^{-3}	2.87×10^{-3}
18×10^4	8.04×10^{-4}	3.73×10^{-5}	4.49×10^{-4}	3.05×10^{-4}
20×10^4	9.13×10^{-4}	4.56×10^{-5}	6.06×10^{-5}	3.26×10^{-5}

values of c_1. The total initialisation time for our approach has been 3.562 sec. Parameter exploration via the standard Gillespie algorithm required 1911 sec, while SA-SSA required only 32 sec.

5.2   Viral Infection Model

We now consider is the viral infection model appeared in [11]. We present the following simplified version of the model which involves three species, the viral template T, the viral genome G, and the viral structural protein S:

$$
\begin{aligned}
T &\xrightarrow{f_1(X)} G + T, & f_1(X) &= k_1 X_T & T &\xrightarrow{f_4(X)} \emptyset, & f_4(X) &= k_4 X_T \\
G &\xrightarrow{f_2(X)} T, & f_2(X) &= k_2 X_G & S &\xrightarrow{f_5(X)} \emptyset, & f_5(X) &= k_5 X_S \\
T &\xrightarrow{f_3(X)} S + T, & f_3(X) &= k_3 X_T & G + S &\xrightarrow{f_6(X)} V, & f_6(X) &= k_6 X_G X_S
\end{aligned}
\tag{5}
$$

The system state is represented as a vector $X = (X_T, X_G, X_S)$. Regarding the model parameters, we follow [11]; for the kinetic constants we have: $k_1 = 1$, $k_2 = 0.025$, $k_3 = 1000$, $k_4 = 0.25$, $k_5 = 1.9985$ and $k_6 = 7.5e - 6$, and initial state $X_0 = (10, 0, 0)$. A random system trajectory can be seen in Fig. 1.

Based on the kinetic constants, we consider the set of fast reactions $\mathcal{R}_{fast} = \{R_3, R_5\}$ and slow reactions $\mathcal{R}_{slow} = \{R_1, R_2, R_4, R_6\}$. Therefore, the fast species will be $Y = (X_S)$, and we have slow species $Z = (X_G, X_T)$, give rise to the following fast and slow subsystems correspondingly:

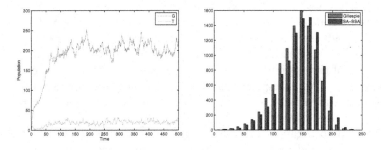

Fig. 1. Left: A random trajectory of the viral infection model, showing the slow species populations. Right: Distribution of the genome population X_G at $t = 50$.

Table 4. Viral infection model: histogram distances for 10^3 simulation runs (estimated self-distance: 0.252).

	G		T	
Time	Nested-SSA	SA-SSA	Nested-SSA	SA-SSA
50	0.988	0.308	0.548	0.242
100	0.244	0.414	0.154	0.226
200	0.388	0.406	0.156	0.204
500	0.346	0.432	0.198	0.238

$$\emptyset \xrightarrow{f_3(\boldsymbol{Y},z)} S, \quad f_3(\boldsymbol{Y},z) = k_3 X_T \qquad T \xrightarrow{f_1(z)} G + T, \quad f_1(z) = k_1 X_T$$
$$S \xrightarrow{f_5(\boldsymbol{Y},z)} \emptyset, \quad f_5(\boldsymbol{Y},z) = k_5 X_S \qquad G \xrightarrow{f_2(z)} T, \quad f_2(z) = k_2 X_G$$
$$T \xrightarrow{f_4(z)} \emptyset, \quad f_4(z) = k_4 X_T$$
$$G \xrightarrow{\hat{f}_6(z)} V, \quad \hat{f}_6(z) = \mathbb{E}_{|z}[f_6(\boldsymbol{Y},z)]$$

The rate of R_6 originally depends on X_G directly, and on X_T indirectly, since the population of T affects the steady-state of the fast process. We consider a random grid of 256 uniformly distributed population values for the genome G and the template T, given upper bounds of 500 and 100 molecules correspondingly. Note that a naïve exploration of the rate expectation would require 50000 evaluations, while we use only 256 for the training set of the GP.

The performance in terms of accuracy for the viral model is summarised in Table 4. We report the histogram distances for slow components, at four timepoints. An example of the histograms generated can be seen in Fig. 1 for the genome G, at time $t = 50$. We see that in all cases the distance from the true distribution is very close to the self-distance estimated for the given number of samples, a fact that implies a very good approximation of the stochastic properties for the slow system. The computational savings are also significant, as can be seen in Table 2.

6 Conclusions

Time-scale separation is a well studied approach to efficiently simulate systems that exhibit stiffness, where systems are partitioned into slow and fast subsystems. Nevertheless, most of the approaches proposed in the literature rely on the structure of the system to produce estimations for the rate expectations for the slow process. We have proposed SA-SSA as a generic approach to simulate the slow-scale subsystems, where these rate expectations are approximated via a machine learning method.

Experiments on examples of stiff systems show that SA-SSA requires a small initialisation cost and results in significant computational savings. For a given level of efficiency, SA-SSA achieved similar or better accuracy than Nested-SSA, whose premise is also a generic simulation framework for stiff systems. Besides any performance comparison, there is a qualitative difference between the two methods. Unlike Nested-SSA, our approach is not transparent with respect to the slow process, since it requires a rough estimate of the reachable state-space. However, the efficiency of SA-SSA is not affected by the complexity of the fast subsystem, in contrast with Nested-SSA, as any relevant cost in only paid during the initialisation phase. Moreover, it has been possible to learn the rate expectations as functions of the model parameters as well; we therefore obtain approximations for a family of systems, provided that these comply with the stiffness assumption.

References

1. Bortolussi, L., Milios, D., Sanguinetti, G.: Smoothed model checking for uncertain continuous time Markov chains. CoRR ArXiv, 1402.1450 (2014)
2. Bortolussi, L., Paškauskas, R.: Mean-field approximation and quasi-equilibrium reduction of markov population models (2014)
3. Bruna, M., Chapman, S.J., Smith, M.J.: Model reduction for slowfast stochastic systems with metastable behaviour. J. Chem. Phys. **140**(17), 174107 (2014)
4. Cao, Y., Gillespie, D.T., Petzold, L.: Multiscale stochastic simulation algorithm with stochastic partial equilibrium assumption for chemically reacting systems. J. Comput. Phys. **206**(2), 395–411 (2005)
5. Cao, Y., Gillespie, D.T., Petzold, L.R.: The slow-scale stochastic simulation algorithm. J. Chem. Phys. **122**(1), 14116 (2005)
6. Cao, Y., Gillespie, D.T., Petzold, L.R.: Accelerated stochastic simulation of the stiff enzyme-substrate reaction. J. Chem. Phys. **123**(14), 144917–12 (2005)
7. Cao, Y., Petzold, L.: Accuracy limitations and the measurement of errors in the stochastic simulation of chemically reacting systems. J. Comput. Phys. **212**(1), 6–24 (2006)
8. Durrett, R.: Essentials of Stochastic Processes. Springer, New York (2012)
9. Gillespie, D.T.: Exact stochastic simulation of coupled chemical reactions. J. Phys. Chem. **81**(25), 2340–2361 (1977)
10. Goutsias, J.: Quasi-equilibrium approximation of fast reaction kinetics in stochastic biochemical systems. J. Chem. Phys. **122**(18), 184102 (2005)
11. Haseltine, E.L., Rawlings, J.B.: Approximate simulation of coupled fast and slow reactions for stochastic chemical kinetics. J. Chem. Phys. **117**(15), 6959 (2002)
12. Mari, L., Bertuzzo, E., Righetto, L., Casagrandi, R., Gatto, M., Rodriguez-Iturbe, I., Rinaldo, A.: Modelling cholera epidemics: the role of waterways, human mobility and sanitation. J. R. Soc. Interface **9**(67), 376–388 (2011)
13. Rao, C.V., Arkin, A.P.: Stochastic chemical kinetics and the quasi-steady-state assumption: Application to the Gillespie algorithm. J. Chem. Phys. **118**(11), 4999 (2003)
14. Rasmussen, C.E., Williams, C.K.I.: Gaussian Processes for Machine Learning. MIT Press, Cambridge (2006)
15. Weinan, E., Liu, D., Vanden-Eijnden, E.: Nested stochastic simulation algorithm for chemical kinetic systems with disparate rates. J. Chem. Phys. **123**(19), 194107 (2005)
16. Zechner, C., Koeppl, H.: Uncoupled analysis of stochastic reaction networks in fluctuating environments. PLoS Comp. Bio. **10**(12), e1003942 (2014)

Approximate Bayesian Computation for Stochastic Single-Cell Time-Lapse Data Using Multivariate Test Statistics

Carolin Loos[1,2], Carsten Marr[1], Fabian J. Theis[1,2], and Jan Hasenauer[1,2(✉)]

[1] Helmholtz Zentrum München - German Research Center for Environmental Health, Institute of Computational Biology, Ingolstädter Landstr. 1, 85764 Neuherberg, Germany
[2] Technische Universität München, Center for Mathematics, Chair of Mathematical Modeling of Biological Systems, Boltzmannstr. 3, 85748 Garching, Germany
jan.hasenauer@helmholtz-muenchen.de

Abstract. Stochastic dynamics of individual cells are mostly modeled with continuous time Markov chains (CTMCs). The parameters of CTMCs can be inferred using likelihood-based and likelihood-free methods. In this paper, we introduce a likelihood-free approximate Bayesian computation (ABC) approach for single-cell time-lapse data. This method uses multivariate statistics on the distribution of single-cell trajectories. We evaluated our method for samples of a bivariate normal distribution as well as for artificial equilibrium and non-equilibrium single-cell time-series of a one-stage model of gene expression. In addition, we assessed our method for parameter variability and for the case of tree-structured time-series data. A comparison with an existing method using univariate statistics revealed an improved parameter identifiability using multivariate test statistics.

Keywords: Parameter estimation · Approximate Bayesian computation · Multivariate test statistics · Single-cell time-series

1 Introduction

Gene expression is known to be affected by different sources of stochasticity [5]. To study these sources, stochastic single-cell time-lapse data are collected [24]. To reveal the underlying mechanisms, models based on continuous time Markov chains (CTMCs) [8] are derived for these data. CTMCs describe the changes in number of molecules and account for intrinsic noise.

The dynamics of CTMCs depend on the process parameters, e.g. reaction rates for the generation of one mRNA. To estimate these parameters, likelihood-based methods can be used. These methods consider all possible paths of the stochastic process by evaluating the transition density, e.g. using the finite state projection [15]. As this is computationally demanding and merely tractable for simple processes and moderate system sizes, likelihood-free approaches have

© Springer International Publishing Switzerland 2015
O. Roux and J. Bourdon (Eds.): CMSB 2015, LNBI 9308, pp. 52–63, 2015.
DOI: 10.1007/978-3-319-23401-4_6

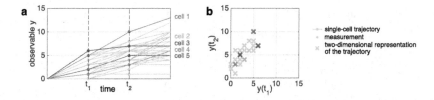

Fig. 1. Multidimensional representation of single-cell trajectories. (a) Single-cell trajectories of observable y. (b) Corresponding representation of the trajectory with time points t_1 and t_2 as scatter plot.

been developed, which are also called approximate Bayesian computation (ABC) methods [14]. ABC methods circumvent the evaluation of the likelihood. A parameter is accepted if the distance between simulated and measured data is sufficiently small, and rejected otherwise. The performance and convergence of ABC depends crucially on the employed distance measure.

Single-cell time-lapse data provide time-resolved information for individual cells. Conventional approaches for stochastic models rely on fitting these trajectories individually [4]. However, this becomes time consuming if the number of cells increases. In this study, we will take a population perspective to estimate parameters based on single-cell time-lapse data collected for instance using fluorescence microscopy. Assuming that one observable y is measured for an individual cell at two time points, the measured trajectories can be viewed as samples from a two-dimensional distribution and depicted as scatter plots (Fig. 1). For multiple observables n_y and/or more time points n_t the single-cell trajectories are samples from a "$n_y n_t$"-dimensional path distribution. Accordingly, distance measures for ABC are provided by multivariate test statistics.

This manuscript is structured as follows: In Sect. 2, we introduce the concept of ABC and present an approach for univariate single-cell snapshot data. For these data, cells are measured once and not tracked over time. In Sect. 3, we introduce two multivariate test statistics and evaluate our novel ABC method on a bivariate normal distribution. In Sect. 4, we apply our method to artificial time-lapse data of a one-stage model of gene expression, accounting for extrinsic cell-to-cell variability as well as for cell division. In Sect. 5, we summarise and discuss our results.

2   Introduction to Approximate Bayesian Computation

ABC is a likelihood-free method for parameter estimation. The method has been introduced by Pritchard et al. [17] in its most basic form, the ABC rejection algorithm. In this manuscript, we use an improved version of the algorithm, ABC with sequential Monte Carlo (ABC SMC), which samples a sequence of distributions with decreasing acceptance threshold ϵ_t [23]:

S1: Initialize ϵ_1 and set the population indicator $t = 1$.
S2.0: Set the particle indicator $i = 1$.

S2.1: If $t = 1$, sample $\boldsymbol{\theta}^{**}$ from the prior distribution $p(\boldsymbol{\theta})$.

Else, sample $\boldsymbol{\theta}^{*}$ from the previous population $\left\{\boldsymbol{\theta}_{t-1}^{(i)}\right\}$ with weights w_{t-1} and draw $\boldsymbol{\theta}^{**} \sim K_t(\boldsymbol{\theta}|\boldsymbol{\theta}^{*})$, with perturbation kernel K_t.
If $p(\boldsymbol{\theta}^{**}) = 0$, return to *S2.1*.
Simulate a candidate data set $\mathcal{D}^{\mathrm{sim}} \sim p(\mathcal{D}|\boldsymbol{\theta}^{**})$.
If $d(\mathcal{D}^{\mathrm{obs}}, \mathcal{D}^{\mathrm{sim}}) \geq \epsilon_t$, return to *S2.1*.

S2.2: Set $\boldsymbol{\theta}_t^{(i)} = \boldsymbol{\theta}^{**}$ and calculate the corresponding weight,

$$w_t^{(i)} = \begin{cases} 1, & \text{if } t = 1 \\ \dfrac{p(\boldsymbol{\theta}_t^{(i)})}{\sum_{j=1}^{P} w_{t-1}^{(j)} K_t(\boldsymbol{\theta}_t^{(i)}|\boldsymbol{\theta}_t^{(j)})}, & \text{if } t > 1. \end{cases}$$

If $i < P$, set $i = i + 1$, go to *S2.1*.

S3: Normalize the weights.
If $\epsilon_t > \epsilon_{\mathrm{end}}$, calculate ϵ_{t+1} e.g. by a quantile selection scheme (see e.g. [2]), set $t = t + 1$, go to *S2.0*.

The evaluation of the likelihood function is replaced by a comparison of observed and simulated data using the distance measure $d(\mathcal{D}^{\mathrm{obs}}, \mathcal{D}^{\mathrm{sim}})$. For the case of $\epsilon_1 = \epsilon_{\mathrm{end}}$ the above described algorithm yields the ABC rejection algorithm.

The generally low probability of observing the data set $p(\mathcal{D}^{\mathrm{obs}})$ yields low acceptance rates and therefore hinders the efficiency of ABC. Thus, lower dimensional summaries $\mathcal{S}$, such as moments, are often used instead of the full data set in the rejection step [3]. If $p(\boldsymbol{\theta}|\mathcal{D}^{\mathrm{obs}}) = p(\boldsymbol{\theta}|\mathcal{S}(\mathcal{D}^{\mathrm{obs}}))$, the summary statistic is sufficient and the true posterior can be obtained for $\epsilon_{\mathrm{end}} \to 0$ [16].

State-of-the-Art: ABC SMC for Single-Cell Snapshot Data

ABC methods have been successfully applied for the analysis of single-cell snapshot data collected e.g. using flow cytometry. The *Inference for Networks of Stochastic Interactions among Genes using High-Throughput data (INSIGHT)* algorithm has already been used for high-dimensional models [13]. Since cells are discarded after being measured in flow cytometry, the measurements at the n_t different time points are independent. The distance between observed and simulated data sets can be calculated in the ABC rejection step using the maximal Kolmogorov-Smirnov (KS) distance over all time points t_k:

$$d_{\mathrm{KS}}(\mathcal{D}^{\mathrm{obs}}, \mathcal{D}^{\mathrm{sim}}) := \max_{k \in \{1, \ldots, n_t\}} \|\hat{F}_{\mathbf{X}_k} - \hat{G}_{\mathbf{Y}_k}\|_{\infty}, \tag{1}$$

with $\mathcal{D}^{\mathrm{obs}} = \{\mathbf{X}_k\}_{k=1}^{n_t}$, $\mathcal{D}^{\mathrm{sim}} = \{\mathbf{Y}_k\}_{k=1}^{n_t}$, and $\hat{F}_{\mathbf{X}_k}, \hat{G}_{\mathbf{Y}_k}$ being the corresponding empirical cumulative distributions. Here, a sample $\mathbf{X}_k$ contains the fluorescence levels of the n single-cells for a time point that is indexed by k and each $\mathbf{Y}_k$ comprises m samples. INSIGHT achieves good results, benefiting from large sample sizes provided by flow cytometry, from using the two-sample Kolmogorov-Smirnov test to compare the data sets, and from exploiting relationships between configurations of the ABC algorithm and the test statistic.

We will adapt the idea of using test statistics for the development of an ABC method for single-cell time-series, which we later will compare with INSIGHT.

3   ABC with Multivariate Test Statistics

In the following, we develop an ABC method for single-cell time-lapse data using hypothesis testing [13,18]. A parameter is accepted as sample of the posterior distribution, if the observed data set and the simulated data set are drawn from the same distribution. This is indicated by a two-sample test.

3.1   Multivariate Test Statistics

For the case of data sets that comprise only one-dimensional samples, tests relying on the KS distance can be used. Since we want to apply ABC with test statistics to multivariate data, we need to find an appropriate multivariate test statistic for the two-sample problem (see [9] for an overview of multivariate two-sample tests).

Cross-match Test (CM). Rosenbaum presented the cross-match test for the multivariate two-sample problem [19]. A complete graph is defined, in which nodes correspond to samples and edge weights correspond to distances, e.g. the euclidean distance, between the samples. To obtain the test statistic a minimum weight non-bipartite matching is performed. The number of cross-matches, i.e., the matched pairs that comprise one observed and one simulated sample, is described by the random variable A_1. The null distribution of A_1 is

$$\Pr(A_1 = a_1) = \frac{2^{a_1}(\frac{n+m}{2})!}{\binom{n+m}{n}a_0!a_1!a_2!}, \tag{2}$$

with a_l being the number of matches with exactly l observed samples. For the case of the total number of samples $n + m$ being uneven see [19]. A higher number of cross-matches indicates a higher similarity of the data sets and we would accept the parameter that has been used to generate the simulated data set. If the number of cross-matches is small, the samples are likely drawn from different distributions and the corresponding parameters are rejected.

We implemented the cross-match test in MATLAB. For this, we integrated a blossom V algorithm[1] [12] to perform the minimum-weight non-bipartite matching, which requires $O((n + m)^3)$ arithmetic operations. The main advantage of the cross-match test is that it is distribution-free and exact, i.e., it does not make assumptions about the underlying distribution and the null distribution is known in closed form.

Maximum Mean Discrepancy Test (MMD). An alternative multivariate test statistic for the two-sample problem is based on the maximum mean discrepancy [9],

$$\mathrm{MMD}[\mathcal{F}, p, q] := \sup_{f \in \mathcal{F}} \left(\mathbb{E}_p[f(x)] - \mathbb{E}_q[f(y)] \right).$$

---

[1] Available at http://pub.ist.ac.at/~vnk/software.html.

If the distributions p and q are equal, the MMD is zero. Moreover, $\mathcal{F}$ is a class of functions $f : \mathcal{X} \to \mathbb{R}$ that is chosen to be the unit ball in a universal reproducing kernel Hilbert space $\mathcal{H}$, to achieve a trade-off between over- and underfitting. If $\mathcal{F}$ comprises not enough functions, the MMD may not be able to detect differences between the distributions p and q. Contrarily, if the class is too powerful, for $p = q$ the MMD may be significantly greater than zero for finite sample sizes. Given samples $\mathbf{X} = (\mathbf{x}_1, \ldots, \mathbf{x}_n)$ and $\mathbf{Y} = (\mathbf{y}_1, \ldots, \mathbf{y}_m)$ of p and q, respectively, an empirical estimate of the MMD is given by

$$\mathrm{MMD}[\mathcal{F}, \mathbf{X}, \mathbf{Y}] := \sup_{f \in \mathcal{F}} \left(\frac{1}{n} \sum_{i=1}^{n} f(\mathbf{x}_i) - \frac{1}{m} \sum_{j=1}^{m} f(\mathbf{y}_j) \right).$$

Using a kernel $k(\mathbf{x}, \mathbf{y}) = \Phi(\mathbf{x})^T \Phi(\mathbf{y})$ with nonlinear feature space mapping $\Phi(\mathbf{x})$, the MMD can be rewritten in terms of the mean embedding $\mu_p := E_p[\Phi(\mathbf{x})]$ as $\mathrm{MMD}[\mathcal{F}, p, q] = \sup_{f \in \mathcal{F}} \langle \mu_p - \mu_q, f \rangle = \|\mu_p - \mu_q\|_{\mathcal{H}}$. With $\mu_{\mathbf{X}} = \frac{1}{n} \sum_{i=1}^{n} \Phi(\mathbf{x}_i)$ and $k(\mathbf{x}, \mathbf{y}) = \langle \Phi(\mathbf{x}), \Phi(\mathbf{y}) \rangle$ the empirical estimate of the MMD is

$$\mathrm{MMD}[\mathcal{F}, \mathbf{X}, \mathbf{Y}] = \left(\frac{1}{n^2} \sum_{i \neq j}^{n} k(\mathbf{x}_i, \mathbf{x}_j) + \frac{1}{m^2} \sum_{i \neq j}^{m} k(\mathbf{y}_i, \mathbf{y}_j) - \frac{2}{nm} \sum_{i,j=1}^{n,m} k(\mathbf{x}_i, \mathbf{y}_j) \right)^{\frac{1}{2}}. \tag{3}$$

In this manuscript, we use a MATLAB implementation of the MMD with an adaptive Gaussian kernel, which has been developed by Gretton et al. [9]. The computational costs for the evaluation of (3) are $O((n + m)^2)$ and the test has shown to perform good even for low sample sizes and high-dimensional data. A connection to summary statistics is given by the fact, that a feature map of a kernel is a sufficient statistic for the exponential family [22].

3.2 Comparison of Test Statistics in ABC SMC for Samples of a Bivariate Normal Random Variable

In the following, we assess the properties of an ABC SMC algorithm using the aforementioned multivariate test statistics. We compare our method with the approach of INSIGHT [13], which neglects connections between different dimensions of the samples. We generate $n = 100$ samples $\mathbf{x}$ of a bivariate normal random variable with mean $\boldsymbol{\mu} = (0, 0)^T$ and a covariance matrix $\boldsymbol{\Sigma} = \left(\begin{smallmatrix} \theta_1 & \theta_2 \\ \theta_2 & \theta_1 \end{smallmatrix} \right)$, with $\theta_1 = 1$ and $\theta_2 = 0.5$ (Fig. 2a). As the covariance matrix has to be positive definite, we use the prior distribution

$$p(\boldsymbol{\theta}) = \begin{cases} \frac{1}{100} & , \text{for } 0 < \theta_1 < 10, 0 \leq \theta_2 < \theta_1 \\ 0 & , \text{otherwise} \end{cases},$$

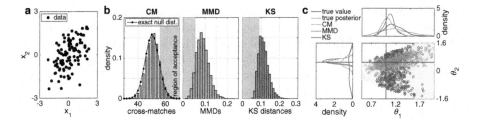

Fig. 2. ABC SMC using test statistics for samples of a bivariate normal distribution. (a) Depiction of 100 samples $\mathbf{x} \sim \mathcal{N}_2(\boldsymbol{\mu}, \boldsymbol{\Sigma})$. (b) Distribution of the CM, MMD and KS statistics for 10000 data sets generated with the true parameters. The gray regions indicate the values for which the parameter is accepted as sample of the final approximation. (c) Posterior approximations for θ_1 and θ_2 using ABC SMC with CM, MMD and KS. The yellow area shows where the prior $p(\boldsymbol{\theta}) > 0$.

for the parameters $\boldsymbol{\theta} = (\theta_1, \theta_2)^T$, which are estimated in the following. Since the efficiency of the algorithm depends on configurations such as the threshold schedule, we only compare the approaches in terms of convergence, i.e., whether it is possible to obtain a reasonable approximation, and not in terms of performance.

The final tolerances, i.e., the maximal allowed MMD $\epsilon_{\mathrm{MMD,end}} = 0.055$ and KS distance $\epsilon_{\mathrm{KS,end}} = 0.99$, and the number of cross-matches that needs to be exceeded $c_{\mathrm{end}} = 56$, are chosen as the 10th percentile of the distances obtained by simulating data and calculating the statistics with the ground truth. Figure 2b shows the distributions of the test statistics. The gray shaded area indicates for which values of the statistic a parameter is accepted in the final population of ABC SMC. For the cross-match test (CM), additionally the exact null distribution is visualized calculated with (2). Note that in contrast to the maximum mean discrepancy (MMD) and the Kolmogorov-Smirnov distance (KS), a high value indicates a good agreement. The results of ABC SMC are visualized in Fig. 2c. ABC SMC with MMD and CM is able to estimate the parameters. The confidence obtained using MMD is much higher than for CM. The KS approach provides an estimation of θ_1 only. The posterior approximation for θ_2 is much wider and only restricted by the relationship $|\theta_2| \leq \theta_1$. The difference can be explained by the lack of information included in the marginal distributions that are examined with KS. Information about θ_1 can only be gained by investigating the correlations among the measurements. The quality of the approximation did not improve significantly for lower tolerances. The CM test requires a higher computation time than the MMD and yields less accurate posterior approximations for the example of a bivariate normal distribution. Although we expect the sampler to converge to the true posterior when using CM and MMD for a tolerance level of zero, we use the MMD for our subsequent studies.

4   Simulation Example: Gene Expression

In this section we apply the ABC SMC scheme with MMD test statistics described before to a one-stage model of gene expression (Fig. 3a). For the generation of artificial data of the gene expression model we use a C implementation of the stochastic simulation algorithm (SSA). We implemented the previously described ABC SMC algorithm in MATLAB and use the following settings: We sample from the posterior distribution of the $\log_{10}$-transformed parameters, for which we use a uniform prior for each parameter, $p(\theta_i) = U[-6, 4]$. To compare observed and simulated data sets we use the maximal KS distance (1), which treats the single-cell time-lapse data as single-cell snapshot data, and the MMD (3), which also considers the tracking information of the data. For the threshold schedule an adaptive quantile approach is used with the 25th percentile. Furthermore, we implemented the k-nearest neighbor perturbation kernel proposed in [7], with $k = P/5$ and $P = 500$ particles per population. We increase the number of particles and repeat the approximation if we do not obtain a similar posterior approximation within three repetitions of the overall ABC SMC sampling. The final threshold ϵ_{end} is chosen in a data-driven fashion. Since we know the true parameters for the simulation study, we generate 1000 data sets using the true values and calculate the corresponding distances. We used the 5th percentile of these distances as final threshold.

4.1   Equilibrium and Non-Equilibrium Time-Series

For an initial evaluation of the proposed ABC SMC using multivariate statistics, we consider two scenarios: In Scenario 1, the initial mRNA number is zero $[\text{mRNA}](0) = 0$. In Scenario 2, the initial mRNA number is sampled from the equilibrium distribution $[\text{mRNA}](0) \sim \text{Poi}(\lambda/\gamma)$ [20]. For both scenarios we generate $n = 10, 100$ and 1000 single-cell time-series for the synthesis rate $\lambda = 5\,\text{h}^{-1}$ and degradation rate $\gamma = 0.3\,\text{h}^{-1}$ using the SSA. We simulate the system for $20\,\text{h}$ and record the mRNA at $n_t = 100$ equidistant time points. The data sets are visualized for the case of $n = 10$ cells in Fig. 3b–c. For the evaluation of our method we assume λ and γ to be unknown and estimate them from the data.

For Scenario 1, in which the population exhibits transient dynamics, both parameters are identifiable with MMD and KS test statistics (Fig. 4a–b). As expected, increasing the number of cells yields a narrower posterior distribution for both statistics. For Scenario 2, ABC SMC using the KS distance cannot

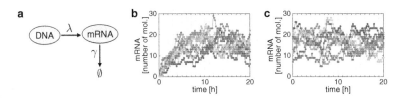

Fig. 3. Illustration of artifical single-cell time-lapse data. (a) Depiction of one-stage model of gene expression with mRNA synthesis rate λ and mRNA degradation rate γ. (b) Scenario 1: Out of steady state time-series of $n = 10$ cells sampled every $\frac{1}{5}$ h. (c) Scenario 2: Steady state time-series of $n = 10$ cells.

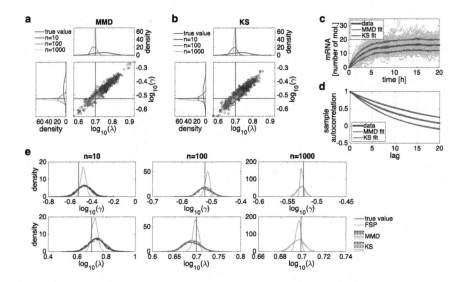

Fig. 4. Results for non-equilibrium time-series (Scenario 1). Posterior approximation obtained by ABC SMC with (a) MMD and (b) KS. (c) Fitted mean and variance of number of molecules for 1000 simulation generated with the MAP estimates. The individual trajectories are illustrated in gray. (d) Fitted mean and variance of the autocorrelation function. (e) Comparison with posteriors obtained using FSP. Different lines indicate different repetitions of ABC SMC.

infer the individual parameters but only the ratio (Fig. 5b). This is explained by the fact that the marginal distributions analysed using the KS distance do not change over time. In contrast, the proposed multivariate method using MMD exploits the temporal fluctuations and can infer both parameters (Fig. 5a).

For the case of 100 cells and 100 measurements, we generate 1000 time-series based on the maximum-a-posteriori (MAP) estimates and compare the mean and variance of the number of molecules for both scenarios (Figs. 4c and 5c) as well as mean and variance of the corresponding sample autocorrelation (Figs. 4d and 5d). The fits and the corresponding properties of the data are almost indistinguishable.

We additionally compare the posterior approximations with those obtained by the finite state projection (FSP) [15]. We sample from the posterior using a FSP-based likelihood (for further details see [1]) and the MCMC toolbox DRAM [10]. The results are shown in Figs. 4e and 5e, for Scenario 1 and 2, respectively. For Scenario 1, the posterior approximations using MMD and KS are similar (Fig. 4e). Both are wider than the approximation obtained with the FSP. For Scenario 2, the posterior distribution obtained by ABC SMC using KS is flat, since no information about the individual parameters can be gained (Fig. 5e). The posterior distribution obtained by ABC SMC using MMD has a higher discrepancy to the FSP than for Scenario 1. This indicates that the MMD can extract more information from the transient dynamics in Scenario 1 than from only the steady state fluctuations in Scenario 2.

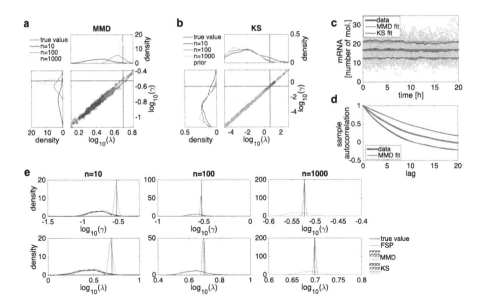

Fig. 5. Results for equilibrium time-series (Scenario 2). Posterior approximation obtained by ABC SMC with (a) MMD and (b) KS. (c) Fitted mean and variance of number of molecules for 1000 simulation generated with the MAP estimates. The individual trajectories are illustrated in gray. (d) Fitted mean and variance of the auto-correlation function. (e) Comparison with posteriors obtained using FSP. Different lines indicate different repetitions of ABC SMC.

4.2 Parameter Variability

Cell-to-cell variability of gene expression can be partitioned into intrinsic and extrinsic noise [5]. In Sect. 4.1, intrinsic noise has been considered, but the proposed approach can in principle also be used to infer extrinsic sources of cell-to-cell variability. In the following example, we model extrinsic noise by assuming variability in the mRNA synthesis and degradation rates. The parameters λ and γ are assumed to be log-normally distributed with means μ_λ, μ_γ and variances σ_λ^2, σ_γ^2. The data comprises 100 time-series measured at 100 time points. The true parameters used for the data generation are $\boldsymbol{\theta} = (\mu_\lambda, \sigma_\lambda^2, \mu_\gamma, \sigma_\gamma^2)^T = (5, 0.1, 0.3, 0.05)^T$. The time-series are depicted in Fig. 6a. The overall variability is higher than in the scenarios without additional variability (Figs. 3a and 4c).

The results obtained for ABC SMC are depicted in Fig. 6b. Here, also the intermediate distributions corresponding to different tolerance values are visualized showing the convergence of the algorithm. It reveals that the posterior distributions of the parameters μ_λ, μ_γ and σ_γ^2 are narrow, indicating identifiability. The posterior distribution for σ_λ^2 is wider and merely an upper bound can be determined. Accordingly, our analysis showed that in principle stochastic and deterministic variability can be reconstructed from single-cell time-lapse data.

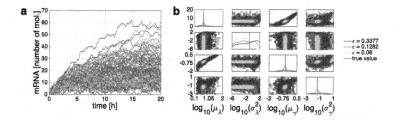

Fig. 6. ABC SMC for cells affected by extrinsic noise. (a) Time-series of cells with parameter variability. (b) Results obtained with MMD. On the diagonal, the marginal posterior distributions for the parameters are shown. The off diagonals provide scatter plots. The colours indicate the population corresponding to different tolerances ϵ and illustrate the convergence with decreasing ϵ.

4.3   Tree Structure

Single-cell time-lapse data often contain information about the ancestors of a cell [6]. We thus propose an approach to include tree-structured data in the ABC SMC sampler. We assume that a simple tree comprises one mother and its two daughter cells. One sample is given by $x_i = (x_{i,\text{mother}}, x_{i,\text{daughter}_1}, x_{i,\text{daughter}_2})$, as visualized in Fig. 7a. Since the samples need to have the same dimension when using MMD, we consider a fixed time interval before and after cell division. This is further motivated by the fact that the time-series exhibit transient dynamics after division, and therefore have a higher information content. Time-series of different lengths could also be interpolated and scaled to the same interval. To assess the quality of our method, we generate $n = 50$ simple trees (Fig. 7b) that each includes one division process. A cell, which is measured at 50 time points, divides after 10 h. The molecules are equally split among the daughter cells. Both daughters are simulated for 10 h and measured at 50 time points.

Figure 7c visualizes the posterior approximations for three repetitions of ABC SMC with MMD. The true value lies within the 90 %-credible interval. This demonstrates the applicability of our method to not only time-series, but also single-cell time-lapse data with additional tree structure. This approach allows to account for connections between the time-series of the mother and the daughter cells and e.g. parameters of the partitioning process could be estimated.

5   Discussion and Outlook

In this paper, we introduced and evaluated an ABC SMC method to infer parameters of CTMCs. Importantly, our method uses multivariate test statistics on the distribution of single-cell trajectories. We studied and compared MMD and CM multivariate statistics, and the univariate KS distance as used in INSIGHT [13]. ABC SMC with MMD provided the best posterior approximations. We found that for equilibrium single-cell time-lapse data the tracking information is important to identify the individual parameters.

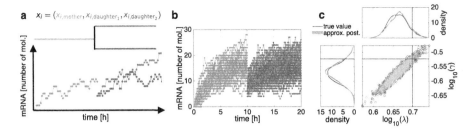

Fig. 7. ABC SMC using multivariate test statistics for tree-structured data. (a) Sample time-series of a mother and its two daughter cells. (b) Simulated data of 50 simple trees. (c) Posterior approximations with MMD. The joint posterior as well as the marginal posteriors of the parameters λ and γ are shown for three repetitions of the sampling with the same final tolerance.

A drawback of the method is the high computation time arising due to computationally expensive stochastic simulations. Thus, efficient simulation methods could be used instead of the SSA [8,11]. These should be combined with appropriate threshold schedules [21] and stopping criteria. So far, we merely used the test static value, but not the acceptance region of the hypothesis test based on a given confidence level. Since INSIGHT [13] benefits from exploiting relationships between configurations of the algorithm and the boundary for the test statistic, this could also be considered for multivariate statistics. This approach could possibly suffer from the low sample sizes of single-cell time-lapse data. In a follow up investigation, it would be worth to study how computation time can be saved by adapting the method to different numbers of observed and simulated samples ($m < n$). Furthermore, as more and more lineage information becomes available, its information content should be evaluated.

In summary, the proposed ABC SMC method using multivariate test statistics seems promising for the analysis of single-cell time-lapse data. It provides a flexible framework, which can easily be extended to similar data types. Using model selection, even sources of cell-to-cell variability might be unraveled.

Acknowledgements. The authors acknowledge financial support from the Postdoctoral Fellowship Program (PFP) of the Helmholtz Zentrum München. Furthermore, the authors thank Dennis Rickert, who provided the C-code implementation of the SSA.

References

1. Andreychenko, A., Mikeev, L., Spieler, D., Wolf, V.: Parameter identification for Markov models of biochemical reactions. In: Gopalakrishnan, G., Qadeer, S. (eds.) CAV 2011. LNCS, vol. 6806, pp. 83–98. Springer, Heidelberg (2011)
2. Beaumont, M.A., Cornuet, J.M., Marin, J.M., Robert, C.P.: Adaptive Approximate Bayesian Computation. Biometrika **96**(4), 983–990 (2009)
3. Beaumont, M.A., Zhang, W., Balding, D.J.: Approximate Bayesian computation in population genetics. Genetics **162**(4), 2025–2035 (2002)

4. Dargatz, C.: Bayesian inference for diffusion processes with application in life sciences. Ph.D. thesis, LMU Munich (2010)
5. Elowitz, M.B., Levine, A.J., Siggia, E.D., Swain, P.S.: Stochastic gene expression in a single cell. Science **297**(5584), 1183–1186 (2002)
6. Etzrodt, M., Endele, M., Schroeder, T.: Quantitative single-cell approaches to stem cell research. Cell Stem Cell **15**(5), 546–558 (2014)
7. Filippi, S., Barnes, C.P., Cornebise, J., Stumpf, M.P.: On optimality of kernels for approximate Bayesian computation using sequential Monte Carlo. Stat. Appl. Genet. Mol. **12**(1), 87–107 (2013)
8. Gillespie, D.T.: Stochastic simulation of chemical kinetics. Ann. Rev. Phy. Chem. **58**, 35–55 (2007)
9. Gretton, A., Borgwardt, K.M., Rasch, M.J., Schölkopf, B., Smola, A.: A kernel two-sample test. J. Mach. Learn. Res. **13**(1), 723–773 (2012)
10. Haario, H., Laine, M., Mira, A., Saksman, E.: DRAM: efficient adaptive MCMC. Stat. Comput. **16**(4), 339–354 (2006)
11. Hasenauer, J., Wolf, V., Kazeroonian, A., Theis, F.J.: Method of conditional moments (MCM) for the Chemical Master Equation: a unified framework for the method of moments and hybrid stochastic-deterministic models. J. Math. Biol. **69**(3), 687–735 (2014)
12. Kolmogorov, V.: Blossom V: a new implementation of a minimum cost perfect matching algorithm. Math. Program. Comput. **1**(1), 43–67 (2009)
13. Lillacci, G., Khammash, M.: The signal within the noise: efficient inference of stochastic gene regulation models using fluorescence histograms and stochastic simulations. Bioinformatics **29**(18), 2311–2319 (2013)
14. Marin, J.M., Pudlo, P., Robert, C.P., Ryder, R.J.: Approximate Bayesian computational methods. Stat. Comput. **22**(6), 1167–1180 (2012)
15. Munsky, B., Khammash, M.: The finite state projection algorithm for the solution of the chemical master equation. J. Chem. Phys. **124**(4), 044104 (2006)
16. Nunes, M.A., Balding, D.J.: On optimal selection of summary statistics for approximate Bayesian computation. Stat. Appl. Genet. Mol. **9**(1), Article 34 (2010)
17. Pritchard, J.K., Seielstad, M.T., Perez-Lezaun, A., Feldman, M.W.: Population growth of human Y chromosomes: a study of Y chromosome microsatellites. Mol. Biol. Evol. **16**(12), 1791–1798 (1999)
18. Ratmann, O., Camacho, A., Meijer, A., Donker, G.: Statistical modelling of summary values leads to accurate approximate Bayesian computations. arXiv preprint arXiv:1305.4283 (2013)
19. Rosenbaum, P.R.: An exact distribution-free test comparing two multivariate distributions based on adjacency. J. R. Stat. Soc. Series B Stat. Methodol. **67**(4), 515–530 (2005)
20. Shahrezaei, V., Swain, P.S.: Analytical distributions for stochastic gene expression. Proc. Natl. Acad. Sci. **105**(45), 17256–17261 (2008)
21. Silk, D., Filippi, S., Stumpf, M.P.: Optimizing threshold-schedules for sequential approximate Bayesian computation: applications to molecular systems. Stat. Appl. Genet. Mol. **12**(5), 603–618 (2013)
22. Song, L.: Learning via Hilbert space embedding of distributions. Ph.D. thesis, University of Sydney (2008)
23. Toni, T., Welch, D., Strelkowa, N., Ipsen, A., Stumpf, M.P.: Approximate Bayesian computation scheme for parameter inference and model selection in dynamical systems. J. R. Soc. Interface **6**(31), 187–202 (2009)
24. Wilkinson, D.J.: Stochastic modelling for quantitative description of heterogeneous biological systems. Nat. Rev. Genet. **10**(2), 122–133 (2009)

Stochastic Analysis of Chemical Reaction Networks Using Linear Noise Approximation

Luca Cardelli[1,2], Marta Kwiatkowska[2], and Luca Laurenti[2(✉)]

[1] Microsoft Research, Cambridge, UK
[2] Department of Computer Science, University of Oxford, Oxford, UK
luca.laurenti@cs.ox.ac.uk

Abstract. Stochastic evolution of Chemical Reactions Networks (CRNs) over time is usually analysed through solving the Chemical Master Equation (CME) or performing extensive simulations. Analysing stochasticity is often needed, particularly when some molecules occur in low numbers. Unfortunately, both approaches become infeasible if the system is complex and/or it cannot be ensured that initial populations are small. We develop a probabilistic logic for CRNs that enables stochastic analysis of the evolution of populations of molecular species. We present an approximate model checking algorithm based on the Linear Noise Approximation (LNA) of the CME, whose computational complexity is independent of the population size of each species and polynomial in the number of different species. The algorithm requires the solution of first order polynomial differential equations. We prove that our approach is valid for any CRN close enough to the thermodynamical limit. However, we show on three case studies that it can still provide good approximation even for low molecule counts. Our approach enables rigorous analysis of CRNs that are not analyzable by solving the CME, but are far from the deterministic limit. Moreover, it can be used for a fast approximate stochastic characterization of a CRN.

1 Introduction

Chemical reaction networks (CRNs) and mass action kinetics are well studied formalisms for modelling biochemical systems. In recent years, CRNs have also been successfully used as a formal programming language for biochemical systems. There are two well established approaches for analyzing chemical networks: deterministic and stochastic. The deterministic approach models the kinetics of a CRN as a system of ordinary differential equations (ODEs) and represents average behaviour, valid in the thermodynamic limit [8]. The stochastic approach, on the other hand, is based on the Chemical Master Equation (CME) and models the CRN as a continuous-time Markov chain (CTMC) [7]. The stochastic behavior can be analyzed by stochastic simulation [9] or by exhaustive probabilistic

This research is supported by a Royal Society Research Professorship and ERC AdG VERIWARE.

model checking of the CTMC, which can be performed, for example, by using PRISM [12].

Exhaustive analysis of the CTMC is able to find the best- and worst-case scenarios and is correct for any population size, but suffers from the state-space explosion problem and can only be used for relatively small systems. In contrast, deterministic methods are much more robust with respect to state-space explosion, but unable to represent stochastic fluctuations, which play a fundamental role when the system is not in thermodynamic equilibrium.

Contributions. In this paper we develop a novel approach for analysing the stochastic evolution of a CRN based on the Linear Noise Approximation (LNA) of the CME. We formulate SEL (Stochastic Evolution Logic), a probabilistic logic for CRNs that enables reasoning about probability, expectation and variance of linear combinations of populations of the species. Examples of properties that can be specified in our logic are shown in Example 1. We propose an approximate model checking algorithm for the logic based on the LNA and implement it in Matlab and Java. We demonstrate that the complexity of model checking is polynomial in the initial number of species and independent of the initial molecule counts, thus ameliorating state-space explosion. Further, we show that model checking is exact when approaching the thermodynamic limit. Though the algorithm may not be accurate for systems far from the deterministic limit, this generally happens when the populations are small, in which case the analysis can be performed by transient analysis of the induced CTMC [11]. Our approach is essential for CRNs that cannot be analyzed by (partial) state space exploration, because of large or infinite state spaces. Moreover, it is useful for a fast (approximate) stochastic characterization of CRNs, since solving the LNA is much faster than solving the CME [6]. We prove asymptotic correctness of LNA-based model checking and show on three examples that it is still possible to obtain very good approximations even for small population systems, comparing with standard uniformisation [11] and statistical model checking implemented in PRISM.

Related Work. Bortolussi *et al.* [1] uses the Central Limit Approximation (CLA) (essentially the same as the LNA) for checking restricted timed automata specifications and they assume fixed population size. Wolf *et al.* [16] develop a sliding window method to approximately verify infinite-state CTMCs, which applies to cases where most of the probability mass is concentrated in a confined region of the state space. This method applies to the induced CTMC, but require at least partial exploration of the state space, and is thus not immune to state-space explosion.

Structure of the Paper. In Sect. 2 we summarise the deterministic and stochastic modelling approaches for CRNs, and in Sect. 3 we describe the Linear Noise Approximation method. Section 4 introduces the logic SEL and the corresponding model checking algorithm based on the LNA. In Sect. 5 we demonstrate our approach on three networks taken from the literature. Section 6 concludes the paper.

2   Chemical Reaction Networks

A *chemical reaction network (CRN)* $C = (\Lambda, R)$ is a pair of finite sets, where Λ is the set of *chemical species* and R the set of reactions. $|\Lambda|$ denotes the size of the set of species. A *reaction* $\tau \in R$ is a triple $\tau = (r_\tau, p_\tau, k_\tau)$, where $r_\tau, p_\tau \in \mathbb{N}^{|\Lambda|}$ and $k_\tau \in \mathbb{R}_{>0}$. r_τ and p_τ represent the stoichiometry of reactants and products and k_τ is the coefficient associated to the rate of the reaction; its dimension is s^{-1}. We often write reactions as $\lambda_1 + \lambda_3 \rightarrow^{k_1} 2\lambda_2$ instead of $\tau_1 = ([1,0,1]^T, [0,2,0]^T, k_1)$, where $\cdot^T$ indicates the transpose of a vector. We define the *net change* associated to a reaction τ by $\upsilon_\tau = p_\tau - r_\tau$. For example, for τ_1 as above, we have $\upsilon_{\tau_1} = [-1, 2, -1]^T$.

We make the assumption that the system is well stirred, that is, the probability of the next reaction occurring between two molecules is independent of the location of those molecules. We consider fixed volume V and temperature; under these assumptions a *configuration* or *state* $x \in \mathbb{N}^{|\Lambda|}$ of the system is given by the number of molecules of each species. We define $[x] = \frac{x}{N}$, the vector of the species *concentration* in x for a given N, where $N = V \cdot N_A$ is the volumetric factor, V is the volume of the solution and N_A is Avogadro's number. The physical dimension of N is $Mol^{-1} \cdot L$, where Mol indicates mole and L is litre. Given $\lambda_i \in \Lambda$ then $\#\lambda_{i}\_x \in \mathbb{N}$ represents the number of molecules of λ_i in x and $[\lambda_i]\_x \in \mathbb{R}$ the concentration of λ_i in the same configuration. In some cases we elide x, and we simply write $\#\lambda_i$ and $[\lambda_i]$ instead of $\#\lambda_{i}\_x$ and $[\lambda_i]\_x$. They are related by $[\lambda_i] = \frac{\#\lambda_i}{N}$. The dimension of $[\lambda_i]$ is $Mol \cdot L^{-1}$.

The propensity $\alpha_{n,\tau}$ of a reaction τ in terms of the number of molecules is a function of the current configuration of the system x such that $\alpha_{n,\tau}(x)dt$ is the probability that a reaction event occurs in the next infinitesimal interval dt. In this paper we assume as valid the stochastic form of the law of mass action, so the propensity rates are proportional to the number of molecules that participate in the reaction. Stochastic models consider the system in terms of numbers of molecules, while deterministic ones, generally, in terms of concentrations, and the relationship is as follows. For a reaction $\tau = (r_\tau, p_\tau, k_\tau)$, given the configuration x and $r_{\tau,i}$, the i-th component of r_τ, then $\alpha_{c,\tau}(x) = k_\tau \prod_{i=1}^{|\Lambda|} \left([\lambda_i]\_x\right)^{r_{\tau,i}}$ is the propensity function expressed in terms of concentrations as given by the deterministic law of mass action. It is possible to show that, for any order of reaction, $\alpha_{n,\tau}(x) \approx N\alpha_{c,\tau}(x)$ if N is sufficiently large. Note that $\alpha_{c,\tau}$ is independent of N. In this paper we are interested only in finite time horizon, because of the problematic character of studying solutions of ODEs for infinite time horizon.

Example 1. Consider the CRN $C = (\{\lambda_1, \lambda_2, \lambda_3\}, R)$, where $R = \{(\lambda_1 + \lambda_2 \rightarrow^{10} \lambda_2 + \lambda_2), (\lambda_2 + \lambda_3 \rightarrow^{10} \lambda_3 + \lambda_3)\}$, with initial conditions $\#\lambda_1 = 98, \#\lambda_2 = 1, \#\lambda_3 = 1$, for a system with $N = 1000$. Figure 1 plots the expectation and standard deviation of population sizes. We may wish to check if the maximum expected value of $\#\lambda_2$ remains smaller than 75 molecules during the first $2\,s$. However, the system is stochastic, so we also need to analyse whether the variance is limited enough when $\#\lambda_2$ reaches the maximum. Sometimes, analysis of first and second moments does not suffice, so it could be of interest to check

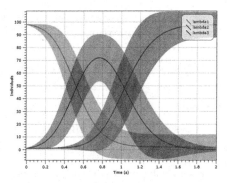

Fig. 1. Expected number and standard deviation of species of the CRN of Example 1 for the given initial conditions, calculated by simulating the CME.

the probability of some events, for instance, is the probability that, between $t_1 = 0.5\,s$ and $t_2 = 1.0\,s$, $\#\lambda_2 - (\#\lambda_1 + \#\lambda_3) > 0$ greater than 0.6?

Deterministic Semantics. Let $C = (\Lambda, R)$ be a CRN. The deterministic model approximates the concentration of the species of the system over time as a set of autonomous polynomial first order differential equations:

$$\frac{d\Phi(t)}{dt} = F(\Phi(t)) \tag{1}$$

$F(\Phi(t)) = \sum_{\tau=(r_\tau,p_\tau,k_\tau)\in R} v_\tau \alpha_{c,\tau}(\Phi(t))$ and $\alpha_{c,\tau}(\Phi(t)) = k_\tau \prod_{i=1}^{|\Lambda|} \Phi_i(t)^{r_{\tau,i}}$.
Function $\Phi : \mathbb{R}_{\geq 0} \rightarrow \mathbb{R}^{|\Lambda|}$ describes the behaviour of the system as a set of deterministic equations assuming a continuous state-space semantics, therefore $\Phi(t) \in \mathbb{R}^{|\Lambda|}$ is the vector of the species concentrations at time t. Assuming $t_0 = 0$, the initial condition is $\Phi(0) = [x_0]$, expressed as a concentration. Note that $F(\Phi(t))$ is Lipschitz continuous, so Φ exists and is unique [7].

Stochastic Semantics. CRNs are well represented by CTMCs, whose transient analysis can be performed via the Chemical Master Equation (CME) [14].

Definition 1. *Given a CRN $C = (\Lambda, R)$ and the volumetric factor N, we define a time-homogeneous CTMC $(X^N(t), t \in \mathbb{R}_{\geq 0})$ with state space $S = \mathbb{N}^{|\Lambda|}$. Given $x_0 \in S$, the initial configuration of the system, then $P(X^N(0)=x_0)= 1$. The transition rate from state x_i to state x_j is defined as $r(x_i, x_j) = \sum_{\{\tau \in R | x_j = x_i + v_\tau\}} N\alpha_{c,\tau}(x_i)$.*

$X^N(t)$ describes the stochastic evolution of molecular populations of each species at time t. For $x \in S$, we define $P^{(t)}(x) = P(X^N(t) = x | X(0) = x_0)$, where x_0 is the initial configuration. The CME describes the time evolution of X^N as:

$$\frac{d}{dt}\left(P^{(t)}(x)\right) = \sum_{\tau \in R}\{N\alpha_{c,\tau}(x - v_\tau)P^{(t)}(x - v_\tau) - N\alpha_{c,\tau}(x)P^{(t)}(x).\} \tag{2}$$

The CME can be equivalently defined in terms of the infinitesimal generator matrix, which admits computing an approximation of the CME using, for example, the sliding window method [16].

We also define the CTMC $(\frac{X^N(t)}{N}, t \in \mathbb{R}_{\geq 0})$ with state space $S = \mathbb{Q}^{|\Lambda|}$. If $[x_0] \in S$ is the initial configuration, then $P(\frac{X^N(0)}{N} = [x_0]) = 1$. The transition rate from state $[x_i]$ to $[x_j]$ is defined as $r([x_i], [x_j]) = \sum_{\{\tau \in R | [x_j] = [x_i] + \frac{v_\tau}{N}\}} N\alpha_{c,\tau}(x_i)$. $\frac{X^N(t)}{N}$ is the random vector describing the system at time t in terms of concentrations. In [7] it is proved that $\lim_{N \to \infty} \sup_{t' \leq t} \| \frac{X^N(t')}{N} - \Phi(t') \| = 0$ almost surely for every time t. This explains the relationship between the two different semantics, where the deterministic solution can be viewed as a limit of the stochastic solution, valid when close enough to the thermodynamic limit.

3   Linear Noise Approximation

The solution of the CME can be computationally expensive, or even infeasible, because the set of reachable states can be huge or infinite. The Linear Noise Approximation (LNA) has been introduced by Van Kampen as a second order approximation of the system size expansion of the CME [14]. Since stochastic fluctuations depend on N, and specifically, for average concentrations, are of the order of $N^{\frac{1}{2}}$ [6], to derive the expansion Van Kampen assumes that:

$$X^N(t) \approx N\Phi(t) + N^{\frac{1}{2}}Z(t) \tag{3}$$

where $Z(t) = (Z_1(t), Z_2(t), \ldots, Z_{|\Lambda|})$ is the random vector, independent of N, representing the stochastic fluctuations, $\Phi(t)$ is given by the solution of Eq. (1) and $X^N(t)$ is the random vector of Definition 1. Using this substitution in the system size expansion and then truncating at the second order, the probability distribution of $Z(t)$ is found to be given by the following linear Fokker-Plank equation [6]:

$$\frac{\partial P(Z, t)}{\partial t} = -\sum_{i=1}^{|\Lambda|}\sum_{j=1}^{|\Lambda|} \frac{\partial F_j(\Phi(t))}{\partial \Phi_i} \frac{\partial (Z_j P(Z, t))}{\partial Z_i} + \frac{1}{2}\sum_{i=1}^{|\Lambda|}\sum_{j=1}^{|\Lambda|} G_{i,j}(\Phi(t)) \frac{\partial^2 P(Z, t)}{\partial Z_i \partial Z_j}$$

$$\tag{4}$$

where $G(\Phi(t)) = \sum_{\tau \in R} v_\tau v_\tau^T \alpha_{c,\tau}(\Phi(t))$ and $F_j(\Phi(t))$ is the j−th component of $F(\Phi(t))$. The solution of Eq. (4) gives a Gaussian process. For every time t, $Z(t)$ has a multivariate normal distribution, whose expected value and covariance matrix are the solution of the following equations [6]:

$$\frac{dE[Z(t)]}{dt} = J_F(\Phi(t))E[Z(t)] \tag{5}$$

$$\frac{dC[Z(t)]}{dt} = J_F(\Phi(t))C[Z(t)] + C[Z(t)]J^T{}_F(\Phi(t)) + G(\Phi(t)) \tag{6}$$

where $J_F(\Phi(t))$ is the Jacobian of $F(\Phi(t))$. We consider as initial conditions $E[Z(0)] = 0$ and $C[Z(0)] = 0$. This means that $E[Z(t)] = 0$ for every t.

It is possible to justify the hypothesis (3) noting that in the lowest order the CME expansion reduces to Eq. (1), and with the following theorem by Kurtz:

Theorem 1. *[7] Consider the subset* $E \subset \mathbb{R}^{|A|}$ *on which are defined the propensity functions* $\alpha_{c,\tau}$. *Let* $Z^N(t)$ *be the random vector given by* $Z^N(t) = N^{\frac{1}{2}}(\frac{X^N(t)}{N} - \Phi(t))$. *Suppose that* $\sum_{\tau \in R} |v_\tau{}^2| \sup_{X \in K} \alpha_{c,\tau}(X) < \infty$ *for each compact* $K \subset E$, *and that, for* $N \to \infty$, $Z^N(0) = Z(0)$, *then* $Z^N(t)$ *converges in distribution to* $Z(t)$.

The LNA thus permits approximation of the probability distribution of $X^N(t)$ with the probability distribution of $Y^N(t) = N\Phi(t) + N^{\frac{1}{2}}Z(t)$. It is easy to show that $Y^N(t)$ has a Gaussian distribution; indeed, $Z(t)$ is Gaussian distributed, and N and $\Phi(t)$ are deterministic.

To compute the LNA it is necessary to solve $O(|A|^2)$ first order differential equations, but the complexity is independent of the initial number of molecules of each species. Therefore, one can avoid the exploration of the state space that methods based on uniformisation rely upon.

Theorem 1 alone only guarantees convergence in distribution. However, in [15], LNA is derived as an approximation of the Chemical Langevin Equation (CLE) [8], rather than system size expansion. This shows that LNA is valid for every real chemical system close enough to the thermodynamical limit, at least for a limited time. Thus, LNA is exact in the limit of high populations, but can also be used for small populations if the behaviour is not too far from the deterministic limit, taking into account the continuous nature of the approximation and Gaussian assumptions on the noise.

3.1 Probabilistic Analysis of CRNs

We have shown that X^N can be approximated by $Y^N(t) = N\Phi(t) + N^{\frac{1}{2}}Z(t)$, where $Y^N(t)$ has a multivariate Gaussian distribution, so it is completely characterized by its expected value and covariance matrix, whose values are respectively $E[Y^N(t)] = N\Phi(t)$ and $C[Y^N(t)] = N^{\frac{1}{2}}C[Z(t)]N^{\frac{1}{2}} = NC[Z(t)]$.

Since Y^N has a multivariate normal distribution then every linear combination of its components is normally distributed. Therefore, given $B = [b_1, b_2, \cdots, b_{|A|}]$ where $b_1, b_2, \ldots, b_{|A|} \in \mathbb{Z}$, we can consider the random variable $BY^N(t)$, which defines a linear combination of the species at time t. For every t, $BY^N(t)$ is a normal random variable, whose expected value and variance are

$$E[BY^N(t)] = BE[Y^N(t)] \tag{7}$$

$$C[BY^N(t)] = BC[Y^N(t)]B^T \tag{8}$$

For a specific time t_k, it is possible to calculate the probability that $BY^N(t_k)$ is within a set I of closed, disjoint real intervals $[l_i, u_i]$, where $l_i, u_i \in \mathbb{R} \cup \{+\infty, -\infty\}$. This probability $\Omega_{Y^N,B,I}(t_k)$ is given by

$$\Omega_{Y^N,B,I}(t_k) = \sum_{[l_i,u_i]\in I} \int_{l_i}^{u_i} g(x|E[BY^N(t_k)], C[BY^N(t_k)])dx \qquad (9)$$

where $g(x|EV, \sigma^2)$ is the Gaussian distribution with expected value EV and covariance σ^2. We recall that it is possible to find numerical solution of Eq. (9) in constant time using the Z table [13].

Example 2. Consider the CRN of Example 1, then we can obtain the probability that $\#\lambda_1 - 2\#\lambda_3$ is at least 10 at time 20 by defining $B' = [1, 0, -2]$, $I' = \{[10, +\infty]\}$ and calculating $\Omega_{Y^N,B',I'}(20)$.

The following theorems are consequences of results in [15], which can be generalized for reactions with a finite number of reagents and products. They show asymptotic pointwise convergence of expected value, variance and probability.

Theorem 2. *Let $C = (\Lambda, R)$ be a CRN. Suppose the solution of Eq. (6) is bounded, then, approaching the thermodynamic limit, for any finite instant of time t_i*

$$\lim_{N\to\infty} \|\Omega_{Y^N,B,I}(t_i) - \widetilde{\Omega}_{X^N,B,I}(t_i)\| = 0, \qquad (10)$$

where $\widetilde{\Omega}_{X^N,B,I}(t_i)$ is the probability that $B(X^N)$ is within I at time t_i.

Theorem 3. *Suppose the solution of Eq. (6) is bounded, then, approaching the thermodynamic limit, for any finite instant of time t_k*

$$\lim_{N\to\infty} \|C[BY^N(t_k)] - C[BX^N(t_k)]\| = 0 \qquad (11)$$

$$\lim_{N\to\infty} \|E[BY^N(t_k)] - E[BX^N(t_k)]\| = 0. \qquad (12)$$

To solve the differential equations (5) and (6), it is necessary to use a numerical method such as adaptive Runge-Kutta algorithm. This yields the solution for a finite set of sampling times $\Sigma = [t_1, \ldots, t_{|\Sigma|}] \in \mathbb{R}^{|\Sigma|}$, where $t_1 \leq \ldots \leq t_k \leq \ldots \leq t_{|\Sigma|}$ and $|\Sigma|$ is the sample size. Assuming Y^N is separable, that is, it is possible to completely define the behavior of Y^N by only considering a countable number of points, we can calculate $\Omega_{Y^N,B,I}$ for any point in Σ and if points are dense enough then this set exhaustively describes the probability that BX^N is within I over time. This restriction is not a limitation since for any stochastic process there exists a separable modification of it [10].

4   Stochastic Evolution Logic (SEL)

Let $C = (\Lambda, R)$ be a CRN with initial state x_0, in a system of size N. We now define the logic SEL (Stochastic Evolution Logic) which enables evaluation of the probability, variance and expectation of linear combinations of populations of the species of C.

The syntax of SEL is given by

$$\eta := P_{\sim p}[B, I]_{[t_1, t_2]} \quad | \quad Q_{\sim v}[B]_{[t_1, t_2]} \quad | \quad \eta_1 \wedge \eta_2 \quad | \quad \eta_1 \vee \eta_2$$

where $Q = \{supV, infV, supE, infE\}$, $\sim = \{<, >\}$, $p \in [0, 1]$, $v \in \mathbb{R}$, $B \in \mathbb{Z}^{|\Lambda|}$, $I = \{[l_i, u_i] \mid l_i, u_i \in \mathbb{R} \cup [+\infty, -\infty] \wedge [l_i, u_i] \cap [l_j, u_i] = \emptyset, i \neq j\}$ and $[t_1, t_2]$ is a closed interval, with the constraint that $t_1 \leq t_2$ and $t_1, t_2 \in \mathbb{R}$. If $t_1 = t_2$ the interval reduces to a singleton.

Formulae η describe global properties of the stochastic evolution of the system. (B, I) specifies a linear combination of the species of C and a set of intervals, where $B \in \mathbb{Z}^{|\Lambda|}$ is the vector defining the linear combination and I represents a set of disjoint closed real intervals. $P_{\sim p}[B, I]_{[t_1, t_2]}$ is the probabilistic operator, which specifies the probability that the linear combination defined by B falls within the range I over the time interval $[t_1, t_2]$. $supE, infE, infV, supV$ respectively yield the supremum and infimum of expected value and variance of the random variables associated to B within the specified time interval.

Example 3. Consider the CRN of Example 1. Checking if the variance of $\#\lambda_1$ remains smaller than K_1 within $[t_j, t_k]$ can be expressed as $supV_{<K_1}[[1, 0, 0]]_{[t_j, t_k]}$. Another example is checking if, in the same interval, $(\#\lambda_1 - \#\lambda_2)$ is at least K_2 or within $[K_3, K_4]$, with $K_3 < K_4 < K_2$, with probability greater than 0.95: $P_{>0.95}[[1, -1, 0], ([K_3, K_4], [K_2, \infty])]_{[t_j, t_k]}$. Equivalently, instead of writing B, we write directly the linear combination it defines. For example, in the latter case we have $P_{>0.95}[(\#\lambda_1 - \#\lambda_2), ([K_3, K_4], [K_2, \infty])]_{[t_j, t_k]}$.

Semantics. Given a CRN $C = (\Lambda, R)$ with initial configuration x_0 in a system of fixed volumetric factor N, its stochastic behaviour is described by the CTMC X^N of Definition 1. We define a path of CTMC X^N as a sequence $\omega = x_0 t_1 x_1 t_1 x_2...$ where x_i is a state and $t_i \in \mathbb{R}_{>0}$ is the time spent in the state x_i. A path is finite if there is a state x_k that is absorbing. $\omega \otimes t$ is the state of the path at time t. $Path(X^N, x_0)$ is the set of all (finite and infinite) paths of the CTMC starting in x_0. We work with the standard probability measure $Prob$ over paths $Path(X^N, x_0)$ defined using cylinder sets [11].

We first define when a path ω satisfies (B, I) at time t

$$\omega, t \models (B, I) \quad \leftrightarrow \quad \exists [l_i, u_i] \in I . l_i \leq B(\omega \otimes t) \leq u_i.$$

Note that $B(\omega \otimes t)$ is well defined because $\omega \otimes t \in \mathbb{N}^{|\Lambda|}$. For η formulas we have

$$X^N, x_0 \models P_{\sim p}[B, I]_{[t_1, t_2]} \quad \leftrightarrow \quad Prob(\omega \in Path(X^N, x_0) \mid \omega, t \models (B, I), t \in [t_1, t_2]) \sim p$$

$$X^N, x_0 \models supV_{\sim v}[B]_{[t_1, t_2]} \quad \leftrightarrow \quad sup(C[B(X^N)], [t_1, t_2]) \sim v$$

$$X^N, x_0 \models infV_{\sim v}[B]_{[t_1,t_2]} \quad \leftrightarrow \quad inf(C[B(X^N)], [t_1, t_2]) \sim v$$

$$X^N, x_0 \models supE_{\sim v}[B]_{[t_1,t_2]} \quad \leftrightarrow \quad sup(E[B(X^N)], [t_1, t_2]) \sim v$$

$$X^N, x_0 \models infE_{\sim v}[B]_{[t_1,t_2]} \quad \leftrightarrow \quad inf(E[B(X^N)], [t_1, t_2]) \sim v$$

$$X^N, x_0 \models \eta_1 \wedge \eta_2 \quad \leftrightarrow \quad X^N, x_0 \models \eta_1 \wedge X^N, x_0 \models \eta_2$$

$$X^N, x_0 \models \eta_1 \vee \eta_2 \quad \leftrightarrow \quad X^N, x_0 \models \eta_1 \vee X^N, x_0 \models \eta_2$$

$inf(\cdot, [t_1, t_2])$ and $sup(\cdot, [t_1, t_2])$ respectively denote the infimum and supremum within $[t_1, t_2]$. $Prob(\omega \in Path(X^N, x_0) \,|\, \omega, t \models (B, I), t \in [t_1, t_2])$ is the probability that the linear combination defined by B falls within I at a time instant t between t_1 and t_2, and is well defined since the probability measure $Prob$ on $Path(X^N, x_0)$ corresponds to transient probability calculated using the CME.

4.1 LNA-based Approximate Model Checking for CRNs

Stochastic model checking of CRNs is usually achieved by transient analysis of the CTMC X^N [11], which involves solving the CME and thus suffers from the state-space explosion problem. We propose an approximate model checking algorithm based on LNA. The inputs are a SEL formula η, the stochastic process X^N induced by the CRN and initial state x_0. The output is $true$ in case the formula is verified, and otherwise $false$.

The algorithm proceeds by induction on the structure of formula η, successively computing whether each subformula is satisfied or not. We assume that Eqs. (5) and (6) are solved numerically where Σ is the finite set of sample points on which their solution is defined and that t_0, initial time, and t_{max}, final time, are always sampling points.

Probabilistic Operator. To evaluate $P_{\sim p}[(B, I)]_{[t_1,t_2]}$ we construct the function $Prob_{(B,I)}(t) = \Omega_{Y^N, B, I}(t_i)$ for $t \in [t_i, t_{i+1}), t_i, t_{i+1} \in \Sigma$ (alternatively, can be constructed as the interpolation of the values of $\Omega_{Y^N, B, I}$ over Σ points).

Lemma 1. $Prob_{(B,I)}$ is integrable on $\mathbb{R}_{\geq 0}$.

Theorem 2 guarantees the pointwise correctness of $Prob_{(B,I)}$ and its integrability allows us to compute the following approximation, then compare to threshold p to decide the truth value. If $t_2 \neq t_1$ then $Prob(\omega \in Path(x_0) \,|\, \omega, t \models (B, I), t \in [t_1, t_2]) \approx \frac{1}{t_2 - t_1} \int_{t_1}^{t_2} Prob_{B,I}(t)dt$ else if $t_1 = t_2$ $Prob(\omega \in Path(x_0) \,|\, \omega, t_1 \models (B, I)) \approx Prob_{B,I}(t_1)$.

Expectation and Variance Operators. To evaluate $sup(C[B(X^N)], [t_1, t_2])$, $inf(C[B(X^N)], [t_1, t_2])$, $sup(E[B(X^N)], [t_1, t_2])$ and $inf(E[B(X^N)], [t_1, t_2])$ we use the LNA, namely, compute the expected value and variance of Eqs. (8) and (7). Theorem 3 guarantees the quality of the approximation. We can now compute the following approximations, then compare to the threshold v:

$$sup(C[B(X^N)], [t_1, t_2]) \approx max\{C[BY^N(t_k)] \,|\, (t_k \in \Sigma \wedge t_1 \leq t_k \leq t_2) \vee (t_k \in L_{[t_1,t_2]})\}$$

$$inf(C[B(X^N)], [t_1, t_2]) \approx min\{C[BY^N(t_k)] | (t_k \in \Sigma \wedge t_1 \leq t_k \leq t_2) \vee (t_k \in L_{[t_1, t_2]})\}$$

and similarly for the expected value. $L_{[t_1, t_2]} = \{t_i | t_i \in \Sigma \wedge \nexists t_j \in \Sigma \text{ such that } |t_1 - t_j| < |t_1 - t_i|\}$ ensures that for any time interval there is at least one sampling point, even if the interval is a singleton.

LNA-based model checking can also be used for systems far from the thermodynamic limit, at a cost of some loss of precision. LNA assumes continuous state space, and it is not possible to justify this assumption for very small populations. However, if the distributions of interest are not multi-modal and the noise term is finite and approximated by a Gaussian distribution, then LNA gives very good approximation even for quite small systems. It is clear that model checking accuracy increases as N grows. We emphasise that the model checking algorithm we have presented is also able to handle CRNs whose stochastic semantics is an infinite CTMC, which occur frequently in biological models.

Complexity of LNA-based Approximate Model Checking. The time complexity for model checking formula η against a CRN $C = (\Lambda, R)$ is linear in $|\eta|$. In the worst case, analysis of a single operator requires the solution of $O(|\Lambda|^2)$ polynomial differential equations for a bounded time. However, an efficient implementation can solve the $O(|\Lambda|^2)$ ODEs only once for the interval $[0, t_{max}]$, and then reuse this result for every operator, where t_{max} is the greatest (finite) time of interest. Note that ODEs are solved in terms of concentrations (a value between 0 and 1 by convention), ensuring independence of the number of molecules of each species, although stiffness can slow down the solution of the LNA.

5 Experimental Results

We implemented the methods in a framework based on Matlab and Java. The experiments were run on an Intel Dual Core $i7$ machine with 8 GB of RAM. To solve the differential equations, we use *Matlab ode45*, a variable step Runge-Kutta algorithm. We employ LNA-based model checking for the analysis of three biological reaction networks: a Phosphorelay Network [5], a Gene Expression Model [16], and the GW network [3]. For every network, the CRN and parameters have been taken from the referenced papers. We coded the same CRNs in PRISM in order to compare accuracy and time of execution with standard uniformisation of the CME [11] and statistical model checking (SMC) techniques (confidence interval method) as implemented in PRISM. For the GW case study, we cannot use global analysis nor SMC, because the state space is too large for direct analysis, and SMC requires many time-consuming simulations to obtain good accuracy. An extended set of experiments can be found in [4].

Phosphorelay Network. We consider a three-layer phosphorelay network whose structure is derived from [5]. Each layer $(L1, L2, L3)$ can be found in phosphorylate form $(L1p, L2p, L3p)$. We consider the initial condition $\#L1p = \#L2p = \#L3p = 0$, $\#L1 = \#L2p = \#L3p = Init$, where $Init \in \mathbb{N}$. Then we analyse the ligand B, whose initial condition is $\#B = 3 * Init$. We are interested in checking the following SEL property:

$$P_{>0.7}[(\#L1p - \#L3p), [0, +\infty]]_{[0,100]} \wedge P_{>0.98}[(\#L3p - \#L1p), [0, +\infty]]_{[300,600]}$$

which is verified if, in the first interval, the probability that $\#L1p$ is greater than $\#L3p$ is > 0.7 and if, between 300 and 600, with probability > 0.98, $\#L3p$ is greater than $\#L1p$. We evaluate this formula in three different initial conditions, firstly $Init = 32$ and $N = 5000$, then $Init = 64$ and $N = 10000$, and finally $Init = 100$ and $N = 15625$, so the same concentration but different numbers of molecules. In all cases, the LNA-based model checking evaluates the formula as true. To understand the quality of the approximation, we check the following quantitative formula $P_{=?}[(\#L3p - \#L1p, [0, +\infty])]_{[T,T]}$ for $T \in [0, 600]$ (in our implementation $=?$ gives the quantity calculated by model checking the operator). We compare the results with the evaluation of the corresponding CSL formula using standard uniformisation (Unif) with error 10^{-7}. The following table shows the results. $MaxErr$ is the maximum error computed by LNA-based approach compared to standard uniformisation and $AvgErr$ is the average error; $Time(\cdot)$ stands for execution time.

Init	Time (LNA)	Time (Unif)	MaxErr	AvgErr
20	0.22 s	2 min	0.0675	0.0519
32	0.23 s	5 min	0.059	0.02
64	0.26 s	> 2 h	0.0448	0.0027
100	0.3 s	> 2 h	0.03	0.0011

Note that as $Init$ increases the error of our method decreases, while the execution time is practically independent of the molecular count. LNA-based algorithms are faster in all cases. Thus our approach can be used even for quite small population systems, giving a fast approximate stochastic characterization.

Gene Expression. We consider a simple CRN that models the transcription of a gene into an mRNA molecule, and the translation of the latter into a protein. The CRN, rates and initial conditions are the same as in [16]. The stochastic semantics of the reaction network is an infinite CTMC, and we use this model to show that our method can handle infinite state-space processes. We consider the quantitative property $supE_{=?}[\#mRNA]_{[T,T]}$, which gives the number of molecules of $mRNA$ in the system at time T. We compare our method with SMC estimation of the same property by using 50000 simulations, for $T = \{300, 600, 900, 1200\}$, and in the following tables we compare the results in terms of execution time ($Time(\cdot)$) and expected value of $\#mRNA$ estimated ($ExpVal(\cdot)$). LNA-based model checking is several orders of magnitude faster without loss of accuracy.

T	Time (LNA)	Time (Simul)	ExpVal (LNA)	ExpVal (Simul)
300	0.52 s	75 s	100.17	100.14 ± 0.1
600	0.54 s	198 s	142.15	142.11 ± 0.1
900	0.54 s	337 s	159.73	159.74 ± 0.1
1200	0.56 s	483 s	167.1	167.1 ± 0.1

DNA Strand Displacement of GW Network. GW is a network related to the G2-M cell cycle switch; under particular initial conditions, it has been shown that GW can emulate the Approximate Majority algorithm [3]. Here, we consider the two-domain DNA strand-displacement implementation of GW [2]. The corresponding CRN is composed of 340 species and 240 reactions. For our analysis the species of interest are R and P, whose initial conditions are $\#R = 90$ and $\#P = 10$; initial conditions of other species are taken from the referenced papers. We check the property $P_{>0.9}[\#R - \#P, [50, +\infty]]_{[6000,35000]}$ for a system of size $N = 45000$, which is verified as true in 28 minutes.

6   Concluding Remarks

We presented a novel probabilistic logic for analysing stochastic behaviour of CRNs and proposed an approximate model checking algorithm based on the LNA of the CME. We have demonstrated on three non-trivial examples that LNA-based model checking enables analysis of CRNs with hundreds of species, and even infinite CTMCs, at a cost of some loss of accuracy. It would be interesting to find bounds on the approximation error when the system is far from the thermodynamic limit. However, the error is not only dependent on the value of N, but also on the structure of the CRN, the rates, and the property. As future work, we plan to improve the accuracy of the method near critical points similarly to the approach of [6], and to extend the logic with more expressive temporal operators. We also intend to release a software tool.

References

1. Bortolussi, L., Lanciani, R.: Model checking Markov population models by central limit approximation. In: Joshi, K., Siegle, M., Stoelinga, M., D'Argenio, P.R. (eds.) QEST 2013. LNCS, vol. 8054, pp. 123–138. Springer, Heidelberg (2013)
2. Cardelli, L.: Two-domain DNA strand displacement. Math. Struct. Comput. Sci. **23**(02), 247–271 (2013)
3. Cardelli, L.: Morphisms of reaction networks that couple structure to function. BMC Syst. Biol. **8**(1), 84 (2014)
4. Cardelli, L., Kwiatkowska, M., Laurenti, L.: Stochastic analysis of chemical reaction networks using linear noise approximation. arXiv preprint (2015). arXiv:1506.07861
5. Csikász-Nagy, A., Cardelli, L., Soyer, O.S.: Response dynamics of phosphorelays suggest their potential utility in cell signalling. J. R. Soc. Interface **8**(57), 480–488 (2011)
6. Elf, J., Ehrenberg, M.: Fast evaluation of fluctuations in biochemical networks with the linear noise approximation. Genome Res. **13**(11), 2475–2484 (2003)
7. Ethier, S.N., Kurtz, T.G.: Markov Processes: Characterization and Convergence, vol. 282. Wiley, New York (2009)
8. Gillespie, D.T.: Deterministic limit of stochastic chemical kinetics. J. Phys. Chem. B **113**(6), 1640–1644 (2009)
9. Gillespie, D.T., Hellander, A., Petzold, L.R.: Perspective: stochastic algorithms for chemical kinetics. J. Chem. Phys. **138**(17), 170901 (2013)

10. Itō, I.: Essentials of Stochastic Processes, vol. 231. American Mathematical Soc., Providence (2006)
11. Kwiatkowska, M., Norman, G., Parker, D.: Stochastic model checking. In: Bernardo, M., Hillston, J. (eds.) SFM 2007. LNCS, vol. 4486, pp. 220–270. Springer, Heidelberg (2007)
12. Kwiatkowska, M., Norman, G., Parker, D.: PRISM 4.0: verification of probabilistic real-time systems. In: Gopalakrishnan, G., Qadeer, S. (eds.) CAV 2011. LNCS, vol. 6806, pp. 585–591. Springer, Heidelberg (2011)
13. Patel, J.K., Read, C.B.: Handbook of the Normal Distribution, vol. 150. CRC Press, Boca Raton (1996)
14. Van Kampen, N.G.: Stochastic Processes in Physics and Chemistry, vol. 1. Elsevier, Amsterdam (1992)
15. Wallace, E., Gillespie, D., Sanft, K., Petzold, L.: Linear noise approximation is valid over limited times for any chemical system that is sufficiently large. IET Syst. Biol. **6**(4), 102–115 (2012)
16. Wolf, V., Goel, R., Mateescu, M., Henzinger, T.A.: Solving the chemical master equation using sliding windows. BMC Syst. Biol. **4**(1), 42 (2010)

Adaptive Moment Closure for Parameter Inference of Biochemical Reaction Networks

Sergiy Bogomolov[1], Thomas A. Henzinger[1], Andreas Podelski[2],
Jakob Ruess[1], and Christian Schilling[2]([✉])

[1] IST Austria, Klosterneuburg, Austria
[2] University of Freiburg, Freiburg, Germany
{sergiy.bogomolov,tah,Jakob.Ruess}@ist.ac.at,
{podelski,schillic}@informatik.uni-freiburg.de

Abstract. Continuous-time Markov chain (CTMC) models have become a central tool for understanding the dynamics of complex reaction networks and the importance of stochasticity in the underlying biochemical processes. When such models are employed to answer questions in applications, in order to ensure that the model provides a sufficiently accurate representation of the real system, it is of vital importance that the model parameters are inferred from real measured data. This, however, is often a formidable task and all of the existing methods fail in one case or the other, usually because the underlying CTMC model is high-dimensional and computationally difficult to analyze. The parameter inference methods that tend to scale best in the dimension of the CTMC are based on so-called moment closure approximations. However, there exists a large number of different moment closure approximations and it is typically hard to say a priori which of the approximations is the most suitable for the inference procedure. Here, we propose a moment-based parameter inference method that automatically chooses the most appropriate moment closure method. Accordingly, contrary to existing methods, the user is not required to be experienced in moment closure techniques. In addition to that, our method adaptively changes the approximation during the parameter inference to ensure that always the best approximation is used, even in cases where different approximations are best in different regions of the parameter space.

Keywords: Stochastic reaction networks · Continuous-time markov chains · Parameter inference · Moment closure

1 Introduction

With the advancement of measurement technologies for biochemical processes in the last decades, quantitative mathematical modeling of biochemical reaction networks has continuously increased in importance [1,14,19]. Chemical reactions inside cells, where some of the reacting species may be present in very low amounts of molecules, are inherently driven by random fluctuations [6,12,20].

© Springer International Publishing Switzerland 2015
O. Roux and J. Bourdon (Eds.): CMSB 2015, LNBI 9308, pp. 77–89, 2015.
DOI: 10.1007/978-3-319-23401-4_8

Accordingly, an accurate mathematical model should take this stochasticity into account. The most widely used class of stochastic models in this context are continuous-time Markov chains (CTMCs) [5]. The advantage of these models is that they are easy to formulate and can be justified based on first principles [4]. The major drawback is that their analytical or computational analysis can be extremely difficult, especially when more than just a few different chemical species play a role for the reaction network. This is because the chemical master equation (CME), which governs the time evolution of the probability distribution of the CTMC, cannot be solved for anything but the simplest systems and even approximation techniques [13,23] tend to fail when the CTMC is high-dimensional. In such cases, an alternative is to focus only on some low-order moments of the probability distribution. Ordinary differential equations that describe the time evolution of these moments can be derived from the CME [2], but their solution typically requires some kind of approximation [18,21]. These approximations, known as moment closure, are usually based on an assumption of the underlying probability distribution and exist in many different varieties [8]. Often, for a given system and given model parameters, some of these approximations provide good results whereas others fail to be sufficiently accurate or fail entirely. Unfortunately, there exists no approach for determining a priori which moment closure technique will provide the best approximation. In general, the only approach that is guaranteed to provide at least statistically exact results is to simply simulate the CTMC using a stochastic simulation algorithm (SSA) [3] and to compute Monte Carlo estimates of the system output of interest based on the simulation results. To obtain precise estimates, however, a large number of simulations may be required, leading to a high computational cost. For the forward analysis of a system, i.e. when the model parameters are known, this is not a serious problem. For the reverse engineering task of identifying the model parameters from measured data, however, the CTMC needs to be analyzed for many different parameter values in order to determine those in best agreement with the measured data. Accordingly, for this task the computational cost of approaches based on stochastic simulation [10] is often prohibitively large.

In this paper, we propose an approach for parameter inference based on moment closure that is complemented by stochastic simulation. In particular, the parameter inference is performed based on the computationally cheap moment closure approximation, whereas the stochastic simulation is employed whenever new regions in the parameter space are explored, either to ensure that the approximation is still sufficiently accurate, or to propose a new approximation that outperforms the previously used one. With this approach we are able to combine the computational advantages of moment closure with the statistical exactness of SSA and obtain a method that is both scalable and does not require a priori knowledge of the performance of different moment closure techniques. Importantly, the method is completely automated and chooses and adapts the approximation from a precomputed library of moment closure methods. Thus, the user only has to specify the model and supply the data and, contrary to previous approaches [9,16,24], no expertise in the analysis of CTMCs is required.

The remaining paper is structured as follows. In Sect. 2, we introduce bio-chemical reaction networks, the chemical master equation and moment closure methods. In Sect. 3, we formulate a maximum-likelihood estimation problem for the model parameters and describe previously published moment-based meth-ods for solving these problems. In Sect. 4, we propose our automated adaptive parameter inference method. In Sect. 5, we study the performance of our method for some benchmark reaction networks. Finally, in Sect. 6, we discuss our results and provide some concluding remarks.

2   Stochastic Modeling of Biochemical Reaction Networks

Consider a biochemical reaction network consisting of m different chemical species $X^1, \ldots, X^m$ that interact according to K different reactions:

$$\nu'_{1k}X^1 + \ldots + \nu'_{mk}X^m \xrightarrow{\theta_k} \nu''_{1k}X^1 + \ldots + \nu''_{mk}X^m, \quad k = 1, \ldots, K, \quad (1)$$

where the coefficients ν'_{ik} and ν''_{ik} determine how many molecules of the i-th species are consumed and produced in the k-th reaction, respectively. Under the assumption that the reaction network is well-stirred and in thermal equi-librium, it can be described by a continuous-time Markov chain $X(t, \theta) = \left[X^1(t, \theta) \cdots X^m(t, \theta)\right]^T$ that takes states $x = [x^1 \cdots x^m]^T \in \mathbb{N}_0^m$ [4]. The tran-sition probabilities of this CTMC are determined by the reaction parameters $\theta = [\theta_1 \cdots \theta_K]^T \in \left(\mathbb{R}_0^+\right)^K$ and the kinetic rate law of the reactions. Here, we restrict our attention to mass action kinetics and elementary chemical reactions (i.e. reactions of order at most 2). These assumptions simplify the computa-tion of moments of the CTMC. It should be noted, however, that they are not strictly necessary for the results of this paper and are mainly imposed because it is very unlikely that, in a three-dimensional space, more than two molecules meet at exactly the same time. Accordingly, any more complicated biochemi-cal reaction can essentially be decomposed into a series of elementary reactions whose reaction rates are governed by the law of mass action. These assumptions lead to transition probabilities of the CTMC that are determined by propensity functions of the form $a_k(x, \theta) = \theta_k h_k(x), k = 1, \ldots, K$, where $h_k(x)$ are at most quadratic polynomials in x. The time evolution of the probability distribution of $X(t, \theta)$ can then be described by the chemical master equation:

$$\dot{p}(x, t) = -p(x, t) \sum_{k=1}^{K} a_k(x, \theta) + \sum_{k=1}^{K} p(x - \nu_k, t) a_k(x - \nu_k, \theta), \quad (2)$$

where $\nu_k = [\nu_{1k} \cdots \nu_{mk}]^T$, $\nu_{ik} = \nu''_{ik} - \nu'_{ik}$, $i = 1, \ldots, m$, and $p(x, t) := P(X(t, \theta) = x)$ is the probability that x molecules of the m chemical species are present at time t.

Since $X(t, \theta)$ has a countably infinite state space, computing the probabilities $p(x, t)$ requires solving an infinite system of coupled ordinary differential equa-tions, which is generally not possible. Approximate solutions can be obtained in some cases, for instance by projection to a finite state space [13,23], but we will not discuss these approaches here.

An alternative is to focus only on some low-order moments of the probability distribution. Ordinary differential equations describing their time evolution can be derived from the CME [2] and written as

$$\dot{\eta}(t) = A(\theta)\eta(t) + B(\theta)\bar{\eta}(t), \tag{3}$$

where $\eta(t)$ is a vector containing the (uncentered) moments up to some desired order L and $\bar{\eta}(t)$ contains moments of order $L+1$. Equation (3) shows that the time evolution of $\eta(t)$ depends on moments of higher order; hence $\eta(t)$ cannot be computed without knowledge of $\bar{\eta}(t)$. Accordingly, the open system of equations Eq. (3) is typically replaced by an approximate closed system of equations

$$\dot{\tilde{\eta}}(t) = A(\theta)\tilde{\eta}(t) + B(\theta)f(\tilde{\eta}(t)), \tag{4}$$

where $\tilde{\eta}(t)$ are approximations of $\eta(t)$. The function f is usually chosen according to an assumption on the underlying probability distribution. Typical examples are to assume that the centered moments (or cumulants) of order $L+1$ are zero [11,22], or to choose f according to a log-normal distribution [21]. In general, the choice of f is made rather arbitrarily without actual knowledge of the underlying distribution. Furthermore, whether a given closure will provide good approximations depends on the system that is being studied, the model parameters, and the order L at which the moment equations are closed. This makes it practically impossible for someone who is not an expert in the use of these methods to choose an appropriate closure. Despite all this, moment closure methods have been successfully applied for analyzing CTMCs, and specifically also for parameter inference [16,24]. The choice of the closure method used in these references, however, was based on trial and error and the success of the performed studies accordingly required a portion of luck.

An alternative approach for analyzing biochemical reaction networks is by using a stochastic simulation algorithm (SSA). It is straightforward to generate statistically exact sample paths $x_1(t), \ldots, x_n(t)$ of $X(t, \theta)$ in this way. From these sample paths, estimators of any system output, for instance some moments or the entire probability distribution at a certain time point, can be constructed. While such an approach is easy to implement and can always be used, it comes with the major drawback that often a large number of sample paths n is required to obtain precise estimates. This can make the use of stochastic simulation for reverse engineering tasks computationally prohibitively expensive.

3   Moment-Based Parameter Inference

In this section, we formulate the parameter inference problem and review previous methods that have been developed to solve it. The goal in this paper is to estimate the reaction rate constants θ from measured data that is of the form

$y = \left\{ x_1^j(t_s), \ldots, x_n^j(t_s), \; s = 1, \ldots, S \right\}$ and corresponds to measuring the number of molecules of the j-th chemical species in n cells at each measurement time point $t_s, s = 1, \ldots, S$ (extension to more than one measured chemical species is straightforward but requires more complicated expressions for the likelihood in Eq. 6 as shown in [17]). We assume that all the collected measurements are statistically independent. This is for instance the case for flow cytometry data where the cells are discarded after being measured so that two different measurements can never come from the same cell. The task of identifying the model parameters from this data can be posed as a maximum-likelihood estimation problem

$$\theta_{\mathrm{MLE}}(y) = \arg \max_\theta \mathcal{L}(y, \theta), \tag{5}$$

where y is the measured data and $\mathcal{L}(y, \theta) = p(y|\theta)$ is the likelihood of the parameters θ, i.e. the probability (density) of the data given that θ are the model parameters. Analytically computing the likelihood is usually impossible, and accordingly, the optimization problem Eq. (5) is typically solved by iterative numerical evaluation of $\mathcal{L}(y, \theta)$ for many different values of θ. Unfortunately, evaluating the likelihood for given parameters θ requires solving the CME with these parameters, which, as discussed in the previous section, is often impossible or computationally expensive itself. For this reason, one option is to use sample moments of the data as measurements instead of the entire data [24]. For instance, one can compute sample means $\hat{\mu}_1(t_s)$ and sample variances $\hat{\mu}_2(t_s), s = 1, \ldots, S$ from the data y and treat the vector $\hat{\mu} := [\hat{\mu}(t_1) \cdots \hat{\mu}(t_S)]^T$, where $\hat{\mu}(t_s) := [\hat{\mu}_1(t_s) \; \hat{\mu}_2(t_s)]$, as new data. In earlier publications [17,24], we have shown that the probability density function $p(\hat{\mu}|\theta)$ of $\hat{\mu}$ is given by

$$p(\hat{\mu}|\theta) = \prod_{s=1}^{S} p(\hat{\mu}(t_s)|\theta), \quad \text{where} \quad p(\hat{\mu}(t_s)|\theta) = \mathcal{N}(M(t_s), \Sigma(t_s)) \quad \text{and} \tag{6}$$

$$M(t_s) = \begin{bmatrix} \mu_1(t_s) \\ \mu_2(t_s) \end{bmatrix} \quad \text{and} \quad \Sigma(t_s) = \frac{1}{n} \begin{bmatrix} \mu_2(t_s) & \mu_3(t_s) \\ \mu_3(t_s) & \mu_4(t_s) - \frac{n-3}{n-1}\left(\mu_2(t_s)\right)^2 \end{bmatrix},$$

where $\mathcal{N}$ stands for the normal distribution, $\mu_1(t_s) = \mu_1(t_s, \theta)$ is the mean and $\mu_i(t_s) = \mu_i(t_s, \theta), i = 2, 3, 4$ are the centered moments of the measured species $X^j(t_s, \theta)$ at time t_s for model parameters θ. Since these moments can be computed from the solution of Eq. (4), we can use this result to approximately compute the likelihood $\mathcal{L}(\hat{\mu}, \theta) = p(\hat{\mu}|\theta)$ without having to solve the CME. Accordingly, we can solve the optimization problem in Eq. (5) using $\hat{\mu}$ instead of y to compute the maximum-likelihood estimator $\theta_{\mathrm{MLE}}(\hat{\mu})$. However, the fact that moments up to order four are required to evaluate the covariance matrices $\Sigma(t_s)$ means that moment closure of order at least $L = 4$ is necessary. To avoid this, one can estimate the covariance matrices $\Sigma(t_s)$ from the data by computing empirical estimates of the moments up to order four and plugging them into the above equation. Throughout this paper, we will follow such a strategy and denote by μ_{data} the moments up to fourth order of the data, i.e. $\mu_{\mathrm{data}} := [\mu_{\mathrm{data}}(t_1) \cdots \mu_{\mathrm{data}}(t_S)]^T$, where $\mu_{\mathrm{data}}(t_s) := [\hat{\mu}_1(t_s) \; \hat{\mu}_2(t_s) \; \hat{\mu}_3(t_s) \; \hat{\mu}_4(t_s)]$ contains the first four centered empirical moments of the data set at time t_s. This strategy is appropriate whenever sufficiently many cells are measured so that the moments up to order four can be estimated with reasonable precision. For flow cytometry data, the number of cells measured per time point typically ranges in the order of thousands or even tens of thousands; hence sufficing precision is always guaranteed.

4   Adaptive Approach for Parameter Inference

The drawback of the approach described in the previous section is that a moment closure method has to be chosen in advance and this closure will be used throughout the entire parameter search. This leads to the problems that, on the one hand, it is a priori very difficult to choose the best closure and, on the other hand, which closure is best may also be different for different parts of the parameter space. The main idea of the method that we propose in the following is to use a small number of simulated trajectories of the system that are generated using a stochastic simulation algorithm (SSA) in order to test different approximations during the parameter space exploration. Specifically, whenever the parameter search leaves a certain area in parameter space, defined as an ϵ-neighborhood around the point at which the last SSA run was carried out, new simulations are performed and all closure methods from a predefined library are evaluated by comparing the different approximations at the current point in parameter space to the simulation results. Importantly, all the approximate moment systems, corresponding to closures of different types and degrees, are precomputed only once, and thus new derivations of the moment equations are not required during the search. To generate these systems we make use of Hespanha's StochDynTools toolbox [7].

Pseudocode of our approach is given in Algorithm 1. The inputs of the algorithm are the CTMC model $X(t, \theta)$, parametrized by the reaction rate constants θ, a set of ODE systems $CL = \{c_1(\theta), \ldots, c_q(\theta)\}$ corresponding to different approximations of the moment dynamics obtained through various closures of different types and degrees, the centered moments up to the fourth order μ_{data} of a measured data set Y, and a maximal number of iterations i_{max} that determines for how many steps in parameter space the search is performed. The algorithm returns the maximum likelihood estimator θ_{MLE}. The core idea of our approach works independently from the actual parameter search technique used in the background. Thus, it can be applied in conjunction with any standard optimization scheme used to minimize some distance between model output and data (for instance simple gradient descent). Accordingly, we focus on the adaptive update of the closure method while abstracting from the actual details of the parameter search for a fixed approximation by the function NEXTPARAMETER (line 18). It takes the current values of the parameters θ_i and the chosen approximate ODE system $c_{\text{best}}(\theta_i)$ and moves the search to the new parameters θ_{i+1} according to some criteria. In our implementation, we instantiate it with a Markov chain Monte Carlo method and a Metropolis-Hastings sampler, based on the likelihood in Eq. (6) [24]. Additionally, this function also takes care of updating the value of the maximum likelihood estimator θ_{MLE} based on the likelihood of the new parameters θ_{i+1}. The remaining pseudocode describes how and when the used closure method is adjusted. We first check whether the current parameter values θ_i are still within the ϵ-neighborhood $N_\epsilon(\theta_{\text{ref}})$, where θ_{ref} are the parameters at which the previous simulation was performed (line 5). In our implementation, we choose a neighborhood in the form of a hyperrectangle of relative size $N_\epsilon(\theta_{\text{ref}}) = \{\theta \mid |\theta - \theta_{\text{ref}}|_k \leq \epsilon \cdot |\theta_{\text{ref}}|_k, k = 1, \ldots, K\}$. If $\theta_i \in N_\epsilon(\theta_{\text{ref}})$, we directly proceed with the standard inference method in line 18, relying on the ODE system $c_{\text{best}}(\theta_i)$ from the most recent evaluation. Otherwise, stochastic simulation is employed with the current parameter values θ_i to compute estimates of the moments $\mu_{\text{SSA}}(\theta_i)$ using the function COMPUTE$_{\text{SSA}}$ (line 6), for which we utilize a standard implementation of Gillespie's SSA in our implementation. These estimates are then compared to the approximations $\mu_{\text{ODE}}(\theta_i)$ obtained with all the different closure methods using the function COMPUTE$_{\text{ODE}}$ which numerically computes the solution of the system of ODEs $c(\theta_i) \in CL$ (lines 8–15). The best approximate system $c_{\text{best}}(\theta_i)$ is chosen as the one that minimizes some distance DIST between estimation and approximation. In general, this distance could be determined in many different ways. In our implementation, we choose DIST as the likelihood of the estimated moments for the measured species X^j (Eq. 6), i.e. we measure the performance of the approximations by evaluating how precise the approximated moments of the system output (not of the entire state) are. Finally, we update the reference point θ_{ref} to θ_i (line 16) and the search continues in the standard way until the next ϵ-neighborhood is left.

5   Case Studies

We applied our inference method to several benchmark stochastic reaction networks. In this section, we report some exemplary results. For all examples, to generate the set of approximate ODE systems CL we used *derivative matching* (dm), *zero cumulants* (zc), *zero variance* (zv) moment closure, each with degree 2, 3, and 4, and *low dispersion* (ld) moment closure with degree 3 and 4 (see [8] for details).

Example 1. The first network is a model that has recently been used to describe agricultural pests [15] but can also be regarded as a model of gene expression in which the produced protein is positively regulated by the current amount of protein and negatively regulated (through an increased degradation rate) by past amounts of protein (i.e. species N could be regarded as an abstraction of a slow process that is activated by C and leads to the production of proteases that degrade C). It is given by the following reactions:

$$\emptyset \xrightarrow{\theta_1} N + C \qquad\qquad N \xrightarrow{\theta_2} 2N + C$$

$$N + C \xrightarrow{\theta_3} C \qquad\qquad C \xrightarrow{\theta_4} \emptyset.$$

Algorithm 1. Adaptive moment-based parameter inference algorithm

---

Input: CTMC $X(t, \theta)$, where $\theta \in (\mathbb{R}_0^+)^K$, set of approximate moment systems $CL = \{c_1(\theta), \ldots, c_q(\theta)\}$ obtained using different closure methods, data μ_{data}, and maximum number of iterations $i_{\max}$
Output: Maximum likelihood estimator θ_{MLE}
1: $\theta_1 :=$ random initial parameter values
2: $\theta_{\text{MLE}} := \theta_1$
3: $\theta_{\text{ref}} := +\infty$
4: **for** $i := 1$ to $i_{\max}$ **do**
5:     **if** $\theta_i \notin N_\epsilon(\theta_{\text{ref}})$ **then**
6:         $\mu_{\text{SSA}}(\theta_i) := \text{COMPUTE}_{\text{SSA}}(X(t, \theta_i))$
7:         $d_{best} := +\infty$
8:         **for all** $c(\theta_i) \in CL$ **do**
9:             $\mu_{\text{ODE}}(\theta_i) := \text{COMPUTE}_{\text{ODE}}(c(\theta_i))$
10:            $d := \text{DIST}(\mu_{\text{SSA}}(\theta_i), \mu_{\text{ODE}}(\theta_i))$
11:            **if** $d < d_{best}$ **then**
12:                $d_{best} := d$
13:                $c_{\text{best}}(\theta_i) := c(\theta_i)$
14:            **end if**
15:         **end for**
16:         $\theta_{\text{ref}} := \theta_i$
17:     **end if**
18:     $\langle \theta_{i+1}, \theta_{\text{MLE}} \rangle := \text{NEXTPARAMETER}(\theta_i, c_{\text{best}}(\theta_i), \mu_{\text{data}}, \theta_{\text{MLE}})$
19: **end for**
20: **return** θ_{MLE}

We assume that $N(0) = C(0) = 0$ and that the true parameters are given by $\theta_1 = 0.03$, $\theta_2 = 0.012$, $\theta_3 = 0.25 \cdot 10^{-4}$ and $\theta_4 = 0.003$, and that 5,000 cells are measured at the time points $t_1 = 10, \ldots, t_{90} = 900$. As settings for our algorithm we used $\epsilon = 0.2$ and performed 200 simulations whenever the search leaves an ϵ-neighborhood, i.e. in line 6 of Algorithm 1.

An exemplary run of our parameter search for $i_{\max} = 1,000$ iterations, started from random initial parameter values, is shown in Fig. 1. It can be seen that all the inferred parameters, i.e. the maximum-likelihood estimates $\theta_{\text{MLE}}(\hat{\mu})$, agree with the true parameter values up to negligible errors with basically no uncertainty. The former is a sign that a precise moment closure method exists for this example, whereas the latter stems from the large number of measurements that we assumed to be available. Figure 2 shows that also the model predictions, computed with the inferred parameters $\theta_{\text{MLE}}(\hat{\mu})$ and the best closure method, agree well both with the data and with SSA estimates of mean and variance obtained with the inferred parameters. We can conclude that the moment closure approximation is very precise and can match the data up to very small errors.

To evaluate on the one hand how important it is to choose a good approximation, and on the other hand whether it is necessary to adaptively change the closure method during the search, we performed the parameter inference

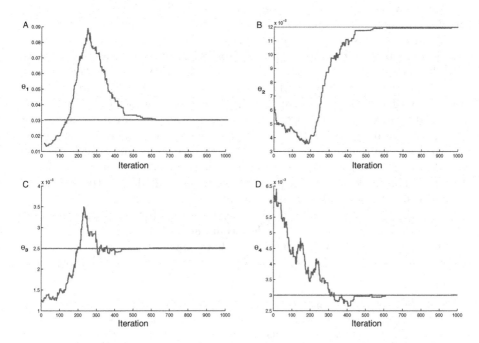

Fig. 1. Parameter search for Example 1. The panels show the values of the parameters in the search as a function of the iteration (blue). It can be seen that after approximately 600 iterations the search is very close to the true values (red lines) for all parameters and retains these values (Color figure online).

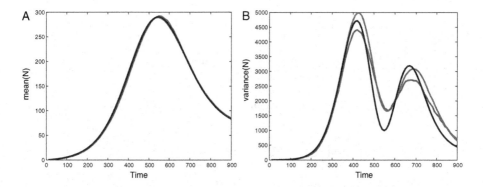

Fig. 2. Model output and data for the inferred parameters. (A) The mean computed with the best closure method (black) and the inferred parameters agrees very well both with the data (red) and the results of stochastic simulation with the inferred parameters (blue). (B) Also all the variances agree very well. The color coding is the same as in (A) (Color figure online).

with the same data and the same algorithm, but fixed an initial closure method and did not allow the search to switch between different approximations (i.e. by choosing $\epsilon = +\infty$). Table 1(a) compares the error in the inferred parameters obtained from our approach to the error in the results when the closure is fixed. It can be seen that for some of the fixed closure approaches the error in the parameter estimates is very large (specifically for all of the zero variance closures). Other methods provide more precise results, but overall all methods with fixed approximation are outperformed by our adaptive approach. Only the fourth order zero cumulants (zc4) closure was more precise than our approach for two of the four parameters. However, for our case study this closure was also computationally the most expensive one and the parameter search with fixed zc4 closure actually took twice as long as the adaptive search, despite the additional stochastic simulations and evaluations of all closure methods needed here.

To further test our results, we investigated how often the approximation was changed during the run of our algorithm and which closure methods were used most often. Table 1(b), column Ex 1, shows how often the different closure methods were chosen as best. It can be seen that some approximations were

Table 1. (a) Relative distance (in percent) between true and inferred parameters obtained from our adaptive algorithm (adapt) and the different closure methods on their own. The smallest distance is marked in bold. (b) Statistics of the used closure methods for the three considered reaction networks. Columns correspond to the different networks (Ex stands for example), rows report in percent how often each of the closure methods was chosen as best in our adaptive search. The bottom block of rows show how often the used approximation was changed as our search progressed through the parameter space (switch), how often stochastic simulation was performed, i.e. how often ϵ-neighborhoods were left and all the closure methods were tested (sim tot), and the total number of iterations in the search (i_{max}).

(a) Example 1

closure	θ_1	θ_2	θ_3	θ_4
adapt	0.44	0.31	**0.65**	**0.29**
dm2	4.45	2.74	2.68	4.32
zc2	11.02	6.11	3.23	2.93
zv2	281.09	74.85	45.72	76.29
dm3	2.54	1.23	1.85	3.55
zc3	9.72	4.80	0.86	2.87
zv3	285.55	79.96	49.01	83.41
ld3	9.08	4.30	6.75	9.63
dm4	3.43	1.33	4.17	9.54
zc4	**0.35**	**0.19**	3.77	9.29
zv4	292.60	78.89	46.60	71.90
ld4	14.44	3.80	12.31	28.06

(b) Search statistics

closure	Ex 1	Ex 2	Ex 3
dm2	15	13	0
zc2	10	23	50
zv2	0	0	0
dm3	10	0	0
zc3	5	16	0
zv3	0	0	0
ld3	0	23	0
dm4	15	0	50
zc4	35	6	0
zv4	0	0	0
ld4	10	19	0
switch	19	30	11
sim tot	23	44	46
i_{max}	1,000	1,000	2,000

never chosen (for instance all of the zero variance closures but also the third order low dispersion closure) whereas derivative matching and zero cumulants closures are chosen most often. Overall, high order closures are preferred over low order closures. This was to be expected, since these usually provide more precise results at the cost of an increased computational effort. Also we highlight that the option to switch the approximation was often used (in 19 out of 23 evaluations), and, compared to a pure simulation-based approach, we needed to employ stochastic simulation only 23 times (instead of 1,000 times).

Further examples. In addition to Example 1 we applied our algorithm to two further reaction networks and performed the same comparisons. Specifically, we considered the model of transient gene expression reported in reference [24] (termed here Example 2) and the first case study in reference [18] (termed here Example 3). The results were overall similar to those obtained for Example 1 and we only report in Table 1(b), columns Ex 2 and Ex 3, how often the different closure methods were used by our adaptive search. It can be seen that in Example 3 the second order zero cumulants and the fourth order derivative matching closure were chosen exclusively, whereas in Example 2, different zero cumulants and low dispersion closures were used most often and there was no noticeable preference for higher order closures.

6   Discussion

Using mathematical models to help in the understanding of complex biological systems is the core idea of systems biology. Up to some years ago, the main bottleneck in the identification of models was the availability of sufficiently precise and abundant data. Recently, measurement technologies have been improving at an amazing pace and nowadays enable us to simultaneously observe the dynamics of many different chemical species at single cell resolution. As these developments continue, we will gain access to data that is sufficiently informative to allow us to infer mathematical models of complex reaction networks from the measurements. However, for stochastic kinetic models that capture the inherent randomness of chemical reactions, this leads to a new bottleneck: the chemical master equation becomes intractable for high-dimensional models and especially the reverse engineering task of identifying model parameters from the measured data quickly becomes computationally infeasible. Parameter inference methods based on moment closure offer a solution to this problem but come with their own drawbacks. The goal of this paper was to address these drawbacks and to provide an automated moment-based inference method that can be used without in-depth knowledge of moment closure. To this end, we interfaced previously proposed approaches with a stochastic simulation algorithm by continuously checking the quality of the approximations and adaptively adjusting the used closure method to the best one available. Accordingly, our approach is generally applicable whenever a sufficiently accurate approximation in the generated library of moment closure methods exists. Importantly, since the approach can adapt the used closure during the exploration of the parameter space, it is not required that a unique closure method provides good approximations for the entire parameter space. Naturally, these benefits come with an increased computational cost compared to most standard moment-based inference approaches. This increase can primarily be attributed to the additional stochastic simulation and the evaluation of all the closure methods that is performed whenever the parameter search leaves an ϵ-neighborhood around the point in parameter space where the last simulation was performed. Accordingly, the parameter ϵ provides a trade-off between computational cost and guarantees that a good approximation is used. For $\epsilon \to \infty$ our approach becomes a standard moment-based inference method, whereas $\epsilon \to 0$ essentially leads to a method akin to those based entirely on stochastic simulation. We believe that this flexibility will prove to be valuable and allow us to investigate a large variety of different reaction networks with one unified inference method.

As future work, we plan to include and test more moment closure methods (e.g. the linear noise approximation), to apply our algorithm to larger and more challenging reaction networks, and to make a complete toolbox for moment-based parameter inference publicly available. In addition to this, in order to speed up our algorithm, we plan to introduce a trade-off between precision and computational cost of the different approximations such that the more expensive high order closure methods are only chosen when the low order closures do not provide acceptable precision.

Acknowledgements. This work was partly supported by the German Research Foundation (DFG) as part of the Transregional Collaborative Research Center "Automatic Verification and Analysis of Complex Systems" (SFB/TR 14 AVACS, http://www.avacs.org/), by the European Research Council (ERC) under grant 267989 (QUAREM) and by the Austrian Science Fund (FWF) under grants S11402-N23 (RiSE) and Z211-N23 (Wittgenstein Award). J.R. acknowledges support from the People Programme (Marie Curie Actions) of the European Union's Seventh Framework Programme (FP7/2007–2013) under REA grant agreement no. 291734.

References

1. Bertaux, F., Stoma, S., Drasdo, D., Batt, G.: Modeling dynamics of cell-to-cell variability in TRAIL-induced apoptosis explains fractional killing and predicts reversible resistance. PLoS Comput. Biol. **10**(10), e1003893 (2014)
2. Engblom, S.: Computing the moments of high dimensional solutions of the master equation. Appl. Math. Comput. **180**(2), 498–515 (2006)
3. Gillespie, D.: A general method for numerically simulating the stochastic time evolution of coupled chemical reactions. J. Comput. Phys. **22**(4), 403–434 (1976)
4. Gillespie, D.: A rigorous derivation of the chemical master equation. Phys. A **188**(1–3), 404–425 (1992)
5. Goutsias, J., Jenkinson, G.: Markovian dynamics on complex reaction networks. Phys. Rep. **529**, 199–264 (2013)
6. Hasty, J., Pradines, J., Dolnik, M., Collins, J.: Noise-based switches and amplifiers for gene expression. Proc. Nat. Acad. Sci. U.S.A. **97**(5), 2075–2080 (2000)

7. Hespanha, J.: StochDynTools - a MATLAB toolbox to compute moment dynamics for stochastic networks of bio-chemical reactions (2006). http://www.ece.ucsb.edu/~hespanha

8. Hespanha, J.: Moment closure for biochemical networks. In: Proceedings of the 3rd International Symposium on Communications, Control and Signal Processing (IEEE), St Julians, Malta, pp. 142–147 (2008)

9. Kügler, P.: Moment fitting for parameter inference in repeatedly and partially observed stochastic biological models. PLoS ONE **7**(8), e43001 (2012)

10. Lillacci, G., Khammash, M.: The signal within the noise: efficient inference of stochastic gene regulation models using fluorescence histograms and stochastic simulations. Bioinformatics **29**(18), 2311–2319 (2013)

11. Matis, T., Guardiola, I.: Achieving moment closure through cumulant neglect. Math. J. **12** (2010). doi:10.3888/tmj.12-2

12. McAdams, H., Arkin, A.: Stochastic mechanisms in gene expression. Proc. Nat. Acad. Sci. U.S.A. **94**(3), 814–819 (1997)

13. Munsky, B., Khammash, M.: The finite state projection algorithm for the solution of the chemical master equation. J. Chem. Phys. **124**, 044104 (2006)

14. Neuert, G., Munsky, B., Tan, R., Teytelman, L., Khammash, M., van Oudenaarden, A.: Systematic identification of signal-activated stochastic gene regulation. Science **339**, 584–587 (2013)

15. Parise, F., Lygeros, J., Ruess, J.: Bayesian inference for stochastic individual-based models of ecological systems: an optimal pest control case study. Front. Environ. Sci. **3**, 42 (2015)

16. Ruess, J., Lygeros, J.: Moment-based methods for parameter inference and experiment design for stochastic biochemical reaction networks. ACM Trans. Model. Comput. Simul. (TOMACS) **25**(2), 8 (2015)

17. Ruess, J., Milias-Argeitis, A., Lygeros, J.: Designing experiments to understand the variability in biochemical reaction networks. J. R. Soc. Interface **10**(88), 20130588 (2013)

18. Ruess, J., Milias-Argeitis, A., Summers, S., Lygeros, J.: Moment estimation for chemically reacting systems by extended Kalman filtering. J. Chem. Phys. **135**, 165102 (2011)

19. Ruess, J., Parise, F., Milias-Argeitis, A., Khammash, M., Lygeros, J.: Iterative experiment design guides the characterization of a light-inducible gene expression circuit. Proc. Nat. Acad. Sci. U.S.A. **112**(26), 8148–8153 (2015)

20. Samoilov, M., Arkin, A.: Deviant effects in molecular reaction pathways. Nat. Biotechnol. **24**(10), 1235–1240 (2006)

21. Singh, A., Hespanha, J.: Lognormal moment closures for biochemical reactions. In: 45th IEEE Conference on Decision and Control, pp. 2063–2068 (2006)

22. Whittle, P.: On the use of the normal approximation in the treatment of stochastic processes. J. Roy. Stat. Soc.: Ser. A (Methodol.) **19**, 268–281 (1957)

23. Wolf, V., Goel, R., Mateescu, M., Henzinger, T.: Solving the chemical master equation using sliding windows. BMC Syst. Biol. **4**, 42 (2010)

24. Zechner, C., Ruess, J., Krenn, P., Pelet, S., Peter, M., Lygeros, J., Koeppl, H.: Moment-based inference predicts bimodality in transient gene expression. Proc. Nat. Acad. Sci. U.S.A. **109**(21), 8340–8345 (2012)

Inferring Executable Models from Formalized Experimental Evidence

Vivek Nigam[1], Robin Donaldson[2], Merrill Knapp[2],
Tim McCarthy[2], and Carolyn Talcott[2(✉)]

[1] Federal University of Paraíba, João Pessoa, Brazil
[2] SRI International, Menlo Park, CA, USA
clt@csl.sri.com

Abstract. Executable symbolic models have been successfully used to analyze networks of biological reactions. However, the process of building an executable model from published experimental findings is still carried out manually. The process is very time consuming and requires expert knowledge. As a first step in addressing this problem, this paper introduces an automated method for deriving executable models from formalized experimental findings called *datums*. We identify the relevant data in a collection of datums. We then translate the information contained in datums to logical assertions. Together with a logical theory formalizing the interpretation of datums, these assertions are used to infer a knowledge base of reaction rules. These rules can then be assembled into executable models semi-automatically using the Pathway Logic system. We applied our technique to the experimental evidence relevant to Hras activation in response to Egf available in our datum knowledge base. When compared to the Pathway Logic model (curated manually from the same datums by an expert), our model makes most of the same predictions regarding reachability and knockouts. Missing information is due to missing assertions that require reasoning about the effects of mutations and background knowledge to generate. This is being addressed in ongoing work.

1 Introduction

Executable models of signal transduction provide insights into how cells work, and a means to understand and predict the effects of perturbations and mutations, key for cellular understanding of disease and therapeutics. For example, using an *executable* model one can apply algorithms to determine how one can prevent a given state from being reached or to compute alternative execution paths that reach a given state. Developing such models is extremely difficult. It requires collecting, organizing and interpreting experimental evidence, and assembling rules representing hypothesized biochemical reactions that make up a signaling network. This is very labor intensive and inferring a rule from experiments requires substantial biological knowledge. Several curated models of signaling and metabolic pathways are available [3,11,17–19]. However, there is a great need for tools to help automate the curation of executable models.

© Springer International Publishing Switzerland 2015
O. Roux and J. Bourdon (Eds.): CMSB 2015, LNBI 9308, pp. 90–103, 2015.
DOI: 10.1007/978-3-319-23401-4_9

The problem of automatically constructing executable models from experimental evidence has several aspects including: (1) formal representation of experimental findings, (2) formal representation of rules as elements of executable models, (3) extracting findings from papers, (4) algorithms for inferring rules from findings and (5) algorithms for assembly of executable models. This paper addresses aspects (1), (2) and (4). The contribution is three fold:

1. We describe a formal representation of experimental evidence called *datums*. Each datum captures relevant information about one or more experiments recording conditions under which a specific state or change in state (modification, activity, location) of a protein or other biochemical happens.
2. We define a language of logical assertions that corresponds to the elements of a datum, and a translation from datum syntax to logical assertions.
3. We define axioms that capture the semantics of datums interpreted as partial information about rules to be used as components of an executable model. The logic is that of Answer Set Programs [9] and we use an existing engine (DLV [12]) to derive minimal models called *answer sets*. Each answer set corresponds to one reaction rule. These models are then parsed into rules of an executable model.

Aspect (3) is being addressed as part of an ongoing DARPA project [7] to advance machine reading and reasoning techniques. We use Pathway Logic (PL) [13] as the formal system for representing and querying executable models of cellular processes. Automated analysis techniques such as forward collection and model-checking are used to assemble executable models and execution pathways by specifying a problem of interest (experimental conditions, targets, ...). The PL algorithms rely crucially on the fact that the rules are curated to work together. For example, rules that connect must use the same level of detail concerning location and modifications of participants. In contrast, automatically inferred rules capture all the relevant available experimental information, resulting in a knowledge base that is more precise and extensible. However, the model assembly process will require automation of the process of transforming rules to work together, without losing information unnecessarily. This is the topic of ongoing work.

We applied our algorithms to a collection of datums supporting a model of activation of Hras in response to Egf. The model is part of the PL collection of models manually curated by an expert. Although this first version of the rule generation logic does not account from some of the information in datums, the resulting model makes the same predictions as the curated model concerning response to Egf stimulation and effects of knockouts, with a small number expected exceptions.

Plan. Section 2 gives a brief overview of Pathway Logic executable models and an informal introduction to datums. Section 3 gives an informal introduction to the rule inference process using an Hras activation rule as an example. Section 4 presents the answer set programming axioms/rules of the datum logic.

Section 4.2 describes the mapping of datums to assertions in the logic. Section 5 presents the Hras case study. Section 6 concludes with related and future work.

2   About Pathway Logic and Datums

2.1   Pathway Logic

Pathway Logic (PL) [13] is a system for modeling and reasoning about cellular processes such as signal transduction, metabolism, and cell-cell communication in the immune system. The PL execution model is based on *rewriting logic* [14,15]. In PL, a cell state is represented as a 'soup' of occurrences, where each occurrence has three components: a protein or other biomolecule (gene, metabolite, ...), a modifier, and a location. The modifier indicates the state of the protein, including binding of small molecules or phosphates, or ability to act on other proteins (enzyme activity). For example, the term < [Hras - GTP] , CLi > is the occurrence of the protein Hras modified by binding to the small molecule GTP (Guanosine-5'-triphosphate), attached to the inside of the cell membrane (CLi). The names used to form occurrences are semantically grounded using meta-data to provide links to standard databases.

Signal transduction steps are formalized as local rewrite rules operating on the relevant part of the cell state. Each rule describes a change in state of a small number of biomolecules (often just one) and the biological context that enables the change. A PL *Rule Knowledge Base* (RKB) consists of symbolic rules containing variables that range over a finite set of proteins, modifications or locations. *STM* (Signal Transduction Model) is a curated PL RKB that constitutes an executable model of signal transduction in the following sense: given an initial state called a (Petri) *dish*, which is a set of occurrences representing an experimental setup, the rules can be applied repeatedly, using the Maude rewrite engine [6], to transform the state. This represents a possible sequence of signaling events in a cell. A set of rule instances that can be applied/fired in some order from an initial state is called an *execution pathway*. Specific model networks can be obtained from an RKB by starting with a dish and using *forward collection*[1] to collect all rule instances that might fire in an execution pathway of this dish. Such models can naturally be viewed as Petri Nets [21].

2.2   Datums: Formal Representation of Experimental Results

The PL *STM* model is an RKB whose rules are inferred from cell culture and test tube experiments. In cell-based experiments, cells are grown under known conditions. The cells may be modified by overexpressing some (possibly mutated) proteins, or knocking out some proteins (preventing expression). The resulting population of cells is treated with a stimulus or stress. Some property of the cells is measured before treatment and at one or more times after treatment to determine change in state, if any. The procedure that measures the property

---

[1] Forward collection in this case is application of rules without removing the premises.

change is called an *assay*. Experiments can also be done in a test tube, and some experiments observe untreated cells.

Every rule in the STM RKB is associated with an evidence file, which contains the collected experimental findings giving evidence supporting the rule. These findings are presented in a formal language called *datums*. A datum describes a collection of experimental findings, all based on the same assay, including a main observation, and effects of perturbation of the experimental system. Technically, the collection consists of separate experiments, but they are intended to be interpreted together, so they are collected in a single datum with *extras*. There are two main types of datum, *state datums* and *change datums*, corresponding to two basic types of biological experiments. *State datums* concern properties of cells in a defined state. *Change datums* summarize the change in the state of something resulting from the addition of a stimulus to cells. Rules are derived from change datums.

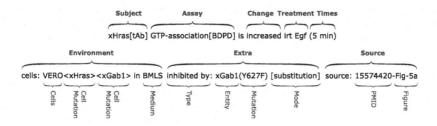

Fig. 1. The elements of a datum.

Datum Structure. The syntax of a datum is designed to be readable by an experimental biologist, but constrained by structure rules and controlled vocabularies so it can be automatically parsed into a formal data structure. The full collection of datums collected for the STM RKB can be accessed via a web query page at light.csl.sri.com/datum. A more detailed description and query examples can be found at pl.csl.sri.com/datumkb.html. The curators notebook (pl.csl.sri.com/CurationNotebook/index.html) contains an intuitive description of datum syntax, catalogs of assays (with their detection methods and other attributes) and cell lines, and a glossary of terms.

The datum in Fig. 1 is a change datum that records an experiment in which the binding of GTP to the protein Hras is increased after addition of Egf (Epidermal Growth Factor) to a cell for 5 min. The first line contains the *subject* (Hras), the *assay* (GTP-association), the *treatment* (Egf) and the *change* (increased). The parenthetical text (*times*) at the end says the measurement was taken 5 min after the treatment. GTP-association is an assay that measures the amount of Hras bound to GTP. The first element of the second line describes the *cellular environment.* In this case VERO cells (a defined cell line) transfected with Gab1 (xGab1), grown in BMLS (Basal Medium Low Serum). The purpose of transfection is that it results in overexpression. The second element is called an "extra".

It records the result of an experiment that is a perturbation of the original experiment. In this case, the cells were transfected with Gab1 with a point mutation (xGab1(Y627F)), in which the tyrosine (Y) at position 627 is replaced by Phenylalanine (F) instead of wild type Gab1 ([substitution]). The third element gives the PubMed identifier of the paper in which the experiment was reported, and the figure where the experimental results were found (15574420-Fig-5a). Source information is not directly used to infer rules, but is crucial for review and updates.

3   Inferring Rules from Datums: An Example

The key ingredients of a datum for rule inference are the subject, assay, treatment, observed change, and cellular environment. Such experimental information is used to constrain the elements of a rule. Specifically, for each assay that measures a change in protein state or location, we associate a rule template that captures the change. The template uses variables for the assay parameters and for additional requirements. The additional requirements can be determined by extras, or by additional experiments. The rule template for a GTP-association assay is

$$
\begin{aligned}
\texttt{TC C < [G - gmods] , Lg >< [P - GDP pmods] , Lp > =>} \\
\texttt{TC C < [G - gmods] , Lg >< [P - GTP pmods] , Lp >}
\end{aligned}
\tag{1}
$$

TC represents the treatment complex that forms to initiate the signal propagation, typically a ligand bound to its activated receptor. C stands for unknown requirements. P is the subject of the assay, Lp is a variable representing the cellular location of P, while G stands for some GEF (Guanosine Exchange Factor) that catalyzes the reaction. pmods and gmods represent the modification state of P and G, respectively. Finally Lp and Lg are the locations of P and G, respectively. Lp, Lg, pmod and gmod must be constrained by additional experiments, or background knowledge.

We can use the datum in Fig. 1 to partially instantiate the GTP-association rule template as follows.

$$
\begin{aligned}
\texttt{EgfTC C < [G - gmods] , Lg >< [Hras - GDP pmods] , CLi > =>} \\
\texttt{EgfTC C < [G - gmods] , Lg >< [HrasP - GDP pmodsd], CLi >}
\end{aligned}
\tag{2}
$$

where EgfTC is the complex that forms when Egf binds to the Egf receptor, which subsequently becomes active and autophosphorylates: < [EgfR - Yphos] : Egf, EgfRC >. We used background knowledge that Hras is anchored to the inside of the plasma membrane to instantiate Lp as CLi.

The next two datums provide evidence that Sos1 is a GEF for Hras.

```
Datum 1:  rHras GDP-dissociation[3H-GDP] is increased by xSos1[tAb]IP
          cells: none, source: 15039778-Fig-2c
Datum 2:  xHras[tAb]IP GTP-association[TLC] is increased itpo xSos1
          cells: HEK293 in BMS, source: 10896938-Fig-1c
```

The first datum says that when you put recombinant Hras (rHras) in a test tube (cells: none) with Sos1 that has been immunoprecipitated (xSos1[tAB]IP) from HEK293 cells, [Hras - GTP] increases. This is direct evidence that Sos1 can act as a GEF in a test tube. We say Sos1 is a *ttGef* (a test tube GEF) for Hras.

Additional evidence that this happens in live cells is needed. The second datum provides such evidence. itpo is a treatment type in which a plasmid for the treatment (Sos1) is introduced into a cell culture and incubated for sufficient time for the treatment protein to become overexpressed. This datum tells us that it is possible that Sos1 can act as a GEF in a cellular environment. We say that Sos1 is an *itpoGef* for Hras. There are datums that report that knocking out Sos1 does not prevent the GDP-GTP exchange. This tells us that there are additional GEFs to be discovered.

Finally, the following datum is evidence for the gabs:GabS requirement.

```
Hras[Ab] GTP-association[BDPD] is increased irt Egf (times)
cells: mEFs in BMLS,   source: 12629518(D) partially reqs: Gab1 [KO]
```

It says that the reaction partially requires Gab1, determined by removing Gab1 from the cellular environment ([KO]). This suggests that Gab1 has a role, but that there may be other proteins that can play the same role as Gab1 in the activation of Hras in response to Egf. To gain confidence in this hypothesis and determine candidate similar proteins, more evidence or background knowledge is needed. This will be the topic of future work and extensions of the datum logic.

4   A Logical Specification for Datums

The interpretation of datums is formalized using *Answer Set Programming* (ASP). We start by briefly explaining ASP before proceeding with the logical specifications of datums.

Answer Set Programs. An ASP program is a collection of clauses of three forms:

$$(1)\, \texttt{D}. \qquad (2)\, \texttt{D :- b1,...,bn}. \qquad (3)\, \texttt{:- b1,...,bn}.$$

where D is either a ground fact, a, or a disjunction of the form a1 v a2, of two ground facts a1 and a2. The symbols b1, ..., bn are ground facts or negated ground facts written not a, where not is negation. The symbol :- should be interpreted as reversed implication and the symbol v as disjunction. Clauses of type (3) are called constraints, specifying that b1, ..., bn should not all be true.

The meaning of an ASP program is a set of ground facts called an *Answer Set*. An answer set of a program $\mathcal{P}$ contains a minimal number of facts that makes each clause of the program $\mathcal{P}$ true. For a formal definition see [9,12].

There are a number of engines that can compute the answer sets of an ASP program. In the present work we have used the DLV engine [12]. Following the usual convention, variables appearing in programs are considered to be shorthand for the set of all possible ground instantiations using the constant and function symbols appearing in the program itself.

4.1   Assertions and Inference Rules for Datums

Some of the main predicates used in the logical theory are given below:

- `subject(S,Dt)` denotes that `S` is the subject of the datum `Dt`.
- `assay(Type,Aux,Dt)` denotes that `Type` is the assay type specified by `Dt`, for example, a phosphorylation or GTP-association. `Aux` is used for assay parameters such as modification sites (`phos!Y627`) or hooks in a binding assay (`none` is used if there are no relevant parameters).
- `treatment(T,Dt)` denotes that `T` is the treatment specified by `Dt`.
- `increased(Dt)`, `irt(Dt)` denote that `Dt` specifies an <u>increase</u> in the changed state of the subject <u>in response to</u> the treatment.

For example, the assertions for the datum of Fig. 1 (Sect. 2) are given below:

```
datum("hras39").                            subject("Hras","hras39").
assay("GTP-Association",none,"hras39").     irt("hras39").
treatment("Egf","hras39").                  increased("hras39").
```

We also have a collection of assertions that are common knowledge, or are implicit in datums collected from experiments by convention. The common knowledge assertions constitute a library used in the inference of the executable rules. An example is the fact that `EgfR` and its modifications are located at `EgfRC`. This is specified by assertions of the form: `location(EgfR, EgfRC, ck)`, where `ck` stands for common knowledge.

Handling Multiple Datums. As described in Sect. 3, some datums contain the evidence for the changes of the subject of a reaction rule. We call these main datums. Other datums, called auxiliary datums, contain evidence about non-subject elements of the reaction, for example, required biomolecules or GEFs. We distinguish these datums using the assertions of the form `useM(Dt)` and `useA(Dt)`, where the former specifies that `Dt` is the main datum and the latter that `Dt` is an auxiliary datum. We specify that an answer set should have exactly one main datum. We do not show the rules here.

Inferred Assertions. We implemented an ASP program that takes the assertions of a datum and generates answer sets, each of which corresponds to a PL rule. In particular, the ASP will derive the following facts:

- `occBf(X1,L1)` denote that before the reaction, `X1` is located at `L1`.
- `occAf(X2,L2)` denotes that after the reaction, `X2` is at location `L2`.
- `occ(X,L)` denotes that the reaction requires `X` at location `L` in order to occur. Such an assertion can be used for a treatment complex or a require composite.
- `moveRule` and `reactRule` denote that the rule to be extracted is either a rule specifying that the subject moves from one location to another without changing its modifications or it is a rule specifying that the subject changes its modifications without changing its location. This separation between move and react rules provides a finer grained specification of a model that simplifies the (meta) reasoning.

These assertions are used to construct rules in our executable model of the form depicted in Eq. 1. Before we explain how these facts are derived, we illustrate how answer sets correspond to rules by example. Consider two answer sets M_1 and M_2, where M_1 contains the set of facts to the left and M_2 contains the set of facts to the right:

$$
\left\{
\begin{array}{c}
\texttt{moveRule,} \\
\texttt{occBf(Hras - mods(Hras),L(Hras)),} \\
\texttt{occAf(Hras - mods(Hras),EgfRC),} \\
\texttt{occ(Egf:EgfR-Yphos,EgfRC)}
\end{array}
\right\}
\quad
\left\{
\begin{array}{c}
\texttt{reactRule,} \\
\texttt{occBf(Hras - mods(Hras) - GDP,L(Hras)),} \\
\texttt{occAf(Hras - mods(Hras) - GTP,L(Hras)),} \\
\texttt{occ(Egf:EgfR-Yphos,EgfRC),} \\
\texttt{occ(Sos1 - mods(Sos1),L(Sos1)),} \\
\texttt{occ(Gab1 - mods(Gab1),L(Gab1))}
\end{array}
\right\}
$$

Here `mods(X)` and `L(X)` are variables that can be instantiated in our executable model by any modifiers and locations, respectively. The answer set M_1 specifies the rule below where `Hras - mods(Hras)` moves from a generic location `L(Hras)` to the location `EgfRC` in the presence of `Egf:EgfR-Yphos` at location `EgfRC`:

$$
\begin{array}{c}
\texttt{< Hras - mods(Hras) , L(Hras) >< Egf:EgfR-Yphos , EgfRC > =>} \\
\texttt{< Hras - mods(Hras) , EgfRC >< Egf:EgfR-Yphos , EgfRC >}
\end{array}
\tag{3}
$$

The answer set M_2 specifies the following rule where the subject `Hras - mods (Hras) - GDP` at a generic location `L(Hras)` is modified to `Hras - mods(Hras) - GTP` in the presence of `Egf:EgfR-Yphos` at location `EgfRC`, `Sos1 - mods(Sos1)` and `Gab1 - mods(Gab1)` at the generic locations `L(Sos1)` and `L(Gab1)`, respectively:

```
< Hras - mods(Hras) - GDP, L(Hras) > < Egf:EgfR-Yphos, EgfRC >
< Sos1 - mods(Sos1), L(Sos1) > < Gab1 - mods(Gab1), L(Gab1) >
=>
< Hras - mods(Hras) - GTP, L(Hras) > < Egf:EgfR-Yphos, EgfRC >
< Sos1 - mods(Sos1), L(Sos1) > < Gab1 - mods(Gab1), L(Gab1) >
```

Specification of Assertion Reasoning. As illustrated above, answer sets specify reaction or move rules. This is specified by the following clauses and constraints:

```
reactRule v moveRule.
:- occBf(X1,L1), occAf(X2,L2), moveRule, X1 <> X2.
:- occBf(X1, L), occAf(X2, L), moveRule.
:- occBf(X1, L1), occAf(X2, L2), reactRule, L1 <> L2.
:- occBf(X, L1), occAf(X, L2), reactRule.
```

The first clause specifies that answer sets must correspond to either move or react rules. The constraints say that in the specification of move rules, the subject should not be modified and it should move. Similarly for react rules, the location of the subject should not change and the subject should be modified. There are other constraints that are omitted, specifying that move rules only make sense when we know where the subject moves to.

We derive `occ`, `occBf` and `occAf` assertions by deriving the corresponding argument, namely the corresponding possibly modified protein and its location. This is done by using the following auxiliary predicates which will be used to infer the elements in a rule of the form in Eq. 1:

- `in(X)` says that there is a possibly modified protein in the rule context, *e.g.*, a treatment complex. `inBf(X)` and `inAf(X)` specify the state of the subject protein before and after the rule, respectively.
- `loc(X,L)` says that a non-subject element `X` is at location `L`. `locBf(X,L)` and `locAf(X,L)` say that the location of the subject `X` is `L` before and after the reaction.

Using these assertions, we derive `occ`, `occBf` and `occAf` assertions using the clauses below:

```
occBf(X,L(X)) :- inBf(X), reactRule.
occAf(X,L(X)) :- inAf(X), reactRule.
occ(X, L(X))  :-  in(X), not hasLocation(X).
occBf(X - mods(X),L) :- subject(X,Dt),useM(Dt),locBf(X,L),moveRule.
occBf(X - mods(X),L(X)) :- subject(X,Dt),useM(Dt),not hasLocBf(X),moveRule.
occAf(modBy(X - mods(X),L) :- subject(X,Dt),useM(Dt),locAf(X,L),moveRule.
```

Here `hasLocation(X)` is an auxiliary assertion (rule omitted), denoting that it is possible to infer a concrete location for `X`.

Datum assertions are used to derive the more basic assertions `in`, `inBf`, `inAf`, `loc`, `locBf` and `locAf`. For example, a GTP-association datum can be used in the following clauses to derive `inBf` and `inAf` facts:

```
inBf(X - mods(X) - GDP) :- irt(Dt), increased(Dt),
        assay(GTP-association, none, Dt), subject(X, Dt), useM(Dt).
inAf(X, mods(X) - GTP) :- irt(Dt), increased(Dt),
        assay(GTP-association, none, Dt), subject(X, Dt), useM(Dt).
in(X) :- tc(X, Dt), useM(Dt).
```

These clauses specify that if the main datum is a GTP-association, then the subject before the reaction should be modified with GDP and after with GTP. Moreover, the treatment complex should be in the dish, specified by the last clause. Similar clauses exists for the other types of datums, such as phosphorylation datums. In a similar way, the location assertions `loc`, `locBf` and `locAf` are derived from datum assertions. Some of them might be derived from common knowledge. We do not show these clauses here.

As described in Sect. 3, other datums provide information about the non-subject elements in a reaction. For example, datums may provide information about GEFs. These are specified by the assertions `ttGEF(Q,S,Dt)` and `itpoGEF(Q,S,Dt)`. Both denote that the datum `Dt` specifies that `Q` could be a GEF for the subject `S`. The former, however, denotes that the experiment was carried out in the test tube, while the latter denotes that the experiment was carried out using cells transfected with `Q`. We infer these assertions from datum assertions as illustrated below.

```
itpoGEF(Q,X,Dt) :- assay(GTP-association,none,Dt), itpo(Dt),
            increased(Dt), subject(X, Dt), treatment(Q,Dt), useA(Dt).
ttGEF(Q,X,Dt) :- assay(GTP-association, none, Dt), by(Dt),
            increased(Dt), subject(X, Dt), treatment(Q,Dt), useA(Dt).
```

4.2   Mapping Datums to Assertions

Each datum is mapped to a set of logical assertions that captures the subject, assay, treatment, treatment type, and change elements of a datum. The mapping algorithm takes as input the JSON representation of datums produced by the datum parser and produces input for the DLV engine as described above.

We ignore datums where the interpretation is complex and often requires specific biological knowledge. We currently ignore any datum with no subject, a mutated subject, a mutated treatment or more than one treatment.

In version 1 of the mapping algorithm, only extras of type "reqs" are captured as their interpretation is relatively straightforward. Extending the mapping algorithm to use "inhibited by" extras is a topic of future work.

Many datums report the same basic experiment, i.e. the same subject, assay and treatment. If these datums also have the same change (result) then the mapping will *merge* them, otherwise the datums are reported to the user as a *conflict* for manual inspection. Conflicts may be particularly troubling because datums span many different cell lines and cell types.

It is then a simple case of mapping each element of the datum (or merged datums) to their logical assertions. For example, the datum from Fig. 1 and the datum from Sect. 3 giving the requirement for Gab1 can be merged, omitting elements not used for generating assertions. The result is

```
xHras[tAb] GTP-association[BDPD] is increased irt Egf
inhibited by: xGab1(Y627F) [substitution] partially reqs: Gab1 [KO]
```

which maps to the following set of assertions:

```
datum("d1-d2").              subject("Hras", "d1-d2").
irt("d1-d2").                assay("GTP-association", none, "d1-d2").
increased("d1-d2").          treatment("Egf", "d1-d2").
reqs("Gab1", "d1-d2").
```

In the case of merged datums, the identifiers of the contributing datums are merged, thus "d1-d2" above. This allows us to track evidence and eventually reason about the quality/quantity of evidence used in generating a rule. The actual merged datum in our case study (Sect. 5) combines 51 datums from the datum knowledge base.

Because we merge all datums for the same change, each set of assertions corresponds to one rule in the model, and contains all information for the set of controls for the rule. Note that auxiliary datums will still be used to find assay specific enzymes such as GEFs or Kinases.

5   Signaling Model of Hras Activation by Egf

To test our rule inference tool, we used a model of Hras activation (GTP binding) in response to Egf derived from the PL STM RKB as a 'gold standard'. The Hras model was derived by generating the subnet relevant to the goal < [Hras − GTP], CLi >. An execution pathway in this model is shown in Fig. 2(a).

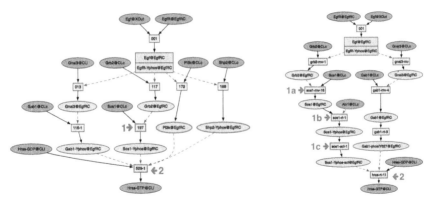

(a) A pathway in the STM Hras model.   (b) An inferred Hras pathway.

Fig. 2. Hras Models

The datums used as input for the inferred model came from the evidence files for these rules together with files containing evidence for Hras GEFs. The JSON datum representation was generated using the datum parser, assertions were generated from the JSON using the assertion mapping tool, and rules were then generated using the logic engine, and automatically converted to Maude syntax.

As discussed in Sect. 1, the final step is assembly of these rules into a model—a connected set of rules that can be executed to reach expected goals, including the activation of Hras. The basic assembly process is carried out using the PL model generation process. We adapted the initial state for the STM Hras subnet to specify the desired model. The abstraction of details to form a connected rule network was carried out by hand, guided by principles developed by the curator of the STM model. Abstracting includes dropping site details from modifications and formalizing knowledge/conjectures such as 'modification implies activation' in specific cases.

The resulting model is more detailed than the STM Hras model. This is expected, due to the separation of modification and translocation rules (the STM model typically collapses these into one step), and the use of location and modification variables that have multiple possible instantiations.

The inferred model answers most of the queries supported by PL in the same way that the STM Hras model does. Examples include reachability of given states, existence of multiple execution paths to the Hras goal, and (RasGrp3, Sos1) as a double knockout pair.

An execution pathway corresponding to the STM model pathway is shown in Fig. 2b. The STM rule 197 for phosphorylation of Sos1 (arrows labeled 1) becomes 3 rules in the inferred model (a move, a modification, and activation). The inferred model has Abl1 (red border) as a requirement for Sos1 phosphorylation. There is a single datum specifying this requirement; the STM curator did not consider one datum showing this requirement as sufficient evidence. Future work includes associating rules with some measure of quantity/quality of evidence, in order to able to assemble models using different criteria for inclusion of rules.

The STM rule 529 for Hras activation (GTP association, arrows labeled 2) includes a requirement for [Shp2 - Yphos] and a requirement for Pi3k (red borders), while the inferred rule does not. These requirements come from extras such as `inhibited by: xPik3r?(mnr)"DN"`... and `inhibited by: xShp2(mnr)"CIA"` that require substantial background knowledge to interpret. For example, `CIA` stands for 'Constitutively InActive'. The inference is that if the endogenous protein is overwhelmed by a mutated form that is lacking some function, then that protein (with that function) is required. Future versions of the assertion generation tool will capture more of these inferences.

6 Related Work and Conclusion

Related work. An excellent survey of executable models of biological processes is given in [8]. There are a number of network reconstruction algorithms based on statistical reasoning techniques such as Bayesian inference [10] or belief propagation [16]. They provide a means of elucidating the networks underlying transcriptomics and proteomics data generated from perturbation experiments. These methods postulate causal relations, but do not capture mechanistic details such as necessary conditions.

Methods more closely related to our approach include the following. Netsynthesis [1,2] is a software for synthesis, inference and simplification of signal transduction networks. The main idea is representing observed indirect causal relationships as network paths, introducing pseudo-vertices for unknown intermediaries of these paths and using techniques from combinatorial optimization to find the most parsimonious graph consistent with all experimental observations. A method based on Petri nets is described in [4]. The reactions of individual proteins are represented as Petri net modules, stored in a database. These modules are similar in spirit to datums. Each place in a module corresponds to a specific functional state of a specific protein domain (e.g. a phosphorylated or unphosphorylated side chain, a catalytically active or inactive domain etc.). For each module, literature references are annotated as part of the modules database entry. Selected modules can be combined to assemble executable Petri net models. The method has been applied to assemble a model of JAK/STAT signaling. In [20], two methods to build signaling models from qualitative data (protein interactions from databases) are proposed, based on analyzing network connectivity and on non-linear optimization. Methods to convert BioPAX models into fully executable models have been proposed, including [5,22]. The work presented here differs from these works in starting from experimental evidence to build knowledge bases and executable models, rather that relying on existing pathway databases.

Conclusion. We have presented an inference system for deriving signal transduction rules from formally represented experimental findings and applied the system to derive rules for a model of Hras activation[2]. Future work includes: extending the mapping of datums to assertions to capture the meaning of experimental perturbations using mutations and fragmentation, extracting formal background knowledge from databases, extending the logic to cover more assays and capture more complex reasoning, such as hypothesizing rule requirements and alternatives by similarity, adding logic to generate *common rules* (rules about protein interactions independent of stimulus), and automating assembly of models from generated rules.

References

1. Albert, R., DasGupta, B., Dondi, R., Sontag, E.: Inferring (biological) signal transduction networks via transitive reductions of directed graphs. Algorithmica **51**(2), 129–159 (2008)
2. Albert, R., DasGupta, B., Sontag, E.: Inference of signal transduction networks from double causal evidence. In: Fenyo, D. (ed.) Methods in Molecular Biology: Topics in Computational Biology. Springer Science+Business Media LLC, New York (2010)
3. Biocyc pathway/genome database collection (2015)
4. Blätke, M., Dittrich, A., Rohr, C., Heiner, M., Schaper, F., Marwan, W.: Jak/stat signalling: an executable model assembled from molecule-centred modules demonstrating a module-oriented database concept for systems and synthetic biology. Mol. Biosyst. **9**(6), 1290–1307 (2012)
5. Blinov, M.L., et al.: Pathway commons at virtual cell: use of pathway data for mathematical modeling. Bioinformatics **30**(2), 292–294 (2014)
6. Clavel, M., Durán, F., Eker, S., Lincoln, P., Martí-Oliet, N., Meseguer, J., Talcott, C. (eds.): All About Maude - A High-Performance Logical Framework. LNCS, vol. 4350. Springer, Heidelberg (2007)
7. DARPA Big Mechanism Project (2015)
8. Fisher, J., Henzinger, T.A.: Executable cell biology. Nat. Biotechnol. **25**(11), 1239–1249 (2007)
9. Gelfond, M., Lifschitz, V.: Logic programs with classical negation. In: ICLP, pp. 579–597 (1990)
10. Hill, S.M., et al.: Bayesian inference of signaling network topology in a cancer cell line. Bioinformatics **28**(21), 2804–2810 (2012)
11. KEGG: Kyoto encyclopedia of genes and genomes (2015)
12. Leone, N., Pfeifer, G., Faber, W., Eiter, T., Gottlob, G., Perri, S., Scarcello, F.: The DLV system for knowledge redlvpresentation and reasoning. ACM Trans. Comput. Logic **7**, 499–562 (2006)
13. Lincoln, P.D., Talcott, C.: Symbolic systems biology and pathway logic. In: Iyengar, S. (ed.) Symbolic Systems Biology, pp. 1–29. Jones and Bartlett, Boston (2010)
14. Meseguer, J.: Conditional rewriting logic as a unified model of concurrency. Theo. Comput. Sci. **96**(1), 73–155 (1992)
15. Meseguer, J.: Twenty years of rewriting logic. J. Logic Algebraic Program. **81**(7–8), 721–781 (2012)

---

[2] The assertion mapping code and logic are currently being extended and improved. We are happy to make the current working version available upon request.

16. Molinelli, E.J., et al.: Perturbation biology: inferring signaling networks in cellular systems. PLoS Comput. Biol. **9**(12), e1003290 (2013). PMID: 24367245, PMCID: PMC3868523
17. Pathway logic (2015)
18. Protein interaction database (2015)
19. Reactome pathway database (2015)
20. Ruths, D.: Deriving Executable Models of Biochemical Network Dynamics from Qualitative Data. Rice University (2009)
21. Talcott, C., Dill, D.L.: Multiple representations of biological processes. In: Priami, C., Plotkin, G. (eds.) Transactions on Computational Systems Biology VI. LNCS (LNBI), vol. 4220, pp. 221–245. Springer, Heidelberg (2006)
22. Willemsen, T., Feenstra, K.A., Groth, P.T.: Building executable biological pathway models automatically from BioPAX. In: Linked Science 2013: Supporting Reproducibility, Scientific Investigations and Experiments, pp. 2–14 (2013)

Symbolic Dynamics of Biochemical Pathways as Finite States Machines

Ovidiu Radulescu[1]([✉]), Satya Swarup Samal[2], Aurélien Naldi[1],
Dima Grigoriev[3], and Andreas Weber[2]

[1] DIMNP UMR CNRS 5235, University of Montpellier, Montpellier, France
`ovidiu.radulescu@univ-montp2.fr`
[2] Institut für Informatik II, Universität Bonn,
Friedrich-Ebert-Allee 144, 53113 Bonn, Germany
`{samal,weber}@cs.uni-bonn.de`
[3] CNRS, Mathématiques, Université de Lille, 59655 Villeneuve d'Ascq, France
`dmitry.grigoryev@math.univ-lille1.fr`

Abstract. We discuss the symbolic dynamics of biochemical networks with separate timescales. We show that symbolic dynamics of monomolecular reaction networks with separated rate constants can be described by deterministic, acyclic automata with a number of states that is inferior to the number of biochemical species. For nonlinear pathways, we propose a general approach to approximate their dynamics by finite state machines working on the metastable states of the network (long life states where the system has slow dynamics). For networks with polynomial rate functions we propose to compute metastable states as solutions of the tropical equilibration problem. Tropical equilibrations are defined by the equality of at least two dominant monomials of opposite signs in the differential equations of each dynamic variable. In algebraic geometry, tropical equilibrations are tantamount to tropical prevarieties, that are finite intersections of tropical hypersurfaces.

1 Introduction

Networks of biochemical reactions are used in computational biology as models of signaling, metabolism, and gene regulation. For various applications it is important to understand how the dynamics of these models depend on internal parameters and environment variables. Traditionally, the dynamics of biochemical networks is studied in the framework of chemical kinetics that can be either deterministic (ordinary differential equations) or stochastic (continuous time Markov processes). Within this framework, problems such as causality, reachability, temporal logics, are hard to solve and even to formalize. Concurrency models such as Petri nets and process algebra conveniently formalize these questions that remain nevertheless difficult. The main source of difficulty is the extensiveness of the set of trajectories that have to be analysed. Discretisation of the phase space does not solve the problem, because in multi-valued networks with m levels (Boolean networks correspond to $m = 2$) the number of the states is m^n and grows exponentially with the number of variables n. An interesting

© Springer International Publishing Switzerland 2015
O. Roux and J. Bourdon (Eds.): CMSB 2015, LNBI 9308, pp. 104–120, 2015.
DOI: 10.1007/978-3-319-23401-4_10

alternative to these approaches is symbolic dynamics which means replacing the trajectories of the smooth system with a sequence of symbols. In certain cases, this could lead to relatively simple descriptions. According to the famous conjecture of Jacob Palis [11], smooth dynamical systems on compact spaces should have a finite number of attractors whose basins cover the entire ambient space. Compactness of ambient space is satisfied by networks of biochemical reactions because of conservation, or dissipativity. For high dimensional systems with multiple separated timescales it reasonable to consider the following property: trajectories within basins of attraction consists in a succession of fast transitions between relatively slow regions. The slow regions, generally called metastable states, can be of several types such as attractive invariant manifolds, Milnor attractors or saddles. Because of compactness of the ambient space and smoothness of the vector fields defining the dynamics, there should be a finite number of such metastable states. This phenomenon, called itinerancy received particular attention in neuroscience [18]. We believe that similar phenomena occur in molecular regulatory networks. A simple example is the set of bifurcations of metastable states guiding the orderly progression of the cell cycle. In this paper we use tropical geometry methods to detect the presence of metastable states and describe the symbolic dynamics as a finite state automaton. The structure of the paper is the following. In the second section we compute the symbolic dynamics of monomolecular networks with totally separated constants. To this aim we rely on previous results [4,12,13]. In the third section we introduce tropical equilibrations of nonlinear networks. Tropical equilibrations are good candidates for metastable states. More precisely, we use minimal branches of tropical equilibrations as proxys for metastable states. In the forth section we propose an algorithm to learn finite state automata defined on these states.

2   Monomolecular Networks with Totally Separated Constants

Monomolecular reaction networks are the simplest reactions networks. The structure of these networks is completely defined by a digraph $G = (V, \mathcal{A})$, in which vertices $i \in V, 1 \leq i \leq n$ correspond to chemical species A_i, edges $(i, j) \in \mathcal{A}$ correspond to reactions $A_i \rightarrow A_j$ with kinetic constants $k_{ji} > 0$. For each vertex, A_i, a positive real variable c_i (concentration) is defined. The chemical kinetic dynamics is described by a system of linear differential equations

$$\frac{dc_i}{dt} = \sum_j k_{ij} c_j - (\sum_j k_{ji}) c_i, \tag{1}$$

where $k_{ji} > 0$ are kinetic coefficients. In matrix form one has : $\dot{c} = Kc$. The solutions of (1) can be expressed in terms of left and right eigenvectors of the kinetic matrix K:

$$c(t) = r^0(l^0, c(0)) + \sum_{k=1}^{n-1} r^k(l^k, c(0)) \exp(\lambda_k t), \tag{2}$$

where r^k, l^k are right and left eigenvectors of K, $Kr^k = \lambda_k r^k$, $l^k K = \lambda_k l^k$.

The system (1) has a conservation law $\frac{d}{dt}(c_1+c_2+\ldots+c_n) = 0$, and therefore there is a zero eigenvalue $\lambda_0 = 0$, $l^0 = (1,1,\ldots,1)$, $(l^0, c(0)) = c_1(0) + c_2(0) + \ldots + c_n(0)$. We say that the network constants are totally separated if for all $(i,j) \neq (i',j')$ one of the relations $k_{ji} \ll k_{j'i'}$, or $k_{ji} \gg k_{j'i'}$ is satisfied.

It was shown in [4,12,13] that the eigenvalues and the eigenvectors of an arbitrary monomolecular reaction networks with totally separated constants can be approximated with good accuracy by the eigenvalues of and the eigenvectors of a reduced monomolecular networks whose reaction digraph is acyclic (has no cycles), and deterministic (has no nodes from which leave more than one edge). Let us denote by $G_r = (V_r, \mathcal{A}_r)$ the reduced digraph, and by κ_i the kinetic constant of the unique reaction that leaves a node $i \in V_r$. The algorithm to obtain G from G_r can be found in [4,12,13] and will not be repeated here. Because G_r is deterministic it defines a flow (discrete dynamical system) on the graph: $\Phi(i) = j$, where j is the unique node following i on the digraph. Reciprocally, we define $\text{Pred}(i) = \phi^{-1}(i)$ as the set of predecessors of the node i in the digraph G_r, namely $\text{Pred}(i) = \{j \in V_r | (j,i) \in \mathcal{A}_r\}$.

We say that a node is a sink if it has no successors on the graph. For the sake of simplicity, we suppose that there is only one sink. For each one of the remaining $n-1$ nodes there is one reaction leaving from it. For a network with totally separated constants we have

$$\kappa_i \ll \kappa_j, \text{ or } \kappa_i \gg \kappa_j \text{ for all } i,j \in [1, n-1], i \neq j \tag{3}$$

For totally separated constants the following lemma is useful

Lemma 1. *If* (3) *is satisfied then, at lowest order, we have*

$$\frac{\kappa_i}{-\kappa_k + \kappa_j} = \begin{cases} 1, & \text{if } i = j \text{ and } \kappa_k < \kappa_i \\ -1, & \text{if } i = k \text{ and } \kappa_j < \kappa_i \\ 0, & \text{if } \kappa_i < \min(\kappa_k, \kappa_j) \\ \pm\infty, & \text{else} \end{cases} \tag{4}$$

The dynamics of the reduced model is given by

$$\frac{dc_i}{dt} = \sum_{j \in \text{Pred}(i)} \kappa_j c_j - \kappa_i c_i, \tag{5}$$

where $\text{Pred}(i)$ is the set of predecessors of the node i in the digraph G_r, namely $\text{Pred}(i) = \{j \in V_r | (j,i) \in \mathcal{A}_r\}$.

As shown in [4] the eigenvectors of the approximated kinetic matrix satisfy

$$\sum_{j \in \text{Pred}(i)} \kappa_j r_j = (\lambda + \kappa_i) r_i \tag{6}$$

$$\kappa_i l_{\Phi(i)} = (\lambda + \kappa_i) l_i, \tag{7}$$

where λ is the eigenvalue, r_i, l_i, $1 \leq i \leq n$ are the components of the right and left eigenvectors, respectively.

Equations (6) and (7) imply that the right and left eigenvectors can be computed by recurrence on the graph, in the direct direction and in the reverse direction, respectively. In order to have non-zero eigenvectors, $\lambda = -\kappa_i$ for some i not a sink, therefore the (non-zero) eigenvalues are $\lambda_k = -\kappa_k, 1 \leq k \leq n-1$. Taking into account the separation conditions (3) we get the following

Proposition 1. *Let us consider that $\kappa_k = 0$ when k is a sink in the graph G_r. Then, the eigenvalues of the kinetic matrix with totally separated constants are $\lambda_k = -\kappa_k$, with $\lambda_k = 0$ when k is a sink. The corresponding left eigenvectors are*

$$l^k_j = \begin{cases} 1, \text{ if } \Phi^m(j) = k \text{ for some } m > 0 \text{ and } \kappa_{\Phi^l(j)} > \kappa_k \text{ for all } l = 0, \dots, m-1 \\ 0, \text{ otherwise} \end{cases},$$

$$(8)$$

and the right eigenvectors are

$$r^k_j = \begin{cases} 1, & \text{if } j = k \\ -1, & \text{if } j = \Phi^m(k) \text{ for some } m > 0 \text{ and } \kappa_{\Phi^m(k)} < \kappa_k < \kappa_{\Phi^l(k)}, \\ & \text{for all } l = 1, \dots, m-1 \\ 0, & \text{otherwise.} \end{cases}$$

The full proof of the Proposition 1 can be found in the appendices.

Let us now discuss the symbolic dynamics of the system. For each eigenvalue $\lambda_k = -\kappa_k$, $\kappa_k > 0$ we associate a transition time $t_k = \kappa_k^{-1}$. Without loss of generality we can consider that $t_1 \ll t_2 \ll \dots \ll t_{n-1}$. Any trajectory of the system is given by (2). At the time t_k one exponential term $exp(\lambda_k t)$ will vanish and the result will be a transition $c \to c - r^k(l^k, c(0))$, provided that $(l^k, c(0)) \neq 0$. In other works, a trajectory can be described as a discrete sequence of states $c(0), c(0) - r^1(l^1, c(0)), \dots$. Let us consider the following normalization $c_1(0) + c_2(0) + \dots + c_n(0) = 1$. Then c_i is the probability of presence in the node i of a particle moving through the reaction network. For monomolecular networks, particles are independent, therefore this simple picture is enough for understanding the dynamics. Let the index i_0 define the initial state of the system $c_{i_0}(0) = 1$, $c_j(0) = 0$ for $j \neq i_0$. i_0 represents the initial position of the particle. According to the Proposition 1 $(l^k, c(0)) = l^k_{i_0} = 1$ if the step κ_k is downstream of i_0 in the graph G_r and if all steps from i_0 to k are faster than κ_k. In this case the jump at t_k is $-r^k$. A jump $-r^k$ has two components different from zero, $-r^k_k = -1$ and $-r^k_j = 1$, where j is the first node downstream of k from which starts a step slower than κ_k. Thus, the jump $-r^k$ corresponds to displacing the particle from k to j. The set of right eigenvectors defines a symbolic flow on the reaction digraph. A particle starting in i_0 first jumps in i_1 where i_1 is the first node such that $\kappa_{i_1} < \kappa_{i_0}$, then continues to i_2 where i_2 is the first node such that $\kappa_{i_2} < \kappa_{i_1}$, and so one and so forth until it gets to the sink. Some nodes have negligible sojourn time, namely nodes such that $\kappa_i > \kappa_j$ for all $j \in \text{Pred}(i)$. This proves the main result of the section.

By transition graph of a finite state machine we mean the digraph $G_{rs} = (V_s, \mathcal{A}_s)$, where V_s is the set of states of the machine and $(i, j) \in \mathcal{A}_s$ if there are transitions from the state i to the state j. We have the following theorem:

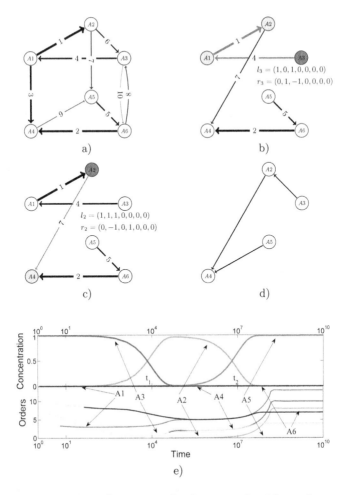

Fig. 1. Symbolic dynamics of a monomolecular network with total separation. The integers γ_i labelling the reactions represent the orders of the kinetic constants, smaller orders meaning faster reactions. The model was reduced using the recipe described in [4,13] (see appendices). (a) full model; (b–c) reduced model with active transitions and corresponding eigenvectors. During a transition the network behaves like a single step: the concentrations of some species (white) are practically constant, some species (yellow) are rapid, low concentration, intermediates, one species (red) is gradually consumed and another (pink) is gradually produced. The net result is the displacement of a particle one or several steps downstream; (d) The transition graph of the finite state machine representing the symbolic dynamics of the network; (e) Trajectory starting from $A3$ (at $t = 0$ the total mass is in $A3$), undergoing two transitions at t_1 and t_2. The simulation has been performed for kinetic constants $\kappa_i = \varepsilon^{\gamma_i}$, with $\varepsilon = 1/50$. On top, concentration of species (concentrations of $A1, A_4, A_6$ are negligible everywhere). At bottom, orders of concentrations (computed as $\log_\varepsilon(x_i)$) with continuous lines if species is tropically equilibrated, dotted lines if not (Color figure online).

Theorem 1. *The symbolic dynamics of a monomolecular network with totally separated constants can be described by a deterministic acyclic finite state machine. The transition graph $G_{rs} = (V_s, \mathcal{A}_s)$ of this machine can be obtained from the graph $G_r = (V_r, \mathcal{A}_r)$ in the following way: $V_s = V_r \setminus \{i \in V_r | \kappa_i > \kappa_j \text{ for all } j \in \mathrm{Pred}(i)\}$, $\mathcal{A}_s = \{(i,j) | i,j \in V_s \text{ and there are } i_0 = i, i_1, \ldots, i_m = j, \text{ such that } i_l \in V_r \setminus V_s, \text{ for } l = 1, \ldots, m-1, \text{ and } (i_l, i_{l+1}) \in \mathcal{A}_r \text{ for } l = 0, \ldots, m-1\}$.*

Remark 1. An example is detailed in Fig. 1.

3   Tropical Equilibrations of Nonlinear Networks with Polynomial Rate Functions

In this section we consider nonlinear biochemical networks described by mass action kinetics

$$\frac{dx_i}{dt} = \sum_j k_j S_{ij} x^{\alpha_j}, \ 1 \le i \le n, \tag{9}$$

where $k_j > 0$ are kinetic constants, S_{ij} are the entries of the stoichiometric matrix (uniformly bounded integers, $|S_{ij}| < s$, s is small), $\alpha_j = (\alpha_1^j, \ldots, \alpha_n^j)$ are multi-indices, and $x^{\alpha_j} = x_1^{\alpha_1^j} \ldots x_n^{\alpha_n^j}$, where α_i^j are positive integers.

For chemical reaction networks with multiple timescales it is reasonable to consider that kinetic parameters have different orders of magnitudes. This can be conveniently formalized by considering that parameters of the kinetic models (9) can be written as

$$k_j = \bar{k}_j \varepsilon^{\gamma_j}. \tag{10}$$

The exponents γ_j are considered to be integer or rational. For instance, the approximation $\gamma_j = \mathrm{round}(\log(k_j)/\log(\varepsilon))$ produces integer exponents, whereas $\gamma_j = \mathrm{round}(d \log(k_j)/\log(\varepsilon))/d$ produces rational exponents, where round stands for the closest integer (with half-integers rounded to even numbers) and d is a strictly positive integer. Kinetic parameters are fixed. In contrast, species orders vary in the concentration space and have to be calculated as solutions to the tropical equilibration problem. To this aim, the network dynamics is first described by a rescaled ODE system

$$\frac{d\bar{x}_i}{dt} = \sum_j \varepsilon^{\mu_j(a) - a_i} \bar{k}_j S_{ij} \bar{x}^{\alpha_j}, \tag{11}$$

where

$$\mu_j(a) = \gamma_j + \langle a, \alpha_j \rangle, \tag{12}$$

and $\langle , \rangle$ stands for the dot product.

The r.h.s. of each equation in (11) is a sum of multivariate monomials in the concentrations. The orders μ_j indicate how large are these monomials, in absolute value. A monomial of order μ_j dominates another monomial of order $\mu_{j'}$ if $\mu_j < \mu_{j'}$.

The tropical equilibration problem consists in the equality of the orders of at least two monomials one positive and another negative in the differential equations of each species. More precisely, we want to find a vector a such that

$$\min_{j,S_{ij}>0} (\gamma_j + \langle a, \alpha_j \rangle) = \min_{j,S_{ij}<0} (\gamma_j + \langle a, \alpha_j \rangle) \tag{13}$$

Computing tropical equilibrations from the orders of magnitude of the model parameters is a NP-hard problem, cf. [17]. However, methods based on the Newton polytope [15] or constraint logic programming [16] exploit the sparseness and redundancy of the system to effectively obtain sets of solutions. The Eq. (13) is related to the notion of *tropical hypersurface*. A *tropical hypersurface* is the set of vectors $a \in \mathbb{R}^n$ such that the minimum $\min_{j,S_{ij}\neq0}(\gamma_j + \langle a, \alpha_j \rangle)$ is attained for at least two different indices j (with no sign conditions). *Tropical prevarieties* are finite intersections of tropical hypersurfaces. Therefore, our tropical equilibrations are subsets of tropical preverieties. The sign condition in (13) was imposed because species concentrations are real positive numbers. Compensation of a sum of positive monomials is not possible for real values of the variables.

Species Timescales. The timescale of a variable x_i is given by $\frac{1}{x_i}\frac{dx_i}{dt} = \frac{1}{\bar{x}_i}\frac{d\bar{x}_i}{dt}$ whose order is

$$\nu_i = \min\{\mu_j | S_{ij} \neq 0\} - a_i. \tag{14}$$

The order ν_i indicates how fast is the variable x_i (if $\nu_{i'} < \nu_i$ then $x_{i'}$ is faster than x_i).

Partial Tropical Equilibrations. It is useful to extend the tropical equilibration problem to partial equilibrations, that means solving (13) only for a subset of species. This is justified by the fact that slow species do not need to be equilibrated. In order to have a self-consistent calculation we compute the species timescales by (14). A partial equilibration is *consistent* if $\nu_i < \nu$ for all non-equilibrated species i. $\nu > 0$ is an arbitrarily chosen threshold indicating the timescale of interest.

Tropical Equilibrations, Slow Invariant Manifolds and Metastable States. In dissipative systems, fast variables relax rapidly to some low dimensional attractive manifold called invariant manifold [3] that carries the slow mode dynamics. A projection of dynamical equations onto this manifold provides the reduced dynamics [8]. This simple picture can be complexified to cope with hierarchies of invariant manifolds and with phenomena such as transverse instability, excitability and itineracy. Firstly, the relaxation towards an attractor can have several stages, each with its own invariant manifold. During relaxation towards the attractor, invariant manifolds are usually embedded one into another (there is a decrease of dimensionality) [2]. Secondly, invariant manifolds can lose local stability, which allow the trajectories to perform large phase space excursions before returning in a different place on the same invariant manifold or on a different one [7]. We showed elsewhere that tropical equilibrations can be used to approximate invariant manifolds for systems of polynomial differential equations [9,10,14]. Indeed, tropical equilibration are defined by the equality of dominant forces acting on the system. The remaining weak non-compensated forces ensure the slow dynamics on the invariant manifold. Tropical equilibrations are thus different from steady states, in that there is a slow dynamics. In this paper we will use them as proxies for metastable states.

Branches of Tropical Equilibrations and Connectivity Graph. For each equation i, let us define

$$M_i(a) = \underset{j}{\mathrm{argmin}}(\mu_j(a), S_{ij} > 0) = \underset{j}{\mathrm{argmin}}(\mu_j(a), S_{ij} < 0), \tag{15}$$

in other words M_i denotes the set of monomials having the same minimal order μ_i. We call *tropically truncated system* the system obtained by pruning the system (11), i.e. by keeping only the dominating monomials.

$$\frac{d\bar{x}_i}{dt} = \varepsilon^{\mu_i - a_i} \Big(\sum_{j \in M_i(a)} \bar{k}_j \nu_{ji} \bar{x}^{\alpha_j} \Big), \tag{16}$$

The tropical truncated system is uniquely determined by the index sets $M_i(a)$, therefore by the tropical equilibration a. Reciprocally, two tropical equilibrations can have the same index sets $M_i(a)$ and truncated systems. We say that two tropical equilibrations a_1, a_2 are equivalent iff $M_i(a_1) = M_i(a_2)$, for all i. Equivalence classes of tropical equilibrations are called *branches*. A branch B with an index set M_i is *minimal* if $M_i' \subset M_i$ for all i where M_i' is the index set B' implies $B' = B$ or $B' = \emptyset$. Closures of equilibration branches are defined by a finite set of linear inequalities, which means that they are polyhedral complexes. Minimal branches correspond to maximal dimension faces of the polyhedral complex. The incidence relations between the maximal dimension faces ($n - 1$ dimensional faces, where n is the number of variables) of the polyhedral complex define the *connectivity graph*. More precisely, minimal branches are the vertices of this graph. Two minimal branches are connected if the corresponding faces of the polyhedral complex share a $n - 2$ dimensional face. In terms of index sets, two minimal branches with index sets M and M' are connected if there is an index set M'' such that $M_i' \subset M_i''$ and $M_i \subset M_i''$ for all i.

Tropical Equilibrations and Monomolecular Networks. Equation (13) have a simpler form in the case of monomolecular networks

$$\min_{j \in \mathrm{Pred}(i)} (\gamma_{ij} + a_j) = \min_{j \in \mathrm{Succ}(i)} (\gamma_{ji} + a_i) \tag{17}$$

where $\mathrm{Pred}(i) = \{j | (j, i) \in \mathcal{A}\}$, $\mathrm{Succ}(i) = \{j | (i, j) \in \mathcal{A}\}$ are the sets of predecessors and successors of the node i in the digraph G.

Let us recall that by min-plus algebra we understand the semi-ring $(\mathbb{R} \cup \{\infty\}, \oplus, \otimes)$ where the two operations are defined as $x \oplus y = \min\{x, y\}$ and $x \otimes y = x + y$. In other words the addition and the min operation play the role of min-plus multiplication and addition, respectively. Therefore Eq. (17) are linear in the unknowns a_i. Computing tropical equilibrations of monomolecular networks boils down to solving linear equations in min-plus algebra. For linear tropical systems there are fast algorithms [5, 6].

We have tested the tropical equilibration conditions (17) for the trajectories of the monomolecular network presented in Fig. 1 by checking if the absolute value of the difference between the r.h.s and l.h.s of (17) is smaller than a threshold. The result is illustrated in Fig. 1(e). For this model, the tropical equilibration solutions are changing along the trajectory. This can been seen by following the orders of the concentrations along the trajectories. These orders change by integers at transition points. Furthermore, at transition points some of the variables that where not previously equilibrated, become equilibrated. The analysis of the tropical equilibrations finds the transitions previously detected in Sect. 2 from the approximated eigenvalues and eigenvectors (t_1 and t_2 for this example) but adds some more. For instance, species $A1$ equilibrates at the timescale $1/\kappa_1 = 10$. This was not taken into account in the description of the automaton in Fig. 1(d) because the species $A1$ is fast and can not accumulate.

4   Learning a Finite State Machine from a Nonlinear Biochemical Network

We are using the algorithm based on constraint solving introduced in [16] to obtain all rational tropical equilibration solutions $a = (a_1, a_2, \ldots, a_n)$ within a box $|a_i| < b$, $b > 0$ and with denominators smaller than a fixed value d, $a_i = p_i/q$, p_i, q are positive integers, $q < d$. The output of the algorithm is a matrix containing all the tropical equilibrations within the defined bounds. A post-processing treatment is applied to this output consisting in computing truncated systems, index sets, and minimal branches. Tropical equilibrations minimal branches are stored as matrices $A_1, A_2, \ldots, A_b$, whose lines are tropical solutions within the same branch. Here b is the number of minimal branches.

Our method computes numerical approximations of the tropical prevariety. Given a value of ϵ, this approximation is better when the denominator bound d is high. At fixed d, the dependence of the precision on ϵ follows more intricate rules dictated by Diophantine approximations. For this reason, we systematically test that the number b and the truncated systems corresponding to minimal branches are robust when changing the value of ϵ.

Trajectories $x(t) = (x_1(t), \ldots, x_n(t))$ of the smooth dynamical system are generated with different initial conditions, chosen uniformly and satisfying the conservation laws, if any. For each time t, we compute the Euclidian distance $d_i(t) = \min_{y \in A_i} \|y - log_\varepsilon(x(t))\|$, where $\|*\|$ denotes the Euclidean norm and $\log_\varepsilon(x) = (\log x_1/\log(\varepsilon), \ldots, \log x_n/\log(\varepsilon))$. This distance classifies all points of the trajectory as belonging to a tropical minimal branch. The result is a symbolic trajectory $s_1, s_2, \ldots$ where the symbols s_i belong to the set of minimal branches. In order to include the possibility of transition regions we include an unique symbol t to represent the situations when the minimal distance is larger than a fixed threshold. We also store the residence times $\tau_1, \tau_2, \ldots$ that represent the time spent in each of the state.

The stochastic automaton is learned as a homogenous, finite states, continuous time Markov process, defined by the lifetime (mean sojourn time) of each state T_i, $1 \leq i \leq b$ and by the transition probabilities $p_{i,j}$ from a state i to another state j. We use the following estimators for the lifetimes and for the transition probabilities:

$$T_i = (\sum_n \tau_n \mathbb{1}_{s_n=i})/(\sum_n \mathbb{1}_{s_n=i}) \tag{18}$$

$$p_{i,j} = (\sum_n \mathbb{1}_{s_n=i,s_{n+1}=j})/(\sum_n \mathbb{1}_{s_n=i}), \ i \neq j \tag{19}$$

As a case study we consider a nonlinear model of dynamic regulation of Transforming Growth Factor beta TGF-β signaling pathway proposed in [1]. This model has a dynamics defined by $n = 18$ polynomial differential equations and 25 biochemical reactions. The paper [1] proposes three versions of the mechanism of interaction of TIF1γ (Transcriptional Intermediary Factor 1 γ) with the Smad-dependent TGF-β signaling. We consider here the version in which TIF1 interacts with the phosphorylated Smad2–Smad4 complexes leading to dissociation of the complex and degradation of Smad4. The results are similar for the other versions of this model. The example was chosen because it is a medium size model based on polynomial differential equations. The computation of the tropical equilibrations for this model shows that there are 9 minimal branches of full equilibrations (in these tropical solutions all variables are equilibrated). The connectivity graph of these branches and the learned automaton are shown in Fig. 2. The study of this example shows that branches of tropical equilibration can change on trajectories of the dynamical system. Furthermore, all the observed transitions between branches are contained in the connectivity graph resulting from the polyhedral complex of the tropical equilibration branches.

The transition probabilities of the automaton are coarse grained properties of the statistical ensemble of trajectories for different initial conditions. Given a state and a minimal branch close to it, it will depend on the actual trajectory to which other branch the system will be close to next. However, when initial data and the full trajectory are not known, the automaton will provide estimates of where we go next and with which probability. For the example studied, the branch B1 is a globally attractive sink: starting from anywhere, the automaton will reach B1 with probability one. This branch contains the unique stable steady state of the initial model. Figure 2 bottom right shows the structure of most probable branches, the ones in which the systems spends most of his time. The branches B1, B3 and B2 correspond to different compositions of the membrane and of the endosome, rich in the receptor RI, rich in the receptor RII and rich in both types of receptors, respectively. Even if this composition is changed on wide domains of orders (planes in the space of orders), the concentrations of effectors are robust (are more constained than the concentrations of receptors).

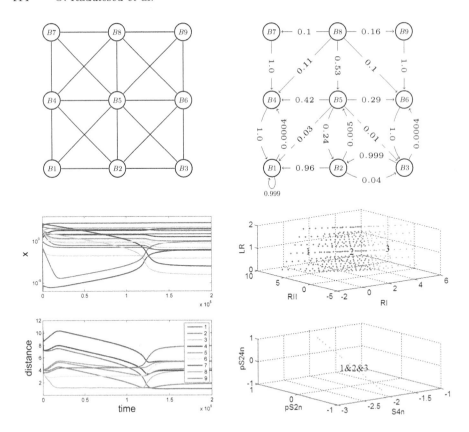

Fig. 2. TGFβ model. Upper left: Connectivity graph of tropical minimal branches; upper right: finite state automaton; bottom left: trajectories with jumps and distances to minimal branches; the closest branch changes with time along the trajectory; bottom right: first three tropical equilibrations minimal branches in various projections in concentration orders space. The variables RI, RII, LR are membrane receptors (signaling input layers) concentration orders, whereas pS2n, S4n, pS24n are nuclear transcription factors and complexes (effectors) concentration orders. The structure tropical branches shows that composition of input layers is more flexible (varies on planes) than the concentrations of effectors (vary on lines).

5   Conclusion

We have presented a method to coarse grain the dynamics of a smooth biochemical reaction network to a discrete symbolic dynamics of a finite state automaton. The coarse graining was obtained by two methods, approximated eigenvectors for mono-molecular networks and minimal branches of tropical equilibrations for more general mass action nonlinear networks. The two methods are compatible one to another, because when applied to monomolecular networks the method based on tropical geometry detects all the transitions indicated by approximated eigenvectors. For both methods the automaton has a small number of states, less than the number of species in the first method and the number of minimal tropical branches in the second method. The coarse grained automaton can be used for studying statistical properties of biochemical networks such as occurrence and stability of temporal patterns, recurrence, periodicity and attainability problems. The coarse graining can be performed in a hierarchical way. For the nonlinear example studied in the paper we computed only the full tropical equilibrations that stand for the lowest order in the hierarchy (coarsest model). As discussed in Sect. 3 we can also consider partial equilibrations when slow variables are not equilibrated and thus refine the automaton. Our approach extends the notion of steady states of a network and propose a simple recipe to characterize and detect metastable states. Most likely metastable states have biological importance because the network spends most of its time in these states. The itinerancy of the network, described as the possibility of transitions from one metastable state to another is paramount to the way neural networks compute, retrieve and use information [18] and can have similar role in biochemical networks.

Acknowledgements. O.R and A.N are supported by INCa/Plan Cancer grant N° ASC14021FSA.

Appendix 1

Proof of Proposition 1. Let us consider that $r_k^k = 1$. Taking $r_k^j = 0$ for all predecessors j of k and for all other nodes that lead to k by the flow Φ satisfy Eq. (6) (main body text) with $\lambda = -\kappa_k$. The same is valid for all the nodes that do not lead to k and are not accessible from k. Remain the nodes that are accessible from k. Let j be such a node. Then $j = \Phi^m(k)$ for some $m > 0$. Equation (6) (main body text) implies that

$$\kappa_{\Phi^{l-1}(k)} r_{\Phi^{l-1}(k)}^k = (-\kappa_k + \kappa_{\Phi^l(k)}) r_{\Phi^l(k)}^k, \text{ for } 1 \leq l \leq m.$$

Thus $r_{\Phi^m(k)}^k = \frac{\kappa_k}{-\kappa_k + \kappa_{\Phi(k)}} \times \frac{\kappa_{\Phi(k)}}{-\kappa_k + \kappa_{\Phi^2(k)}} \times \ldots \times \frac{\kappa_{\Phi^{m-1}(k)}}{-\kappa_k + \kappa_{\Phi^m(k)}}$. Suppose that $\kappa_k <$ $\kappa_{\Phi^l(k)}$ for $l = 1, \ldots, m - 1$ and $\kappa_{\Phi^m(k)} < \kappa_k$. Using Lemma 1 (main body text) it follows $r_{\Phi^m(k)}^k = -1$. If any of the previous inequality does not hold then at least one factor in the expression of $r_{\Phi^m(k)}^k$ vanishes and the remaining factors are finite, thus $r_{\Phi^m(k)}^k = 0$. Consider now that $l_k^k = 1$. Taking $l_k^j = 0$ for all the nodes j that can be obtained from k and for all other nodes that do not lead to k by the flow Φ satisfy Eq. (7) (main body text) with $\lambda = -\kappa_k$. The remaining nodes are all leading to k. Let j be such a node. Then $k = \Phi^m(j)$ for some $m > 0$. Equation (7) (main body text) implies that

$$\kappa_{\Phi^{l-1}(j)} l_{\Phi^l(j)}^k = (-\kappa_k + \kappa_{\Phi^{l-1}(j)}) l_{\Phi^{l-1}(j)}^k, \text{ for } 1 \leq l \leq m.$$

Hence $l_j^k = \frac{\kappa_j}{-\kappa_k + \kappa_j} \times \frac{\kappa_{\Phi(j)}}{-\kappa_k + \kappa_{\Phi(j)}} \times \ldots \times \frac{\kappa_{\Phi^{m-1}(j)}}{-\kappa_k + \kappa_{\Phi^{m-1}(j)}}$. Suppose that $\kappa_{\Phi^l(j)} > \kappa_k$, for all $l = 0, \ldots, m - 1$. Using Lemma 1 (main body text) it follows $l_j^k = 1$. If one of these inequalities is not satisfied for a $l = 0, \ldots, m-1$ then the corresponding factor in the expression of l_j^k vanishes and $l_j^k = 0$.

The above formulas cover the zero eigenvalue case if we consider that $\kappa_k = 0$ for k being the sink. It follows that $r_k^0 = 1$ and $r_j^0 = 0$ elsewhere. Furthermore, $l_j^0 = 1$ for all j.

Appendix 2

Algorithm for reduction of monomolecular networks with total separation. This algorithm consists of three steps.

I. Constructing of an Auxiliary Reaction Network: Pruning.

For each A_i branching node (substrate of several reactions) let us define κ_i as the maximal kinetic constant for reactions $A_i \to A_j$: $\kappa_i = \max_j\{k_{ji}\}$. For correspondent j we use the notation $j = \phi(i)$: $\phi(i) = \arg\max_j\{k_{ji}\}$.

An auxiliary reaction network $\mathcal{V}$ is the set of reactions obtained by keeping only $A_i \to A_{\phi(i)}$ with kinetic constants κ_i and discarding the other, slower reactions. Auxiliary networks have no branching, but they can have cycles and confluences. The correspondent kinetic equation is

$$\dot{c}_i = -\kappa_i c_i + \sum_{\phi(j)=i} \kappa_j c_j, \tag{20}$$

If the auxiliary network contains no cycles, the algorithm stops here.

II. Gluing Cycles and Restoring Cycle Exit Reactions.

In general, the auxiliary network $\mathcal{V}$ has several cycles $C_1, C_2, \ldots$ with periods $\tau_1, \tau_2, \ldots > 1$.

These cycles will be "glued" into points and all nodes in the cycle C_i, will be replaced by a single vertex A^i. Also, some of the reactions that were pruned in the first part of the algorithm are restored with renormalized rate constants. Indeed, reaction exiting a cycle are needed to render the correct dynamics: without them, the total mass of the cycle is conserved, with them the mass can also slowly leave the cycle. Reactions $A \to B$ exiting from cycles ($A \in C_i$, $B \notin C_i$) are changed into $A^i \to B$ with the rate constant renormalization: let the cycle C^i be the following sequence of reactions $A_1 \to A_2 \to \ldots A_{\tau_i} \to A_1$, and the reaction rate constant for $A_i \to A_{i+1}$ is k_i (k_{τ_i} for $A_{\tau_i} \to A_1$). For the limiting (slowest) reaction of the cycle C_i we use notation $k_{\lim i}$. If $A = A_j$ and k is the rate reaction for $A \to B$, then the new reaction $A^i \to B$ has the rate constant $k k_{\lim i}/k_j$. This rate is obtained using quasi-stationary distribution for the cycle. If kinetic constants are expressed as powers of a small positive parameter ϵ, i.e., if $k = \epsilon^\gamma$, then the order of the constant has to be changed according to the rule $\gamma \to \gamma + \gamma_{lim} - \gamma_j$, where γ, $\gamma_{lim\,i}$, γ_j are the orders of the constants k, $k_{\lim i}$ and k_j, respectively.

The new auxiliary network $\mathcal{V}^1$ is computed for the network of glued cycles. Then we decompose it into cycles, glue them, iterate until a acyclic network is obtained $\mathcal{V}^n$ (Fig. 3).

III. Restoring Cycles.

The dynamics of species inside glued cycles is lost after the second part. A full multi-scale approximation (including relaxation inside cycles) can be obtained by restoration of cycles. This is done starting from the acyclic auxiliary network $\mathcal{V}^n$ back to $\mathcal{V}^1$ through the hierarchy of cycles. Each cycle is restored according to the following procedure:

For each glued cycle node A_i^m, node of $\mathcal{V}^m$,

- Recall its nodes $A_{i1}^{m-1} \rightarrow A_{i2}^{m-1} \rightarrow ...A_{i\tau_i}^{m-1} \rightarrow A_{i1}^{m-1}$; they form a cycle of length τ_i.
- Let us assume that the limiting step in A_i^m is $A_{i\tau_i}^{m-1} \rightarrow A_{i1}^{m-1}$
- Remove A_i^m from $\mathcal{V}^m$
- Add τ_i vertices $A_{i1}^{m-1}, A_{i2}^{m-1}, ...A_{i\tau_i}^{m-1}$ to $\mathcal{V}^m$
- Add to $\mathcal{V}^m$ reactions $A_{i1}^{m-1} \rightarrow A_{i2}^{m-1} \rightarrow ...A_{i\tau_i}^{m-1}$ (that are the cycle reactions without the limiting step) with correspondent constants from $\mathcal{V}^{m-1}$

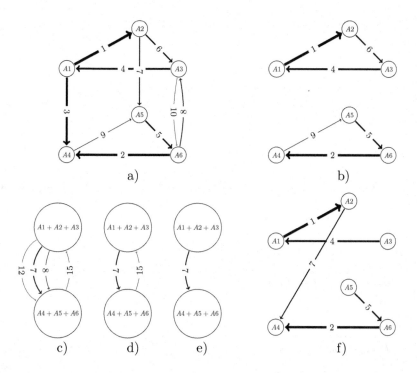

Fig. 3. The successive steps of the reduction algorithm, illustrated for the prism model used in the paper. (a) is the initial model; (b) is the auxiliary network resulting from step I, pruning; (c) is the result of gluing 3 species cycles and renormalizing the exit reactions (the constants of orders 3, 7, 10, 8 are renormalized to 3+6−1 = 8, 7+6−6 = 7, 10+6−4 = 12, and 8+9−2 = 15, respectively); (d) is the auxiliary network after one more iteration; (e) results from gluing and then restoring the 3 species cycles without the limiting step (constant of order 15); (f) results from restoring the single species cycles without their limiting steps.

- If there exists an outgoing reaction $A_i^m \rightarrow B$ in $\mathcal{V}^m$ then we substitute it by the reaction $A_{i\tau_i}^{m-1} \rightarrow B$ with the same constant, i.e. outgoing reactions $A_i^m \rightarrow \dots$ are reattached to the beginning of the limiting steps
- If there exists an incoming reaction in the form $B \rightarrow A_i^m$, find its prototype in $\mathcal{V}^{m-1}$ and restore it in $\mathcal{V}^m$
- If in the initial $\mathcal{V}^m$ there existed a "between-cycles" reaction $A_i^m \rightarrow A_j^m$ then we find the prototype in $\mathcal{V}^{m-1}$, $A \rightarrow B$, and substitute the reaction by $A_{i\tau_i}^{m-1} \rightarrow B$ with the same constant, as for $A_i^m \rightarrow A_j^m$ (again, the beginning of the arrow is reattached to the head of the limiting step in A_i^m)

Appendix 3

Description of the TGFb model used in this paper. The model is described by the following system of differential equations

$$\frac{dx_1}{dt} = k_2 x_2 - k_1 x_1 - k_{16} x_1 x_{11}$$

$$\frac{dx_2}{dt} = k_1 x_1 - k_2 x_2 + k_{17} k_{34} x_6$$

$$\frac{dx_3}{dt} = k_3 x_4 - k_3 x_3 + k_7 x_7 + k_{33} k_{37} x_{18} - k_6 x_3 x_5$$

$$\frac{dx_4}{dt} = k_3 x_3 - k_3 x_4 + k_9 x_8 - k_8 x_4 x_6$$

$$\frac{dx_5}{dt} = k_5 x_6 - k_4 x_5 + k_7 x_7 + 2 k_{11} x_9 - 2 k_{10} x_5^2 - k_6 x_3 x_5 + k_{16} x_1 x_{11}$$

$$\frac{dx_6}{dt} = k_4 x_5 - k_5 x_6 + k_9 x_8 + 2 k_{13} x_{10} - 2 k_{12} x_6^2 - k_{17} k_{34} x_6 + k_{31} k_{36} x_8 - k_8 x_4 x_6$$

$$\frac{dx_7}{dt} = k_6 x_3 x_5 - x_7 (k_7 + k_{14})$$

$$\frac{dx_8}{dt} = k_{14} x_7 - k_9 x_8 - k_{31} k_{36} x_8 + k_8 x_4 x_6$$

$$\frac{dx_9}{dt} = k_{10} x_5^2 - x_9 (k_{11} + k_{15})$$

$$\frac{dx_{10}}{dt} = k_{15} x_9 - k_{13} x_{10} + k_{12} x_6^2$$

$$\frac{dx_{11}}{dt} = k_{23} x_{14} - k_{30} x_{11}$$

$$\frac{dx_{12}}{dt} = k_{18} - x_{12} (k_{20} + k_{26}) + k_{30} x_{11} + k_{27} x_{15} - k_{22} k_{35} x_{12} x_{13}$$

$$\frac{dx_{13}}{dt} = k_{19} - x_{13} (k_{21} + k_{28}) + k_{30} x_{11} + k_{29} x_{16} - k_{22} k_{35} x_{12} x_{13}$$

$$\frac{dx_{14}}{dt} = k_{22} k_{35} x_{12} x_{13} - x_{14} (k_{23} + k_{24} + k_{25})$$

$$\frac{\mathrm{d}x_{15}}{\mathrm{d}t} = k_{26}x_{12} - k_{27}x_{15}$$

$$\frac{\mathrm{d}x_{16}}{\mathrm{d}t} = k_{28}x_{13} - k_{29}x_{16}$$

$$\frac{\mathrm{d}x_{17}}{\mathrm{d}t} = k_{31}k_{36}x_8 - k_{32}x_{17}$$

$$\frac{\mathrm{d}x_{18}}{\mathrm{d}t} = k_{32}x_{17} - k_{33}k_{37}x_{18}$$

These variables are as follows:

- Receptors on membrane: x_{12} = RI, x_{13} = RII, x_{14} = LR.
- Receptors in the endosome: x_{11} = LRe, x_{15} = RIe, x_{16} = RIIe.
- Transcription factors and complexes in cytosol: x_1 = S2c, x_3 = S4c, x_5 = pS2c, x_7 = pS24c, x_9 = pS22c, x_{18} = S4ubc.
- Transcription factors and complexes in the nucleus: x_2 = S2n, x_4 = S4n, x_6 = pS2n, x_8 = pS24n, x_{10} = pS22n, x_{17} = S4ubn.

References

1. Andrieux, G., Fattet, L., Le Borgne, M., Rimokh, R., Théret, N.: Dynamic regulation of Tgf-B signaling by Tif1γ: a computational approach. PloS One **7**(3), e33761 (2012)
2. Chiavazzo, E., Karlin, I.: Adaptive simplification of complex multiscale systems. Phys. Rev. E **83**(3), 036706 (2011)
3. Gorban, A., Karlin, I.: Invariant Manifolds for Physical and Chemical Kinetics. Lecture Notes in Physics, vol. 660. Springer, Heidelberg (2005)
4. Gorban, A., Radulescu, O.: Dynamic and static limitation in reaction networks, revisited. In: Guy B. Marin, D.W., Yablonsky, G.S. (eds.) Advances in Chemical Engineering - Mathematics in Chemical Kinetics and Engineering. Advances in Chemical Engineering, vol. 34, pp. 103–173. Elsevier, Amsterdam (2008)
5. Grigoriev, D.: Complexity of solving tropical linear systems. Comput. Complex. **22**(1), 71–88 (2013)
6. Grigoriev, D., Podolskii, V.V.: Complexity of tropical and min-plus linear prevarieties. Comput. Complex. **24**, 31–64 (2015)
7. Haller, G., Sapsis, T.: Localized instability and attraction along invariant manifolds. SIAM J. Appl. Dyn. Syst. **9**(2), 611–633 (2010)
8. Maas, U., Pope, S.B.: Simplifying chemical kinetics: intrinsic low-dimensional manifolds in composition space. Combust. Flame **88**(3), 239–264 (1992)
9. Noel, V., Grigoriev, D., Vakulenko, S., Radulescu, O.: Tropical geometries and dynamics of biochemical networks application to hybrid cell cycle models. In: Feret, J., Levchenko, A. (eds.) Proceedings of the 2nd International Workshop on Static Analysis and Systems Biology (SASB 2011). Electronic Notes in Theoretical Computer Science, vol. 284, pp. 75–91. Elsevier (2012)
10. Noel, V., Grigoriev, D., Vakulenko, S., Radulescu, O.: Tropicalization and tropical equilibration of chemical reactions. In: Topical and Idempotent Mathematics and Applications, vol. 616. American Mathematical Society (2014)
11. Palis, J.: A global view of dynamics and a conjecture on the denseness of finitude of attractors. Astérisque **261**, 339–351 (2000)

12. Radulescu, O., Gorban, A.N., Zinovyev, A., Lilienbaum, A.: Robust simplifications of multiscale biochemical networks. BMC Syst. Biol. **2**(1), 86 (2008)
13. Radulescu, O., Gorban, A.N., Zinovyev, A., Noel, V.: Reduction of dynamical biochemical reactions networks in computational biology. Front. Genet. **3**(131) (2012)
14. Radulescu, O., Vakulenko, S., Grigoriev, D.: Model reduction of biochemical reactions networks by tropical analysis methods. Mathematical Model of Natural Phenomena (2015, in press)
15. Samal, S.S., Radulescu, O., Grigoriev, D., Fröhlich, H., Weber, A.: A tropical method based on newton polygon approach for algebraic analysis of biochemical reaction networks. In: 9th European Conference on Mathematical and Theoretical Biology (2014)
16. Soliman, S., Fages, F., Radulescu, O.: A constraint solving approach to model reduction by tropical equilibration. Algorithms Mol. Biol. **9**(1), 24 (2014)
17. Theobald, T.: On the frontiers of polynomial computations in tropical geometry. J. Symbolic Comput. **41**(12), 1360–1375 (2006)
18. Tsuda, I.: Chaotic itinerancy as a dynamical basis of hermeneutics in brain and mind. World Futures: J. Gen. Evol. **32**(2–3), 167–184 (1991)

Feature Learning Using Stacked Autoencoders to Predict the Activity of Antimicrobial Peptides

Francy Camacho[1,2,3]([⊠]), Rodrigo Torres[2], and Raúl Ramos-Pollán[3]

[1] School of Computer Science, Universidad Industrial de Santander (UIS),
Carrera 27 calle 9, Bucaramanga, Colombia
francy.camacho1@correo.uis.edu.co
[2] Grupo de Investigación en Bioquímica y Microbiología (GIBIM),
School of Chemistry, UIS, Carrera 27 calle 9, Bucaramanga, Colombia
rtorres@uis.edu.co
[3] Center for High Performance and Scientific Computing,
School of Computer Science, Universidad Industrial de Santander,
Carrera 27 calle 9, Bucaramanga, Colombia
rramosp@uis.edu.co

Abstract. In recent years, pattern recognition methods have been applied to determine the activity of biological molecules, including the prediction of antimicrobial activity of synthetic and natural peptides where Quantitative Structure-Activity Relationship methodologies are widely used. Traditionally, works focused on designing descriptors for sequences to yield better correlations with the biological activity and improve predictors performance. Albeit there have been remarkable results, the small size of available datasets leave large room for improvement. In this work, rather than hand-crafting new descriptors, our approach consists in automatically learning them from existing ones. We use stacked autoencoders (a class of unsupervised neural networks), and the descriptors learnt are fed to a support vector regression task to predict biological activity. This method improves results in existing literature by roughly 12% simultaneously in different metrics, providing interesting insights into the nature of descriptors learnt and suggesting its applicability in other areas in protein properties prediction.

Keywords: Autoencoder · Stacked autoencoder · Antimicrobial peptides · Support vector regression

1 Introduction

Recently, the development of new antibiotics has become a necessity due to the emergence and spread of resistant strains [1]. Few drugs can face this problem and, together with the reduction of pharmaceutical industries researching new antibacterial agents, this has become a threat to public health [2]. Antimicrobial peptides are a promising alternative to traditional antibiotics due the broad spectrum of biological activity and low probability to produce resistance in bacteria, although the design and synthesis of new peptides have been limited, inter

© Springer International Publishing Switzerland 2015
O. Roux and J. Bourdon (Eds.): CMSB 2015, LNBI 9308, pp. 121–132, 2015.
DOI: 10.1007/978-3-319-23401-4_11

alia the huge number of possible sequences that we can obtain if we take the twenty natural amino acids [3].

For this reason, methodologies such as QSAR (Quantitative Structure-Activity Relationship) are being widely used to predict the activity of peptides (Minimal Inhibitory Concentration, MIC), using regression methods to try to find peptides with high MIC [4–6]. QSAR is based on the idea that a sequence or peptide structure can be described through physico-chemical properties (descriptors), and these are correlated with biological activity present in the peptide through a mathematical function [7].

Albeit results are promising there is still plenty of room for improvement mostly due to the small size of existing datasets. In this work, we take the approach of using machine learning methods to learn new descriptors from the existing ones rather than further devising new ones [8–10]. We use stacked autoencoders (a class of unsupervised neural networks) to learn new descriptors which are then fed to a support vector regression task to predict biological activity from them. Our results were satisfactory reducing the $RMSE_{ext}$ from 0.96 to 0.84 and improving R^2_{ext} from 0.72 to 0.85 compared to literature and suggest that this method can be considered in different application areas in protein prediction.

This paper is structured as follows. Section 2 describes the datasets we used and provides a general overview on how stacked autoencoders work. Section 3 explains the experiment setups we devised. Section 4 describes the results we obtained and provides some insights on their interpretation. Finally, Sect. 5 draws the conclusions.

2   Materials and Methods

2.1   Dataset and Descriptors

We use the dataset CAMELs, which is made of 101 sequences of peptides of the same length (15 aminoacids). Each peptide has been tested against several strains of microorganisms and its activity was reported measured as the mean antibiotic potency against these [6].

From the aminoacid sequences of peptides it is possible to compute descriptors representing quantitatively several physico-chemical properties. There is a wide range of descriptors and in this work we started off from the properties described by Zhou et al. [4], where different descriptor groups were extracted from the primary structure of peptides, using a web tool called PROFEAT [11]. It computes ten groups of properties as shown in Table 1, where *AllDesc* is full set of available descriptors. Due to technical problems on PROFEAT's web site we used instead propy (available as a Python library [12]) to compute the ten groups of descriptors just mentioned. Additionally, we verified that propy source code implements the same equations for each descriptor according to PROFEAT's user manual.

Table 1. Ten groups of descriptors compute for the dataset. The initial and final columns represent the number of descriptors before and after of preprocessing, respectively

Descriptors	Initial	Final
Dipeptide Composition (Ddcd)	400	106
Normalized MoreauBroto autocorrelation (Dnmba)	240	112
Moran autocorrelation (Dmad)	240	112
Geary autocorrelation (Dgad)	240	112
Composition, transition and distribution (Dctd)	147	147
Sequence order coupling number (Dsoc)	20	20
Quasi sequence order (Dqso)	50	46
Pseudoaminoacid composition type I (Dpaac)	30	23
Pseudoaminoacid composition type II (Dapaac)	30	23
All Descriptors (AllDesc)	1517	730

2.2 Autoencoders (AEs)

Autoencoders (AEs) [13,17] are a special class of neural networks that are used in an unsupervised manner. Typically, supervised machine learning methods (such as neural networks) are provided with input data and the expected predictions to generate a predictive model from input data (such as for predicting antimicrobial activity from peptide descriptors). Unlike that, unsupervised methods, such as AEs, only use input data to learn a new representation without using the expected predictions.

An AE is a symmetric neural network with one hidden layer (Fig. 1(a)), i.e. the number of neurons in the input and output layers is the same. For each input vector, the expected output is set to be the same input vector and training happens similarly to a neural network, approximating the output of the network to the input data, minimizing the error between both. This way, if training succeeds to reconstruct the input data at the output layer, the hidden layer will contain a new representation of the input data which will be more compact if the hidden layer has less neurons than the input layer, or more sparse if it has more neurons.

The activation of neurons in the hidden layer is the result of lineal combination of the input vector x:

$$a^{(2)} = f\left(W^{(1)} * x + b_1\right) \tag{1}$$

where $W^{(1)}$ is weight vector, b_1 is the bias o intercept term and f is the sigmoid function, where $f = \frac{1}{1+e^a}$. Likewise, at the output layer, the activation is given by

$$h_{W,b}(x) = f\left(W^{(2)} * a + b_2\right). \tag{2}$$

where $W^{(2)}$ is weight vector in the output layer, b_2 is the bias o intercept term and f is the sigmoid function. The error the network incurs when reconstructing the input at the output is given by the cost function $J(W, b)$ which is typically minimized through a gradient descent method:

$$J(W, b) = \frac{1}{2} \| h_{W,b}(x) - y \|^2 \qquad (3)$$

An interesting feature of AEs is that the number of neurons in the hidden layer can be smaller or greater than the input layer. If it is less, it will force the network compress the information, similarly to Principal Component Analysis [14]. If it is greater, the network will learn a more distributed representation in the sense that more neurons will be used to represent the same information at the input layer. In this case, we are interested in forcing the network to activate at small number of neurons from the hidden layer at each input producing a sparse representation of the data and, thus, forcing each neuron to specialize to detect a different input pattern.

In order to achieve this a sparsity restriction is included in the cost function $J(W, b)$ that controls how many neurons are activated:

$$J_{sparse}(W, b) = J(W, b) + \beta \sum_{i=1}^{c2} KL(\rho \parallel \hat{\rho}_j) \qquad (4)$$

where β is the weight that penalize the sparsity, $c2$ is the number of neurons in the hidden layer, ρ is sparsity parameter (in this work, $\rho = 0.05$), $\hat{\rho}_j = \frac{1}{m} \sum_{i=1}^{m} \left[a_j^{(2)}(x^{(i)}) \right]$ is the average of activation of neurons in the hidden layer. KL is Kullback-Leibler divergence:

$$KL(\rho \parallel \rho_j) = \rho \, log \, \frac{\rho}{\hat{\rho}_j} + (1 - \rho) log \, \frac{1 - \rho}{1 - \hat{\rho}_j} \qquad (5)$$

The parameters $W^{(1)}, b_1, W^{(2)}, b_2$ are optimized so that $J(W, b)$ is minimized through back-propagation and L-BFGS [15].

2.3   Stacked Autoencoder (SAE)

A stacked autoencoder is a neural network with two or more layers of autoencoders that are used in an unsupervised manner. The main idea with SAE is to capture high order features from the data. Training is conducted using the approach called greedy-wise, i.e. each hidden layer is trained separately and the output of each one is used as input for the next layer [17]. For instance, to train a stacked autoencoder with two hidden layers, we first create and train an autoencoder with one hidden layer and keep only the primary feature activations $h^{(1)}$ (see Fig. 1a) after training. Next, we feed the data to this first autoencoder and, for each input instance, we obtain the values at the output layer $h^{(1)}$ as a new representation of the data. Instead of directly using this new data, we feed it to a second autoencoder and perform the training process again (Fig. 1b).

At the end, the output of this second layer at $h^{(2)}$ is the final representation of our data.

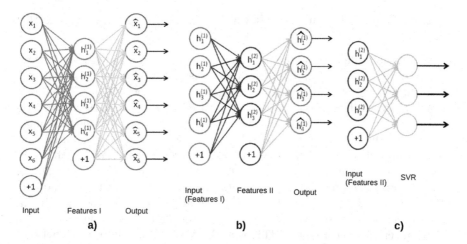

Fig. 1. Stacked autoencoder with 2 hidden layer. (a) contains 6 input neurons and four neurons in the hidden layer. Note this is a compressing autoencoder. It contains $(6+1)*4+(4+1)*6 = 58$ connections. Figure taken from [17]

2.4 Processing Workflow

In all our processing workflow is then composed of three stages:

1. **Preprocessing:** all descriptors are preprocessed by (1) standardizing their values so that for each descriptor its mean was zero and its standard deviation was one; and (2) removing the ones with the same value in all peptides (its standard deviation was zero).
2. **Unsupervised Feature Learning:** Different configurations of AEs and SAEs are trained and run on the preprocessed data producing a new representation.
3. **Supervised Prediction:** Different configurations of a Support Vector Regression task are run on the new representation obtained in the previous phase to effectively predict antimicrobial activity for the initial peptides. See (Fig. 1c)

3 Experimental Setup

3.1 Experimental Configurations

Starting off from the 10 groups of descriptors for each peptide obtained with propy, we devised four general experimental setups, and run each descriptor group through each setup after preprocessing as described above. The four setups where the following:

Original: We performed a Support Vector Regression directly on each group of descriptors without further processing or feature learning. The purpose of this setup is to give us a baseline against which to measure the behavior of further setups.

AE: We trained different configurations of AEs to learn a new set of features which were then fed to a Support Vector Regression task. In each AE configuration we vary the number of neurons in the hidden layer from 20 to 1000 neurons. This allows for configurations producing both compact and sparse representations with respect to the number of descriptors in each group. When the number of neurons in the hidden layer was between 20 and 500 and it was varied with a step of 20, and when it was between 500 and 1000 it was varied with a step of 50. This resulted in 35 AE configurations which were used for each groups of descriptors. Each one of these configurations contains several thousand connections that need to be trained. For instance an AE with 500 neurons in the hidden layer for descriptor group *Ddcd* with 106 descriptors contains around 106 K connections (see Fig. 1a). This way the size of our AEs ranged between 800 connections (for the AE with the 20 descriptors of the Dsoc group and 20 neurons in the hidden layer) and 1.46 million connections (for AllDesc and 1000 neurons in the hidden layer).

SAE2: For each AE configuration we created a two layer stacked autoencoder by adding an additional hidden layer with half the neurons, producing therefore another 35 configurations. As explained, each configuration was trained layer-wise. Sizes of SAE2 configurations ranged between 800 connections and 1.6 million.

SAE4: Likewise SAE2 but the number of neurons in the second hidden layer was obtained by dividing the number of neurons in the first hidden layer by 4 yielding, again, another 35 configurations. Sizes of SAE4 configurations ranged between 600 connections and 1.1 million.

As we have 10 descriptor groups, in total we run 1060 experimental configurations (350 for **AE**, **SAE2** and **SAE4** and 10 for **Original**). Deeper stacked autoencoder configurations (with more layers) were not yet considered due to their computational cost as the purpose of this paper is to validate the general utility of the method.

3.2   Validation and Supervised Training

For supervised training, we split the data in a subset for training and another one for validation according to Zhou et al. [4] as strictly as possible (using the same validation split). Then, we optimized the free parameters of the Support Vector Regression task. For this, we created a grid varying (C, γ, ϵ) and for each combination of parameters we used the train data split to train a SVR with 5 fold cross-validation and with the average score of R^2_{ext} we choose the parameters yielding the maximum score. Our parameter grid resulted from varying the ranges of the free parameters as follows C (10 to 72.5 with a step of 2.5), γ ($10^{-1.5}$ to $10^{0.5}$ varying the potency with a step of 0.25), ϵ (0.1 to 0.9 with a step of 0.1). The grid therefore contained 1872 parameter combinations which were run with each configuration described in Sect. 3.1. Therefore, we trained 1'984'320 Support Vector Regression cross validation processes and selected one for each one of the 1060 **Original, AE, SAE2** and **SAE4** configurations.

Finally, with the best combination of parameters (C, γ, ϵ) for each configuration we trained a SVR with the full training split (no cross-validation) and tested it with the validation data split for obtaining the final performance. The performance metrics used for the validation set were Root Mean Square Error ($RMSE_{ext}$), Correlation coefficient of multiple determination (R^2_{ext}), Pearson correlation coefficient (R) and R^2_{pred} (R^2 predictive) [16]. The subscript ext represents that these metrics were used with validation set (or external validation set).

4   Results and Discussion

Our approach differs from most of the studies used in the prediction of antimicrobial peptides [4,5,8–10,18,19] in that descriptors are learnt automatically in an unsupervised machine learning task. Table 2 summarizes the results we obtained and those of the referenced literature. Our results are shown in the four bottom lines of the table together with the descriptor group and AE or SAE configuration with which they were obtained.

Details can be found in Figs. 2, 3 and 4 where we plotted the performance of each descriptor group with R^2_{ext}, $RMSE_{ext}$ and R_{ext} respectively, together with the performance reported in the literature as shown in Table 2 as dashed lines. Experiments with setup **SAE2** were not plotted as they were not significantly better than **SAE4**. In all, the complete set of experiments took some 40 compute hours. The compute time for each set of descriptors and AE or SAE configuration varies greatly depending on the number of connections of the specific configuration.

Table 2. Comparative results for different algorithms used for prediction of antimicrobial peptides

Method	R_{ext}	$RMSE_{ext}$	R^2_{ext}	R^2_{pred}	Ref
GA-SVM	0.78	1.39	-	-	[4]
PSO-GA-SVM	0.9	0.96	-	-	[4]
STR-MLR	-	-	0.326	-	[18]
G/PLS	0.8	-	0.67	0.64	[5]
ANN	-	-	0.72	-	[19]
Setup Original (Dqso+SVR)	0.87	1.10	0.73	0.74	This work
Setup AE (Dctd(900)+SVR)	0.9	1.10	0.739	0.74	This work
Setup SAE2 (Dqso(140,70)+SVR)	0.96	0.864	0.841	0.842	This work
Setup SAE4 (Dqso(800,200)+SVR)	0.97	0.845	0.848	0.849	This work

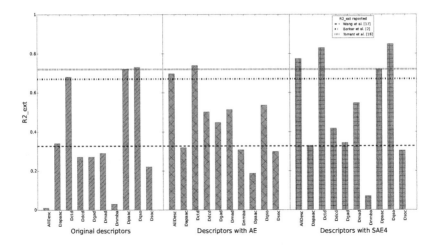

Fig. 2. Best performance for R^2_{ext} for each group of descriptors in tree experimental setup. R^2_{ext} higher is better

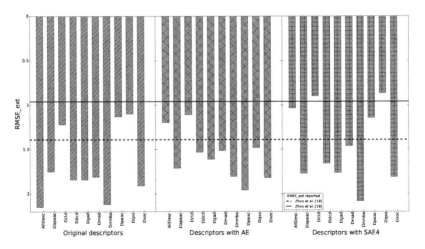

Fig. 3. Best performance for $RMSE_{ext}$ for each group of descriptors in tree experimental setup. $RMSE_{ext}$ closer to zero is better

For **AE** configurations the best group of original descriptors were consistently *Dctd* (with 147 original descriptors) obtained with 900 hidden neurons, performing better than literature only in the R^2_{ext} metric. This is an AE with over 265 K connections. However **SAEs** perform consistently better than results in the literature with different sets of descriptor groups, mostly *Dqso* and *Dctd*.

For further detail, Fig. 5 shows the results for all autoencoder and stacked autoencoder configurations for each dataset for metric R^2_{ext} for each variation of number of neurons in the first hidden layer. Recall that for **SAE2** the second hidden layer contains half the neurons of the first layer and for **SAE4** it contains

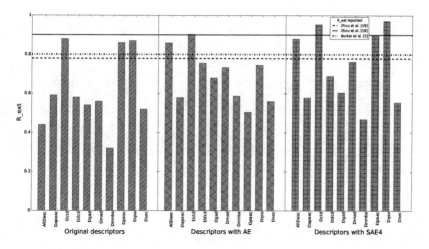

Fig. 4. Best performance for R_{ext} for each group of descriptors in tree experimental setup. R_{ext} higher is better

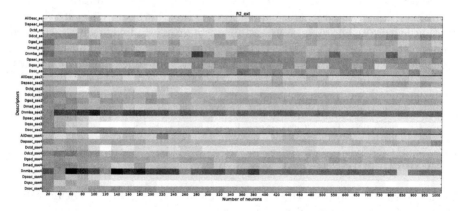

Fig. 5. Graphical representation of results for autoencoder and stacked autoencoder for each variation of number of neurons in the hidden layer. The best results for AE was Dctd with neuron 900, SAE2 was Dqso with neurons 140 and 70 and SAE4 was Dqso with neurons 800 and 200

one fourth. Higher R^2_{ext} is represented lighter with white being the best score and black the worse.

It can be observed how descriptor groups *Dqso* and *Dpaac* behave consistently well with both **SAE2** and **SAE4** and *Dmmba* behave consistently worse than others. The behavior with **AE** seems somewhat different with descriptor group *Dctd* working better overall.

Moreover, it can as well be noted that configurations with more neurons in the hidden layers seem to work better (for each row, scores on the right tend to

Table 3. Correlation among descriptors within each descriptor group for SAE configurations compared with the original representation. If pixel is darker, the correlation is closer to -1, while if it is lighter, the correlation is closer to 1

-	AllDesc	Dapaac	Dctd	Ddcd	Dgad	Dmad	Dnmba	Dpaac	Dqso	Dsoc
Original										
SAE										
Neurons	100,25	460,230	120,30	220,110	100,50	180,90	40,10	320,160	800,200	60,15

be lighter). This seems to favor AEs and SAEs that learn sparse representations as opposed to the ones learning more compact (compressing) ones.

Finally, in order to shed some light on the interpretation on the learnt features, we compare the intercorrelations among the original features of each descriptor group and those obtained with the best SAE configuration starting from that original dataset. This can be seen in Table 3 where we picture the correlation matrix among n variables as an $n \times n$ grayscale image with each pixel representing the correlation between the corresponding variables. Thus, complete independence among variables is represented by a white diagonal surrounded by a black background.

It can be observed that, in general, the new features obtained through SAEs generally enhance the independence of the original descriptors as the backgrounds in row 2 in Table 3 are generally darker.

5   Conclusions

In this work, we approach the task of predicting the activity of antimicrobial peptides by using autoencoders and stacked autoencoders to learn new descriptors rather than hand-crafting them, in an unsupervised manner, without using the known activity as measured in the laboratory. When feeding the new features to a supervised machine learning method, we show how learnt representations consistently provide satisfactory results as compared with recent works.

Besides we also show how, among the learnt representations, sparse ones seem to be preferable to more compact ones as they probably give a better chance for data separability for the supervised prediction task later on. Moreover, we also show how the learnt representations also enhance the independence of the initial descriptors reducing the correlation among them.

We believe this approach to be worthwhile exploring in other areas in prediction of properties protein sharing data characteristics and problem complexity. Moreover we have identified descriptor groups which consistently behave better. This could help design better candidate peptides in the future.

However, we also observed the importance of the selection of the original set of descriptors from which the learning process starts. This suggests probably hybrid approaches where specialists hand-craft a base collection of descriptors and the unsupervised learning process complements them with automatically learnt ones. Future work is expected to continue in this direction.

Acknowledgments. The authors thank the support of the High Performance and Scientific Computing Centre at Universidad Industrial de Santander (www.sc3.uis.edu. co). This project was funded by COLCIENCIAS (Project number: 1102-5453-1671) and Vicerrectoría de Investigación y Extensión (VIE) from UIS.

References

1. Amábile-Cuevas, C.F.: Antimicrobial resistance in developing countries. In: Sosa, A.d.J., Byarugaba, D.K., Amábile-Cuevas, C.F., Hsueh, P.R., Kariuki, S., Okeke, I.N. (eds.) Antimicrobial Resistance in Develoving Countries, Chap. 1, pp. 15–27. Springer, New York (2010)
2. Projan, S.J.: Why is big Pharma getting out of antibacterial drug discovery? Curr. Opin. Microbiol. **6**(5), 427–430 (2003)
3. Fjell, C.D., Hiss, J., Hancock, R.E.W., Schneider, G.: Designing antimicrobial peptides: form follows function. Nature Rev. Drug Discov. **11**(1), 37–51 (2012)
4. Zhou, X., Li, Z., Dai, Z., Zou, X.: QSAR modeling of peptide biological activity by coupling support vector machine with particle swarm optimization algorithm and genetic algorithm. J. Mol. Graph. Model. **29**(2), 188–196 (2010)
5. Borkar, M.R., Pissurlenkar, R.R.S., Coutinho, E.C.: HomoSAR: bridging comparative protein modeling with quantitative structural activity relationship to design new peptides. J. Comput. Chem. **34**(30), 2635–2646 (2013)
6. Cherkasov, A., Jankovic, B.: Application of 'inductive' QSAR descriptors for quantification of antibacterial activity of cationic polypeptides. Molecules **9**(12), 1034–1052 (2004). (Basel, Switzerland)
7. Taboureau, O.: Methods for building quantitative structure-activity relationship (QSAR) descriptors and predictive models for computer-aided design of antimicrobial peptides. In: Giuliani, A., Rinaldi, A.C. (eds.) Antimicrobial Peptides, Methods in Molecular Biology, Methods in Molecular Biology, Chap. 6, vol. 618, pp. 77–86. Humana Press, Totowa (2010)
8. Shu, M., Yu, R., Zhang, Y., Wang, J., Yang, L., Wang, L., Lin, Z.: Predicting the activity of antimicrobial peptides with amino acid topological information. Med. Chem. **9**(1), 32–44 (2013)
9. Hemmateenejad, B., Yousefinejad, S., Mehdipour, A.R.: Novel amino acids indices based on quantum topological molecular similarity and their application to QSAR study of peptides. Amino Acids **40**(4), 1169–1183 (2011)
10. Lin, Z., Long, H., Bo, Z., Wang, Y., Wu, Y.: New descriptors of amino acids and their application to peptide QSAR study. Peptides **29**(10), 1798–1805 (2008)
11. Li, Z.R., Lin, H.H., Han, L.Y., Jiang, L., Chen, X., Chen, Y.Z.: PROFEAT: a web server for computing structural and physicochemical features of proteins and peptides from amino acid sequence. Nucleic Acids Res. **34**(Web Server issue), W32–W37 (2006)
12. Cao, D.S., Xu, Q.S., Liang, Y.Z.: propy: a tool to generate various models of Chous PseAAC. Bioinform. Appl. Note **29**(7), 960–962 (2013)

13. Shin, H., Orton, M.R., Collins, D.J., Doran, S.J., Leach, M.O.: Stacked autoencoders for unsupervised feature learning and multiple organ detection in a pilot study using 4D patient data. IEEE Trans. Pattern Anal. Mach. Intell. **35**(8), 1930–1943 (2013)
14. Wold, S., Esbensen, K., Geladi, P.: Principal component analysis. Chemometr. Intell. Lab. Syst. **2**(1–3), 37–52 (1987)
15. Liu, D.C., Nocedal, J.: On the limited memory BFGS method for large scale optimization. Math. Program. **45**, 503–528 (1989)
16. Kiralj, R., Ferreira, M.M.C.: Basic validation procedures for regression models in QSAR and QSPR studies: theory and application. J. Braz. Chem. Soc. **20**(4), 770–787 (2009)
17. Ng, A., Ngiam, J., Foo, C.Y., Mai, Y., Suen, C.: Unsupervised feature learning and deep learning. http://ufldl.stanford.edu/wiki/index.php/UFLDL_Tutorial
18. Wang, Y., Ding, Y., Wen, H., Lin, Y., Hu, Y., Zhang, Y., Xia, Q., Lin, Z.: QSAR modeling and design of cationic antimicrobial peptides based on structural properties of amino acids. Comb. Chem. High Throughput Screen. **15**(4), 347–353 (2012)
19. Torrent, M., Andreu, D., Nogués, V.M., Boix, E.: Connecting peptide physicochemical and antimicrobial properties by a rational prediction model. PLoS One **6**(2), e16968 (2011)

Structural Simplification of Chemical Reaction Networks Preserving Deterministic Semantics

Guillaume Madelaine[1,2]($\boxtimes$), Cédric Lhoussaine[1,2], and Joachim Niehren[1,3]

[1] CRIStAL, UMR 9189, 59650 Villeneuve-d'ascq, France
guillaume.madelaine@ed.univ-lille1.fr
[2] University of Lille, Villeneuve-d'ascq, France
[3] INRIA Lille, Lille, France

Abstract. We study the structural simplification of chemical reaction networks preserving the deterministic kinetics. We aim at finding simplification rules that can eliminate intermediate molecules while preserving the dynamics of all others. The rules should be valid even though the network is plugged into a bigger context. An example is Michaelis-Menten's simplification rule for enzymatic reactions. In this paper, we present a large class of structural simplification rules for reaction networks that can eliminate intermediate molecules at equilibrium, without assuming that all molecules are at equilibrium, i.e. in a steady state. We prove the correctness of our simplification rules for all contexts that preserve the equilibrium of the eliminated molecules. Finally, we illustrate at a concrete example network from systems biology that our simplification rules may allow to drastically reduce the size of reaction networks in practice.

1 Introduction

In systems biology [18], reaction networks are used to represent biological systems. They enable formal analyses [9], simulations with several semantics [7], parameter estimations and identifications [1], etc. With bigger and bigger networks, in order to keep the analyses as simple as possible, or to have quick simulations (in particular in the context of real-time control [29]), we need to be able to simplify reaction networks. Indeed, the reactions of many metabolic reaction networks are often motivated by simplifications of concrete chemical reactions, see e.g. [21], but these simplifications are always done in informal manner without any semantical guarantees. An exception is Michaelis-Menten's simplification rule of enzymatic reactions, which is properly justified under quasi-steady-state assumption [27].

One usual approach is to simplify the ordinary differential equation (ODE) systems, that describe the deterministic semantics of reaction networks, but not the reaction networks themselves. In [17], authors presented a method based on the structure of enzyme-catalysed reactions to compute a simplified ODE

This work has been funded by the French National Research Agency research grant Iceberg ANR-IABI-3096.

O. Roux and J. Bourdon (Eds.): CMSB 2015, LNBI 9308, pp. 133–144, 2015.
DOI: 10.1007/978-3-319-23401-4_12

system at steady-state. In [6], authors used dependency analysis of rule-based models to obtain a simplified ODE system. Many other simplification methods use the distinction between slow and fast reactions, as for instance methods based on invariant manifolds [11], quasi-steady state [3,27], quasi-equilibrium approximation [12] or tropicalization [28]. Other methods reduce the number of parameters, for instance by using Lie symmetries [19]. However, most of those methods require the parameter values, or at least their magnitudes, and those data are often unknown. Moreover, it is useful to preserve the reaction network and not just its ODE system, and transforming an ODE system back to a reaction network is a difficult issue, since not always possible, or not possible in a unique manner [8].

Another approach is to consider reaction networks as programs [5,16,25], and to apply simplification rules directly to such programs, similarly to what is done in compiler construction [24,26]. This means to directly simplify the reaction network and not the corresponding ODE system, or even while ignoring the kinetics all over. Such structural simplification methods are usually based on a small-step semantics, saying how chemical solutions may evolve non-deterministically. They are often contextual, i.e. the simplification rules remain correct when the network is plugged into a bigger context. In our own previous work [20], we proposed to simplify reaction networks while preserving the reachability of final components, called attractors. However, those methods do not fit well with the deterministic semantics, even though the simplification rules obtained seem sensible for biological systems. Previous structural simplification methods were presented in [10], where subgraph epimorphisms are used to reduce reaction networks. Similar works had been done in Petri Nets [2,23], preserving its usual properties (liveness, deadlock, termination, etc.). In [4], Cardelli presented morphisms that preserve the deterministic semantics, but does not give simplification rules for them.

In this article, we aim at finding a new approach for simplifying reaction networks that preserves the deterministic semantics, i.e. the evolution of concentrations of molecular species over time. The approach should be structural in that it applies to reaction networks directly without computing the ODE system. It should be contextual, so that we can easily simplify modules or subnetworks in a larger context while preserving the overall dynamics. Therefore, we propose a collection of simplification rules that eliminate intermediate molecules while preserving the dynamics of all others. Some simplification rules are based on partial equilibrium conditions on the intermediate molecules (but a general steady-state is not assumed). Such conditions were already assumed to justify Michaelis-Menten's exact simplification for enzymatic reactions [22] which is widely accepted. There the intermediate complex needs to be at equilibrium; when it is only close to the equilibrium, then a small error is made which can be estimated. A network obtained by applying a simplification rule has the same deterministic semantics than the original one, in all contexts that preserve the equilibrium conditions on intermediate molecules. For applying a simplification rule, the corresponding ODE system is not needed, and the kinetic parameters may be unknown. We illustrate the usefulness of the simplification by applying it to biological examples, where it allows to drastically reduce the size of reaction networks.

Outline. We first illustrate the basic ideas and motivations at an example in Sect. 2. We recall the formal definitions of reaction networks with their deterministic semantics in Sect. 3. In Sect. 4, we contribute a contextual equivalence relation for reaction networks, and in Sect. 5 a set of simplification axioms, that we prove correct with respect to this equivalence relation. In Sect. 6, we illustrate at a biological example, how much reaction networks can be simplified in practice. We finally conclude and discuss future work in Sect. 7.

2   Preliminary Example

We first present a preliminary example, to illustrate our simplification.

Consider the reaction network *Gene* in Fig. 1 on the left. It has four species: a gene G, an inhibitor *Inh*, some *mRNA*, and a protein P. The reaction r_1 describes a transcription, the production of *mRNA* in presence of a gene G. This gene is required to apply the reaction, but its amount is not modified by it. This reaction has also a modulator, *Inh*, indicated by a dashed arrow. A modulator influences the speed rate of a reaction, but is not required to apply it. Here, *Inh* slows down the reaction r_1. The reaction r_2 is the translation of *mRNA* into the protein P, while the reaction r_3 (resp. r_4) describes the degradation of *mRNA* (resp. P). Aside from the first one, every reaction has a simple mass-action kinetic.

In order to simplify the network, we first need to specify how the environment interacts with the network: this is indicated by pending dotted arrows in Fig. 1. We consider here that G and *mRNA* are internal molecules, that is, they can not be modified by the context. Then, the context can be any set of reactions that

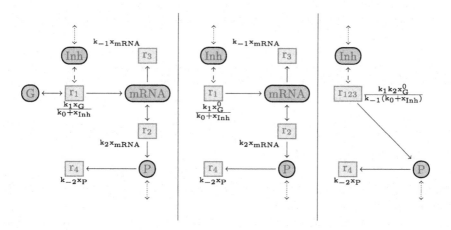

Fig. 1. Reaction graphs of the *Gene* network on the left, and its two simplifications. Molecules are represented by circles, and reactions by squares. In the kinetic expressions near the reactions, the k_i are parameters while x_A is a variable representing the concentration of a molecule A. x_G^0 denotes the initial concentration of G. A dash arrow means that the molecule acts as a modulator in the reaction, while a dot arrow means that the molecule can be modified by the context.

does not contain G and $mRNA$. It can for instance transform P into another protein, or produces something else in presence of Inh, etc.

In this network, we are especially interested in the protein P, and on the contrary we want to eliminate the intermediate $mRNA$. To do that, we will assume that $mRNA$ is at equilibrium, i.e. its concentration, x_{mRNA}, is constant over time.

Let us simplify our network. First, notice that the gene G is not modified by any reaction. It is used in reaction r_1, but only as an activator, i.e. on both sides of the reaction. Moreover, G is an internal molecule, that can not be modified by the context. Therefore its concentration, x_G, is constant over time: $x_G = x_G^0$. Then we make this modification in the kinetic expression of reaction r_1, and remove completely G from the network. The new network is pictured in Fig. 1 (middle).

Now, consider the intermediate $mRNA$. It is an internal molecule, and its (complete) ordinary differential equation is:

$$\frac{dx_{mRNA}}{dt} = \frac{k_1 x_G^0}{k_0 + x_{Inh}} - k_{-1} x_{mRNA}$$

Since we assumed that $mRNA$ is at equilibrium, i.e. $\frac{dx_{mRNA}}{dt} = 0$, we deduce:

$$x_{mRNA} = \frac{k_1 x_G^0}{k_{-1}(k_0 + x_{Inh})}$$

Therefore we remove $mRNA$ from the network, and replace the variable x_{mRNA} in the kinetics of reaction r_2, by the expression computed above. We obtain the simplified network in Fig. 1 (right) where r_1, r_2 and r_3 are merged into the new reaction r_{123}.

As we will see in this paper, the simplification rules used above preserve the deterministic semantics of reaction networks, in every context. So the simplified network is contextual equilibrium-equivalent to the first one. Note that we can not simplify the network anymore, since both Inh and P can be modified by the context.

3   Reaction Networks

We introduce reaction networks and define their deterministic semantics in terms of ordinary differential equations.

Let $Spec$ be a set of molecular species ranging over by A, B, C. We define a *(chemical) solution* $s \in Sol : Spec \to \mathbb{N}_0$ as a function from molecular species to natural numbers. Given natural numbers $n_1, \ldots, n_k$, we denote by $n_1 A_1 + \ldots + n_k A_k$ the solution that contains n_i molecules of species A_i for all $1 \leq i \leq k$ and 0 molecule of all other species.

A *kinetic reaction* $r = (s_1 \to s_2; e)$ is a pair composed of a *reaction* $s_1 \to s_2$ and a *kinetic expression* e. The reaction transforms the solution s_1, called *reactants*, into the solution s_2, called *products*. The molecules present in the same amount in both reactants and products are called *activators*. They are not modified by the reaction, but are required to apply it. Kinetic expressions are symbolic functions defined from *concentration variables*, $Vars_{Spec} = \{x_A \mid A \in Spec\}$, symbols of *initial concentrations*, $Const_i = \{x_A^0 \mid A \in Spec\}$, and symbols of *kinetic parameters*, $Const_k = \{k_0, k_1, \ldots\}$:

$$e, f, \ldots ::= x \mid x^0 \mid k \mid e + f \mid e - f \mid e \times f \mid e/f \mid -e \mid (e)$$

where $x \in Vars_{Spec}$, $x^0 \in Const_i$ and $k \in Const_k$. As usual, we also simply denote ef for $e \times f$.

The *(chemical) concentration* of a chemical species is a function from time to positive numbers $\mathbb{R}_+ \to \mathbb{R}_+$. Kinetic expressions are interpreted as actual kinetic functions by means of an *assignment* α that maps concentration variables to concentrations (α_c), initial concentrations to non negative real values (α_0) and kinetic parameters to non negative real values (α_k):

$$\alpha_c : Vars_{Spec} \to (\mathbb{R}_+ \to \mathbb{R}_+) \qquad \alpha_0 : Const_i \to \mathbb{R}_+ \qquad \alpha_k : Const_k \to \mathbb{R}_+$$

We only consider assignments α consistent on initial concentrations, that is for any species A, $\alpha_c(x_A)(0) = \alpha_0(x_A^0)$. Given an assignment α, the interpretation $[e]_\alpha$ of a kinetic expression e is thus defined as follows:

$$[x]_\alpha(t) = \alpha_c(x)(t) \quad [x^0]_\alpha(t) = \alpha_0(x^0) \quad [k]_\alpha(t) = \alpha_k(k) \quad [(e)]_\alpha(t) = [e]_\alpha(t)$$

$$[-e]_\alpha = -[e]_\alpha \quad [e \ op \ f]_\alpha(t) = [e]_\alpha(t) \ op \ [f]_\alpha(t) \text{ where } op \in \{+, -, \times, /\}$$

Given a set of kinetic reactions, we only consider assignments α such that for any kinetic expression e occurring in this network, its interpretation $[e]_\alpha : \mathbb{R}_+ \to \mathbb{R}_+$ is a continuously differentiable function from time to non negative real numbers, standing for the actual reaction rate. Kinetic reactions $(s_1 \to s_2; e)$ also have to respect the following *coherence property*: the actual rate given by any assignment α is equal to zero if and only if one of the reactants is not present: $\forall \alpha$. $[e]_\alpha(t) = 0$ iff $\exists A \in s_1.[x_A]_\alpha(t) = 0$. Note that a kinetic expression can contain concentration variables of molecules that are not present in the reactants of the reaction; such molecules, called *modulators*, are not required to apply the reaction, but modify its rate.

Definition 1. *A reaction network is a pair $\langle I, R \rangle$, composed of a set of internal molecules I, which specifies that some molecules can not interact with the context, and a set of kinetic reactions R.*

From any network $N = \langle I, R \rangle$ and from its kinetic expressions, we can infer a *system of ordinary differential equations* defined by

$$ODE(N) = \left[\frac{dx_A}{dt} = \sum_{(s_1 \to s_2; e) \in R} (s_2(A) - s_1(A))e \right]_{A \in Spec}$$

Given any assignment α_0 of the initial concentrations and any assignment α_k of the kinetic parameters, by the Cauchy-Lipschitz theorem, the system $ODE(N)$ has a unique differentiable solution α_c, defined on a maximal interval including 0. Moreover, we only consider solutions α_c defined on (at least) $[0, +\infty[$. Otherwise, we say that N has no *valid solution* for these assignments.

An *equilibrium condition* e is defined similarly to kinetic expressions and interpreted as function from time to positive numbers. It is satisfied by an assignment α iff α_c satisfies $\dfrac{de}{dt} = 0$ given the initial concentration and parameter assignments α_0 and α_k. An equilibrium condition can for instance impose the equilibrium of a particular molecule (for instance $e = x_A$), a solution ($e = \sum_{A \in s} s(A)x_A$), or a reaction ($e = f$ for the reaction $(r ; f)$). We denote by **E** a set of equilibrium conditions. Given a network N and equilibrium conditions **E**, the deterministic dynamics of N that satisfies E is defined as

$$sol(N, \mathbf{E}) = \{\alpha \mid \alpha_c \text{ satisfies } E \text{ and is a valid solution of } ODE(N)$$
$$\text{for initial concentrations } \alpha_0 \text{ and parameter assignments } \alpha_k\}$$

Since we are particularly interested in the molecules that are not at equilibrium, we say that two assignments α and α' are *equal modulo equilibrium conditions*, denoted $\alpha_{\mathbf{E}}\alpha'$, if they are equal on those molecules.

4   Contextual Equilibrium-Equivalence

We present here a notion of weak equilibrium-equivalence between reaction networks, then the definition of contexts, and finally the contextual equilibrium-equivalence.

Definition 2 (Weak Equilibrium-Equivalence). *Two networks N and M are* weakly equilibrium-equivalent *for **E**, denoted $N \sim^{\mathbf{E}} M$, if they have the same solutions modulo equilibrium conditions* $sol(N, \mathbf{E}) =_{\mathbf{E}} sol(M, \mathbf{E})$.

A *context* $\mathcal{C}$ is itself a reaction network. Given a set of internal molecules I, we say that a context $\mathcal{C}$ is *compatible* with I if $\forall A \in I$, A has no occurrence in $\mathcal{C}$. We denote by $Context(I)$ the set of compatible contexts with I. Given a network $N = \langle I, R \rangle$ and a compatible context $\mathcal{C} = \langle I', R' \rangle \in Context(I)$, we denote by $\mathcal{C}[N] = \langle I \cup I', R \cup R' \rangle$ the network placed into the context.

Definition 3 (Contextual Equilibrium-Equivalence). *Let **E** be an equilibrium, the reaction networks $N = \langle I, R \rangle$ and $M = \langle I', R' \rangle$ are* contextually equilibrium-equivalent *for **E**, denoted $N \equiv^{\mathbf{E}} M$, if they are weakly equilibrium-equivalent in any compatible context, i.e.* $\forall \mathcal{C} \in Context(I \cup I'). \mathcal{C}[N] \sim^{\mathbf{E}} \mathcal{C}[M]$.

5   Simplification Axioms

In this section, we present some simplification axioms, that transform a network into a contextually equilibrium-equivalent network. The soundness proofs of those axioms are given in the annex[1]. These simplification axioms reduce the size of a reaction network, either by completely removing a molecule from the set of reactions, by decreasing the number of reactions, or by simplifying a reaction.

We first present 2 simple simplification axioms, followed by 4 instances of a more general axiom, based on the presence of an intermediate molecule. Finally, we present this general axiom. Notice that the axioms are quite similar to the ones we presented for the attractor equivalence with a qualitative and observational semantics in [20].

The first 2 simplification axioms are given in Fig. 2. The first one, (USELESS), deletes a reaction $s \to s$ that does not impact the network dynamics. The axiom (ACTIVATOR) removes an internal molecule A only used as an activator in the reactions (i.e. is always present in the same amount in both sides of the reaction). It is for instance the case for the gene G in the *Gene* network in Sect. 2.

The next four axioms in Fig. 3 are instances of the more general axiom (INTERMEDIATE). These axioms aim at eliminating an internal and intermediate molecule which is at equilibrium.

In the first one, (INTER), the intermediate molecule A is only used in two reactions, one time as the unique product, and the other as the unique reactant.

$$\frac{}{\langle \emptyset, \{(s \to s \; ; \; e)\}\rangle \equiv^{\mathbf{E}} \langle \emptyset, \emptyset \rangle} \text{ (USELESS)}$$

$$\frac{\forall (s_1 \to s_2 \; ; \; e) \in R, s_1(A) = s_2(A)}{\langle \{A\}, R \rangle \equiv^{\mathbf{E}} \langle \{A\}, \{(r \backslash A \; ; \; e[x_A^0/x_A]) \mid (r \; ; \; e) \in R\} \rangle} \text{ (ACTIVATOR)}$$

Fig. 2. Simple simplification axioms.

$$\frac{x_A \in \mathbf{E} \qquad A \notin s, s'}{\langle \{A\}, \{(s \to A \; ; \; e), (A \to s' \; ; \; k_2 x_A)\}\rangle \quad \equiv^{\mathbf{E}} \quad \langle \{A\}, \{(s \to s' \; ; \; e)\}\rangle} \text{ (INTER)}$$

$$\frac{x_C, x_E \in \mathbf{E}}{\begin{array}{l} \langle \{E, C\}, \\ \{(E + S \to C \; ; \; k_1 x_E x_S), \\ (C \to E + S \; ; \; k_{-1} x_C), \\ (C \to E + P \; ; \; k_2 x_C))\}\rangle \end{array} \equiv^{\mathbf{E}} \begin{array}{l} \langle \{E, C\}, \\ \{(S \to P \; ; \; k_2(x_E^0 + x_C^0) \dfrac{x_S}{x_S + \dfrac{k_2 + k_{-1}}{k_1}})\}\rangle \end{array}} \text{ (MICHAELIS-MENTEN)}$$

$$\frac{x_A \in \mathbf{E} \qquad A \notin s, s'}{\begin{array}{l} \langle \{A\}, \{(s \to s + A \; ; \; e), \\ (A \to \emptyset \; ; \; k_{-1} x_A), (A \to s' \; ; \; k_2 x_A)\}\rangle \end{array} \equiv^{\mathbf{E}} \langle \{A\}, \{(s \to s+s' \; ; \; \dfrac{k_2}{k_{-1} + k_2} e)\}\rangle} \text{ (CASCADE}_1\text{)}$$

$$\frac{x_A \in \mathbf{E} \qquad A \notin s, s'}{\begin{array}{l} \langle \{A\}, \{(s \to s + A \; ; \; e), \\ (A \to \emptyset \; ; \; k_{-1} x_A), (A \to A + s' \; ; \; e')\}\rangle \end{array} \equiv^{\mathbf{E}} \langle \{A\}, \{(s \to s + s' \; ; \; e'[\dfrac{e}{k_{-1}}/x_A])\}\rangle} \text{ (CASCADE}_2\text{)}$$

Fig. 3. Instances of intermediate molecule axiom.

---

[1] www.cristal.univ-lille.fr/~guillaume.madelaine/doc/2015_structural_simplification.pdf.

(INTERMEDIATE)

$$\frac{x_A \in \mathbf{E} \quad \forall t.(\sum_l e_l')(t) \neq 0 \quad A \notin s^{(1)}, s^{(2)}, s_l^{(2')}, s_m^{(1'')}, s_m^{(2'')} \quad x_A \notin e_l'}{\begin{array}{l} \langle \{A\}, \{(s^{(1)} \rightarrow s^{(2)} + A \; ; \; e)\} \cup \\ \{(A \rightarrow s_l^{(2')} \; ; \; x_A e_l')\}_l \cup \\ \{(s_m^{(1'')} + A \rightarrow s_m^{(2'')} + A \; ; \; e_m'')\}_m \rangle \end{array} \equiv_{\mathbf{E}} \begin{array}{l} \langle \{A\}, \{(s^{(1)} \rightarrow s^{(2)} + s_l^{(2')} \; ; \; e_l' \dfrac{e}{\sum_l e_l'})\}_l \cup \\ \{(s_m^{(1'')} + s^{(1)} \rightarrow s_m^{(2'')} + s^{(1)} \; ; \; e_m''[\dfrac{e}{\sum_l e_l'} / x_A])\}_m \rangle \end{array}}$$

Fig. 4. General intermediate molecule axiom.

Since A is at equilibrium, the kinetic expressions of these reactions have to be equal, i.e. $e = k_2 x_A$. The axiom removes A and merges both reactions into one, keeping only the kinetic expression e. The parameter k_2 is eliminated.

The second axiom, (MICHAELIS-MENTEN), simplifies a three-steps enzyme-catalyzed transformation. A substrate S binds to an enzyme E to form the complex C. Then the complex either transforms back to $S + E$, or produces the product P while releasing E. Assuming that the enzyme E and the complex C are at equilibrium, the axiom merges the reactions into a unique one, that directly transforms S into P. The equilibrium of C imposes that the simplified reaction has a Michaelis-Menten kinetics of the form $V\dfrac{x_S}{x_S + K}$ [22].

The last two, (CASCADE$_1$) and (CASCADE$_2$), concern a cascade of reactions, where the intermediate molecule A, at equilibrium, is produced in presence of some activators s, and then is either degraded or used to produce some s'. The axioms eliminate A, so the simplified networks directly produced s' in presence of s. The simplified kinetic expressions are obtained by computing the value of x_A at equilibrium, and by replacing it in the third kinetic reaction.

We finally present in Fig. 4 the general axiom (INTERMEDIATE). In this axiom, we consider an intermediate internal molecule A, at equilibrium. It simplifies a model with one reaction that can produce A, with a (non-empty) set of reactions that has only A as reactant and whose kinetic expressions are linear in x_A, and possibly a set of reactions with A as activator. Then the axiom eliminates A, and merges two-by-two the reactions. The linearity of the kinetic expression of some reactions is necessary to easily compute the expression of x_A at equilibrium, that is in this case $x_A = \sum_j e_j / \sum_l e_l'$.

6   Simplification of the *Tet-On* Reaction Network

We present here the simplification of the *Tet-On* system [13–15] using our axioms. The initial *Tet-On$_{detailed}$* reaction network, depicted in Fig. 5 (left), has 10 reactions and 11 parameters. We simplify it into the contextually equilibrium-equivalent *Tet-On$_{simple}$* network, depicted on Fig. 5 (right), with only two reactions and 3 parameters.

The *Tet-On* system [13–15] describes how the production of activated green fluorescent proteins (GFP_a) in a cell can be stimulated by the presence of doxy-cycline (Dox) outside the cell. The detailed network is *Tet-On$_{detailed}$* $= \langle I, R \rangle$

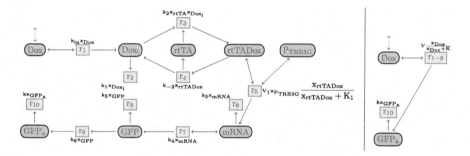

Fig. 5. Reaction graphs of the detailed (left) and simplified (right) *Tet-On* networks. Molecules are represented by circles, and reactions by squares. In the kinetic expressions near the reactions, the k_i are parameters while x_A is a variable representing the concentration of a molecule A. A dash arrow means that the molecule acts as a modulator in the reaction, while a dot arrow means that the molecule can be modified by the context. In the right network, the parameters are $V = x^0_{P_{TRE3G}} V_1 k_4 k_6 / k_3 (k_5 + k_6)$ and $K = k_1 k_{-2} K_1 / x^0_{rtTA} k_{in} k_2$.

$$Dox \rightarrow Dox + Dox_i \; ; \; k_{in} x_{Dox} \qquad (1)$$

$$Dox_i \rightarrow \emptyset \; ; \; k_1 x_{Dox_i} \qquad (2)$$

$$rtTA + Dox_i \rightarrow rtTADox \; ; \; k_2 x_{rtTA} x_{Dox_i} \qquad (3)$$

$$rtTADox \rightarrow rtTA + Dox_i \; ; \; k_{-2} x_{rtTADox} \qquad (4)$$

$$mRNA \rightarrow \emptyset \; ; \; k_3 x_{mRNA} \qquad (6)$$

$$mRNA \rightarrow mRNA + GFP \; ; \; k_4 x_{mRNA} \qquad (7)$$

$$GFP \rightarrow \emptyset \; ; \; k_5 x_{GFP} \qquad (8)$$

$$GFP \rightarrow GFP_a \; ; \; k_6 x_{GFP} \qquad (9)$$

$$GFP_a \rightarrow \emptyset \; ; \; k x_{GFP_a} \qquad (10)$$

$$P_{TRE3G} + rtTADox \rightarrow P_{TRE3G} + rtTADox + mRNA \quad ; \quad V_1 x_{P_{TRE3G}} \frac{x_{rtTADox}}{x_{rtTADox} + K_1} \quad (5)$$

Fig. 6. Reactions of the detailed *Tet-On$_{detailed}$* network.

where every molecule is internal except for *Dox* (i.e. $I = Spec \backslash Dox$), and R is the set of reactions from Fig. 6, inspired by the *Tet-On* model from [15].

In the network, the doxycycline *Dox* moves into the cell and becomes *Dox$_i$* by reaction (1). We assume here that the amount of *Dox* is controlled by the environment (for instance by a microfluidics device [30]), and therefore the network can not modify its concentration. Then *Dox$_i$* is either degraded by reaction (2), or binds to the artificial transcription factor *rtTA* by reaction (3). The complex *rtTADox* either dissociates (4), or activates the transcription of the gene *P$_{TRE3G}$*, producing *mRNA* (5). *mRNA* either degrades (6) or is translated into *GFP* (7). Finally, *GFP* needs to be activated into *GFP$_a$* (9) in order to become fluorescent and thus observable by a microscope. Both *GFP* and *GFP$_a$* can also be degraded (8, 10).

We are particularly interested by *GFP$_a$*, since it is the only experimentally observable molecule. Therefore we assume that all other molecules are at equilibrium, i.e. $\mathbf{E} = \{x_X \mid X \in Spec \backslash GFP_a\}$. The simplification follows the axioms from Figs. 2, 3 and 4, so that will prove that the two networks are contextually equilibrium-equivalent for $\mathbf{E}$. Note that in the following simplification, for the sake of readability, some kinetic expressions were sometimes slightly rewritten into equivalent expressions.

Let us first remark that the gene P_{TRE3G} is only used as an activator, in the reaction 5. So we apply the axiom (ACTIVATOR), removing P_{TRE3G} from this reaction, while replacing $x_{P_{TRE3G}}$ by $x^0_{P_{TRE3G}}$ in its kinetic function. Then $rtTADox$ is an internal molecule at equilibrium, present in three reactions: one that produces it (3), one that consumes it (4), and one that uses it as an activator (5). Then we use the axiom (INTERMEDIATE) on it, followed directly by (USELESS), and merge the three reactions into:

$$rtTA + Dox_i \rightarrow rtTA + Dox_i + mRNA \; ; \; x^0_{P_{TRE3G}} V_1 \frac{x_{rtTA} x_{Dox_i}}{x_{rtTA} x_{Dox_i} + k_{-2} K_1/k_2} \quad (11)$$

$rtTA$ is only used as activator, so we apply (ACTIVATOR) and simplify (11) into:

$$Dox_i \rightarrow Dox_i + mRNA \; ; \; x^0_{P_{TRE3G}} V_1 \frac{x_{Dox_i}}{x_{Dox_i} + k_{-2} K_1/x^0_{rtTA} k_2} \quad (12)$$

Apply axiom (CASCADE)$_1$ on GFP, replacing the reactions (7), (8) and (9) by:

$$mRNA \rightarrow mRNA + GFP_a \; ; \; (k_4 k_6/(k_5 + k_6)) x_{mRNA} \quad (13)$$

Also, apply (CASCADE)$_2$ on Dox_i, and replace reactions (1), (2), and (12) by:

$$Dox \rightarrow Dox + mRNA \; ; \; x^0_{P_{TRE3G}} V_1 \frac{x_{Dox}}{x_{Dox} + k_1 k_{-2} K_1/(x^0_{rtTA} k_{in} k_2)} \quad (14)$$

Finally we use the axiom (INTERMEDIATE) followed by (USELESS) on $mRNA$, and merge the reactions (6), (13) and (14) into:

$$Dox \rightarrow Dox + GFP_a \; ; \; \frac{x^0_{P_{TRE3G}} V_1 k_4 k_6}{k_3(k_5 + k_6)} \frac{x_{Dox}}{x_{Dox} + k_1 k_{-2} K_1(x^0_{rtTA} k_{in} k_2)} \quad (15)$$

Defining two new parameters $V = x^0_{P_{TRE3G}} V_1 k_4 k_6/(k_3(k_5 + k_6))$ and $K = k_1 k_{-2} K_1/(x^0_{rtTA} k_{in} k_2)$, we eventually obtain the following reaction network:

$$Dox \rightarrow Dox + GFP_a \; ; \; V \frac{x_{Dox}}{x_{Dox} + K} \qquad GFP_a \rightarrow \emptyset \; ; \; k x_{GFP_a}$$

Notice that, aside from the kinetics, the simplified network is equal to the one we obtained with our qualitative simplification in [20].

7   Conclusion

We presented a new structural simplification of reaction networks, that preserved the deterministic semantics. The simplification is contextual, and is based on equilibrium conditions on intermediate molecules. We shown the usefulness of the simplification by applying it on two biological networks.

We are currently implementing the simplification algorithm, with a more complete set of axioms and compatible with the SBML format. This axioms include variants of the axioms presented here, for instance with different equilibrium conditions, but also other types of axioms, using for instance symmetries in the network. We plan to apply the simplification more systematically to biological systems. It would also be interesting to compare in depth the power of our structural simplification rules to that of the King-Altman method on ODE system [17]. On the theoretical side, as future work, we want to investigate an approximated equivalence, with approximated equilibrium conditions, and to compute the maximal error of a simplification. A similar simplification method with a stochastic semantics will also be considered.

Acknowledgment. The authors would like to thank Michel Petitot for its useful discussions as well as members of the *PalBioSys* research network.

References

1. Ashyraliyev, M., Fomekong-Nanfack, Y., Kaandorp, J.A., Blom, J.G.: Systems biology: parameter estimation for biochemical models. Febs J. **276**(4), 886–902 (2009)
2. Berthelot, G., Roucairol, G.: Reduction of petri-nets. In: Mazurkiewicz, A. (ed.) Mathematical Foundations of Computer Science, vol. 45, pp. 202–209. Springer, Heidelberg (1976)
3. Bodenstein, M.: Eine theorie der photochemischen reaktionsgeschwindigkeiten. Z. Phys. Chem. **85**(329), 0022–3654 (1913)
4. Cardelli, L.: Morphisms of reaction networks that couple structure to function. BMC Syst. Biol. **8**(1), 84 (2014)
5. Cardelli, L., Zavattaro, G.: On the computational power of biochemistry. In: Horimoto, K., Regensburger, G., Rosenkranz, M., Yoshida, H. (eds.) AB 2008. LNCS, vol. 5147, pp. 65–80. Springer, Heidelberg (2008)
6. Danos, V., Feret, J., Fontana, W., Harmer, R., Krivine, J.: Abstracting the differential semantics of rule-based models: exact and automated model reduction. In: Logic In Computer Science (LICS), pp. 362–381. IEEE (2010)
7. De Jong, H.: Modeling and simulation of genetic regulatory systems: a literature review. J. Comput. Biol. **9**(1), 69–105 (2002)
8. Fages, F., Gay, S., Soliman, S.: Inferring reaction models from ODEs. In: Gilbert, D., Heiner, M. (eds.) CMSB 2012. LNCS, vol. 7605, pp. 370–373. Springer, Heidelberg (2012)
9. Fages, F., Soliman, S.: Formal Cell Biology in Biocham. In: Bernardo, M., Degano, P., Zavattaro, G. (eds.) SFM 2008. LNCS, vol. 5016, pp. 54–80. Springer, Heidelberg (2008)
10. Gay, S., Soliman, S., Fages, F.: A graphical method for reducing and relating models in systems biology. Bioinformatics **26**(18), 575–581 (2010)
11. Gorban, A.N., Karlin, I.V.: Method of invariant manifold for chemical kinetics. Chem. Eng. Sci. **58**(21), 4751–4768 (2003)
12. Gorban, A.N., Karlin, I.V., Ilg, P., Öttinger, H.C.: Corrections and enhancements of quasi-equilibrium states. J. Non-newton. Fluid Mech. **96**(1), 203–219 (2001)

13. Gossen, M., Bujard, H.: Tight control of gene expression in mammalian cells by tetracycline-responsive promoters. Proc. Natl. Acad. Sci. U.S.A. **89**(12), 5547–5551 (1992)

14. Gossen, M., Freundlieb, S., Bender, G., Müller, G., Hillen, W., Bujard, H.: Transcriptional activation by tetracyclines in mammalian cells. Science **268**(5218), 1766–1769 (1995)

15. Huang, Z., Moya, C., Jayaraman, A., Hahn, J.: Using the Tet-On system to develop a procedure for extracting transcription factor activation dynamics. Mol. BioSyst. **6**(10), 1883–1889 (2010)

16. John, M., Lhoussaine, C., Niehren, J., Versari, C.: Biochemical reaction rules with constraints. In: Barthe, G. (ed.) ESOP 2011. LNCS, vol. 6602, pp. 338–357. Springer, Heidelberg (2011)

17. King, E.L., Altman, C.: A schematic method of deriving the rate laws for enzyme-catalyzed reactions. J. Phys. Chem. **60**(10), 1375–1378 (1956)

18. Kitano, H.: Systems biology: a brief overview. Science **295**(5560), 1662–1664 (2002)

19. Lemaire, F., Sedoglavic, A., Urguplu, A.: Moving frame based strategies for reduction of ordinary differential/recurrence systems using their expanded lie point symmetries (2008)

20. Madelaine, G., Lhoussaine, C., Niehren, J.: Attractor equivalence: an observational semantics for reaction networks. In: Fages, F., Piazza, C. (eds.) FMMB 2014. LNCS, vol. 8738, pp. 82–101. Springer, Heidelberg (2014)

21. Mäder, U., Schmeisky, A.G., Flórez, L.A., Stülke, J.: SubtiWiki-a comprehensive community resource for the model organism Bacillus subtilis. Nucleic Acids Res. **40**, 1278–1287 (2012)

22. Michaelis, L., Menten, M.L.: Die kinetik der invertinwirkung. Biochem. z **49**(333–369), 352 (1913)

23. Murata, T., Koh, J.: Reduction and expansion of live and safe marked graphs. IEEE Trans. Circuits Syst. **27**(1), 68–70 (1980)

24. Pitts, Andrew M.: Operational semantics and program equivalence. In: Barthe, Gilles, Dybjer, Peter, Pinto, Luís, Saraiva, João (eds.) APPSEM 2000. LNCS, vol. 2395, p. 378. Springer, Heidelberg (2002)

25. Regev, A., Shapiro, E.: Cellular abstractions: cells as computation. Nature **419**(6905), 343 (2002)

26. Schmidt-Schauss, M., Sabel, D., Niehren, J., Schwinghammer, J.: Observational program calculi and the correctness of translations. J. Theor. Comput. Sci.(TCS) **577**, 98–124 (2015)

27. Segel, L.A., Slemrod, M.: The quasi-steady-state assumption: a case study in perturbation. SIAM Rev. **31**(3), 446–477 (1989)

28. Soliman, S., Fages, F., Radulescu, O., et al.: A constraint solving approach to tropical equilibration and model reduction. In: WCB-Ninth Workshop on Constraint Based Methods for Bioinformatics, Colocated with CP **2013**, (2013)

29. Uhlendorf, J., Bottani, S., Fages, F., Hersen, P., Batt, G.: Towards real-time control of gene expression: controlling the hog signaling cascade. In: Pacific Symposium On Biocomputing, pp. 338–349. World Scientific (2011)

30. Uhlendorf, J., Miermont, A., Delaveau, T., Charvin, G., Fages, F., Bottani, S., Batt, G., Hersen, P.: Long-term model predictive control of gene expression at the population and single-cell levels. Proc. Natl. Acad. Sci. **109**(35), 14271–14276 (2012)

Automating the Development of Metabolic Network Models

Robert Rozanski[1][(✉)], Stefano Bragaglia[2], Oliver Ray[2], and Ross King[1]

[1] School of Computer Science, University of Manchester, Manchester M13 9PL, UK
[2] Department of Computer Science, University of Bristol, Bristol BS8 1TH, UK
rozanskr@cs.man.ac.uk

Abstract. Although substantial progress has been made in the automation of many areas of systems biology, from data processing and model building to experimentation, comparatively little work has been done on integrated systems that combine all of these aspects. This paper presents an active learning system, "Huginn", that integrates experiment design and model revision in order to automate scientific reasoning about Metabolic Network Models. We have validated our approach in a simulated environment using substantial test cases derived from a state-of-the-art model of yeast metabolism. We demonstrate that Huginn can not only improve metabolic models, but that it is able to both solve a wider range of biochemical problems than previous methods, and to utilise a wider range of experiment types. Also, we show how design of extended crucial experiments can be automated using Abductive Logic Programming for the first time.

1 Introduction

Biological systems are extremely complicated. Even the model cellular systems of *Escherichia coli* and *Saccharomyces cerevisiae* consist of thousands of genes, proteins, small molecules, *etc.*, all interacting in complicated spatiotemporal ways. In addition, as biological systems have evolved through Darwinian evolution, Ockham's razor is not as effective as it is in the physical sciences.

Currently, although many computational tools are used to build systems biology models, the evaluation and analysis of these models is still mostly done by humans, who identify conflicting results, suspicious or low-confidence elements of models, ask specific questions to test the models, and run manual experiments. However, humans can only investigate small parts or aspects of models, because of their typical size and complexity. This bottleneck could be overcome by automating model development, *i.e.* the process of asking specific questions, running tailored experiments to answer them, and revising models if needed.

Huginn is an open-source software, available at:
github.com/robaki/huginnCMSB2015.
All figures included in this paper are in public domain; files can be downloaded from:
github.com/robaki/huginnCMSB2015.

© Springer International Publishing Switzerland 2015
O. Roux and J. Bourdon (Eds.): CMSB 2015, LNBI 9308, pp. 145–156, 2015.
DOI: 10.1007/978-3-319-23401-4_13

1.1   Adam, a Robot Scientist

King *et al.* [10] created an automated system that investigated the problem of orphan enzymes in metabolic models of yeast. The system, "Adam", was able to propose initial hypothetical models, and then design two-factor growth experiments to test them. The experiments were run using automated laboratory equipment. The data were then analysed to determine which models to refute. Adam, although successful, has multiple limitations. Its methods of proposing hypotheses were specific to the problem of orphan enzymes. Its experiment design and hypothesis testing algorithms were limited to only one type of experiment, and could not be easily extended. It also lacked general revision capabilities. These limitations make Adam unsuitable candidate for a general-purpose metabolic model development system.

1.2   Huginn

We have developed Huginn[1], to overcome some of the limitations of Adam. In doing this we have drawn from Machamer's, Darden's and Craver's (MDC) theory of discovering mechanisms. We have adopted MDC concept of mechanism to represent Metabolic Network Models (MNM) in a way suitable for automated system. We have also used their characterisation of the final stage of the model development process as a guide to the design of Huginn. We used Logic Programming, and Abductive Logic Programming (ALP) (Gringo [9], Clasp [8] and XHAIL [15]) to automate model construction and revision, as well as testing consistency of models with experiments. We have also used them to automate experiment design in a novel way.

1.3   Metabolic Networks as Biological Mechanisms

A significant amount of research in biology is concerned with development of models of mechanisms (*e.g.* of DNA replication). By representing what is happening in biological systems these models provide a way to predict and explain their behaviour in a way understandable to humans. Recently the notion of mechanisms in biology has attracted the attention of philosophers of science, who have tried to specify what these mechanisms are, and how they are discovered [2,5,6].

In this study we have adopted the notion of mechanism proposed by MDC [14]. They characterise mechanisms as collections of entities and activities organised in such ways that they can produce regular changes from setup to termination conditions. For example, a model of cellular respiration would show how cells produce ATP from glucose through a series of chemical reactions and transport processes.

The core qualitative information about metabolism are the chemical reactions and other processes that can occur in an organism, as well as chemical substances involved in them. MNM represents these processes in a form of hypergraphs. MNMs typically abstract away not only the concentration and dynamics

---

[1] From the Norse mythology – one of two ravens scouting the world for Odin.

of the system, but also some of the conditions, *e.g.* certain enzymes are expressed only under specific conditions. MNMs can be understood as MDC-type descriptions of mechanism. MNM show how certain chemicals are produced from other chemicals by representing continuous chemical paths from the former to the latter. Initial and termination conditions are the presence of specific species (*e.g.* metabolites) and genes in specific compartments (*e.g.* cytosol). Activities like chemical reactions, transport, gene expression and complex formation connect these conditions through intermediate steps.

2 Methods

2.1 Discovery of Mechanisms

The MDC concept of biological mechanism was developed to better understand the discovery of mechanisms. Discovery should not be understood here as an event, but as an extended iterative process of exploration, specification, building, testing and revision. According to MDC [4,6], the process starts with exploring and characterising the phenomenon of interest, *i.e.* one that is to be explained by description of mechanism. Then, incomplete and often abstract sketches of mechanisms are formulated, taking into account clues such as the nature of the phenomenon, its context (*e.g.* evolutionary), its spatial and time characteristics. These sketches show how the phenomenon could possibly be produced. Through specification and initial evaluation sketches are turned into schemata: these still may be to some extend incomplete or abstract, but contain enough information to allow production of fully specified models. Then, through further instantiation (if required) and searching for direct experimental evidence, final descriptions of mechanisms are produced. The transition between each of these stages involves construction, evaluation and anomaly resolution (revision), which is guided by specific strategies.

In this paper we focus on the latter stages of the discovery process, where phenomenon is fully characterised and models of mechanisms are composed entirely of non-abstract elements, *i.e.* they are constructed from specific reactions, proteins, metabolites and not from place-holder elements. We implement a number of strategies proposed by MDC in the design of Huginn, specifically:

- continuity and productivity are taken into account in construction, consistency testing, and revision
- generation and elimination of rival hypotheses (using crucial experiments)
- searching for direct evidence for hypotheses by *in vivo* and *in vitro* experiments:
  - entity and activity detection
  - characterising entities *in vitro* (enzymes' properties and complex formation)
  - disrupting mechanisms, and studying changes (gene deletions and changes in medium composition).

The model development process used in Huginn (see Fig. 1) is initialised in a number of steps. First, initial models and experiment results are recorded. Then models are checked for consistency with the results, as well as other criteria, like ability to produce termination conditions (*i.e.* synthesize final compounds) and presence of disconnected activities (*e.g.* reactions which substrates are not present in the model). Models that failed are revised. If the pool of initial models is smaller than user-specified threshold, then additional models are produced to fill that gap and the system is ready to enter its proper development cycle.

The first step in the development cycle is to design an experiment to test current working models. Then, the experiment is executed (simulated) and results used to test working models. Refuted models are then revised. If there is no way to make a model logically consistent with the results, then one or more of them will be ignored. This ability to ignore results is important for dealing with limitations of the Knowledge Representation method, as well as factors such as experimental noise, and the open world problem. The quality scores of models are then recalculated based on the number of covered and ignored results.

Huginn stops development process if at least one of three conditions is true. The first condition is lack of progress. If there were new models produced recently or if the best (highest quality score) model has recently changed, then development continues. The second condition is running out of experiments to execute, which happens when working models become empirically equivalent. In this case Huginn tries to redesign models at random, but if it fails 10 times, it stops. The last condition is running out of time or exceeding maximum number of cycles: both values are specified by the user.

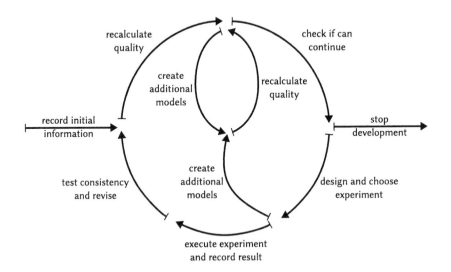

Fig. 1. Model development process

2.2   Abductive Logic Programming

The development process relies on four core operations: consistency checking, revision, production of additional models and experiment design. We use ALP for these operations. Abductive inference is typically understood as inference to the best explanation. In ALP abduction is defined as constructing a hypothesis H, that together with background knowledge B, entails a set of examples E:

$$B \cup H \models E$$

Unlike deduction, abduction is a defeasible form of inference, *i.e.* given true background knowledge and examples (observations), it may produce false hypotheses. However, it has the advantage of being able to produce novel knowledge. ALP tools have been used previously for completion [3,11] and revision of metabolic networks [16]. Thanks to optimisation capabilities of existing tools one can generate theories that not only satisfy hard logical constraints, but are also optimal with respect to user-specified criteria.

2.3   Representing Models Using Logic

MNM can be formalised and translated into datalog-style logic programs. Entities are defined by their type, identifier and version. Huginn currently supports four types of entities: metabolite, protein, complex or gene. Versions enable one to represent uncertainty regarding an entity's properties. Two currently supported properties are *catalyses* and *transports*.

Huginn supports five types of activities: chemical reaction, complex formation, expression, transport or growth. Substrate and product predicates are used for all types of activities, and these specify not only what entities are required and produced, but also in what compartments. Apart from substrates, chemical reactions and transport may need catalysts or transporters respectively.

Models are defined by specifying which setup conditions and activities they contain.

All these facts describe the elements involved in the MNM. In order to determine which metabolites are synthesizable, simulation rules are added to this description. A group of rules marks as *active* activities which all substrates are either initially present or synthesizable (in appropriate compartment) and which catalyst/transporter requirements are met. An additional rule marks all products of active reactions as synthesizable.

2.4   Experiment Types and Predictions

Model descriptions need to be supplemented with prediction and consistency rules to support the use of empirical information. Predictions describe what outcome models predict w.r.t. description of experiment. Outcome is binary: true or false. In addition, model can be indifferent w.r.t. experiment (it does not predict any outcome). Model is inconsistent with a result of experiment if outcome of the experiment is different from the predicted one. Prediction rules determine predicted outcomes of experiments. There are seven types of experiments currently used in Huginn and each of them has its separate set of prediction rules:

Entity Detection: detection of metabolites, proteins or complexes.

Entity Localisation: as above, but in a specified compartment.

Activity Detection: used for detecting growth.

Activity Reconstruction: checks if activities can be reconstructed without enzymes or transporters.

Reconstruction Enzymatic Reaction: checks whether given entity can catalyse specific reaction.

Reconstruction Transporter Required: as above, but for transporters.

Two Factor Growth Experiment: used previously to test candidate parent genes of orphan enzymes [10]. It tests whether decreased growth rate after gene deletion can be offset by addition of a particular metabolite.

Some types of experiments can include interventions: addition or substraction of a specific entity from specific compartment. In our study we have restricted interventions to manipulation of the growth medium (addition/substraction of nutrients) and gene deletions. The way the interventions are handled differs depending on the nature of the task (revision, experiment design, *etc.*).

2.5   Automating Crucial Tasks

As mentioned above, the four essential tasks in the model development cycle are: consistency check, revision, construction of additional models and experiment design. All of these tasks were automated using Logic Programming (LP) techniques.

Consistency Check: This step consist of checking whether models are consistent with all known results as well as additional structural criteria. Specifically, models must synthesize all compounds specified in the termination conditions, they must not contain any activities that are missing substrates, and they cannot contain two versions of the same entity (that situation would be equivalent to having inconsistent beliefs about the entity's properties).

Revision: The models are revised by supplementing requirements from consistency check with mode declarations that specify what activities can be added and removed. XHAIL then tries to minimise a number of changes to the model. In cases where more than one optimal solution is found, one of them is chosen at random to keep the population of working models at a constant size.

For example, metabolite met_8 was detected in cells with deleted gene $g26$. This outcome is in conflict with predictions of model (a) (Fig. 2). In that model met_8 can be synthesized from input metabolites met_7 and met_11 (marked

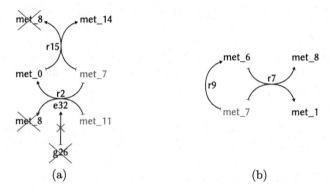

Fig. 2. Revision example: (a) deletion of $g26$ disrupts reactions $r2$ (lack of enzyme) and $r15$ (lack of substrate: met_0) and thus prevents the model from producing met_8, contrary to experimental results. Consistency with the results can be restored by adding two additional reactions (b) which can produce met_8 independently from $g26$ (Color figure online).

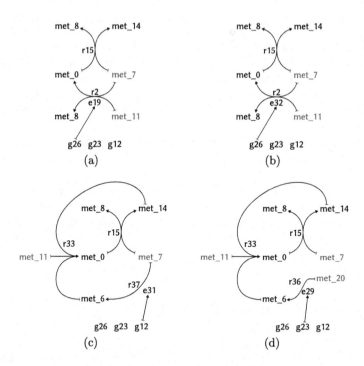

Fig. 3. Experiment design example: models (a) and (b) rely on gene $g26$ to produce met_0 and met_14, while models (c) and (d) rely on genes $g12$ and $g23$ respectively. Thus experiment consisting in deleting $g26$ and detecting either met_0 or met_14 will split these models into two groups: one predicting that the metabolite will be synthesised despite deletion, the other that it will not be.

green) in reaction $r2$, which requires enzyme coded by $g26$. Alternatively, it can be synthesised in $r15$, but that requires some source of substrate met_0. Since the only source of met_0 is $r2$, deletion of $g26$ disrupts both reactions and met_8 is not produced. Consistency with the experimental result can be restored by adding reaction(s) that can synthesise met_8 independently from $g26$, *e.g.* reactions $r9$ and $r7$ (Fig. 2(b)).

Construction of Additional Models: Additional models are constructed using almost the same approach as in revision, but with the addition of a requirement that the resulting models must be different (contain different set of activities) from any of the working models.

Experiment Design: The idea behind our approach to experiment design is to design experiment that will split the working models into two groups of equal size: one predicting that outcome of experiment is true, the other that it is false. This can be understood as an extension of the concept of crucial experiment. The same principle was used before as a strategy for choosing experiments from pre-generated sets [11].

For example, lets consider four models from Fig. 3. The input metabolites are met_7, met_11 and met_20 (marked green, only shown where relevant), and the output metabolite is met_14. All models synthesize met_14 in $r15$, but differ in ways they produce required substrate for this reaction: met_0. Models (a) and (b) rely on $r2$ and gene $g26$, while models (c) and (d) use $r37$ (needs gene $g12$) and $r36$ (needs $g23$) respectively. Therefore, if $g26$ is deleted models (a) and (b) will predict that met_14 is not produced, while (c) and (d) will predict that it is produced. One of plausible experiments for this group of models is then a *detection entity* experiment, detecting met_14 and involving one gene deletion (of $g26$).

Since some models may be considered to be better and therefore more probably correct in a subjective sense, we split not raw numbers of models, but rather their total quality score. Since designing experiment that will split scores into equal groups is not always possible, this task was implemented as optimisation problem. The system tries to minimise total penalty, which is calculated as follows:

$$P = |0.5 * \sum_{m} q(m) - \sum_{m \in T} q(m)| + |0.5 * \sum_{m} q(m) - \sum_{m \in F} q(m)| + \sum_{m \in I} q(m)$$

where m is model, $q(m)$ is model's quality, T, F and I are sets of models that predict that the outcome is true, false or indifferent respectively. Due to the complicated nature of this task it was implemented using Gringo/Clasp directly, not through XHAIL.

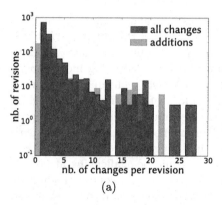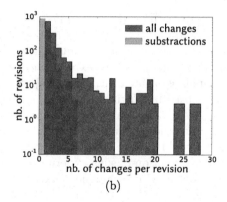

(a)　　　　　　　　　　　　　(b)

Fig. 4. Size of model revisions: each model revision may consist of multiple changes (additions or substractions of activities). The histograms compare distributions of additions (a) and substractions (b) with all changes. Note the log scale on y axis.

3   Results and Conclusions

The goal of our study was to evaluate whether the proposed system can be used in model development. To answer this question we supplied Huginn with initial models containing errors and run the development process to see whether the models would be improved. At this initial stage of evaluation the use of real biochemical experiments is not necessary, and would not be cost effective. Instead we ran simulations using reference models, which are fragments of the yeast consensus metabolic model 7.11 [1], between 14 and 54 activities in size. The knowledge bases containing the activities and entities for model development were created by mixing elements from a given reference model with additional, erroneous elements the role of which is to make the development process harder. The initial models were created by randomly selecting a set of activities from these knowledge bases.

The improvement of models consisted of removing and adding activities so that working models resemble the reference model. To quantify the difference between a model and the reference model we use the symmetric difference between the sets of activities involved in the models.

The results of our simulations show that Huginn can successfully improve initial models, with an average reduction in initial error of 76 %. For the smaller test-cases the working models tended to quickly become empirically equivalent and attempts to recover from it through generation of randomised models would fail. For the larger test-cases, Huginn tended to continue development until timeout. The final unsuccessful attempt to construct new models was not associated with any quality change[2] ($p = 0.13$). However, since in many cases constructing random models allowed Huginn to recover and continue development process, including this ability was beneficial (see Footnote 2) ($p = 0.0003$).

---

[2] Tested using pair-wise comparison of improvement and then a binomial test.

One of the significant differences between Adam and Huginn is Huginn's ability to use more types of experiments. To test whether this difference is beneficial we ran additional simulations while limiting available experiment types to only *two-factor growth* experiments. The results show that using more experiments is associated with larger improvements[3] (p = 0.047). The main experiment types used by Huginn were *two-factor growth* (48 % of experiments) and *entity detection* (51 %). A portion of the latter involved multiple interventions: gene deletions and medium manipulations (19 %). In the most extreme case an experiment would use 4 gene deletions on top of manipulating the medium composition. While execution of such experiment in practice would be challenging at best, it shows that ALP techniques used in Huginn can successfully cope with complex experiment design problems. The rest of the experiments used in development were *entity localisation* experiments.

Another significant difference between Adam and Huginn is in their revision abilities. Adam can only add individual missing expression activities. While, thanks to XHAIL, Huginn can handle a wider range of activities, also remove them, and introduce multiple changes in one revision. Therefore, it should be able to make more substantial changes to MNM structures. Our simulations show that it is indeed the case: 50 % of the revisions involved more than one change (addition/substraction), while the largest involved as many as 28 changes (Fig. 4). Many of these revisions combined the addition and substraction of activities (29 %). The majority of revisions (60 %) involved changing elements other than expression activities. These results show that Huginn takes advantage of its enhanced revision abilities to introduce larger changes to the models, and is therefore capable of solving wider range of biochemical problems than Adam – not only the problem of orphan enzymes, but also other structural problems in the metabolic networks.

We conclude that Huginn qualitatively improves on Adam by using more types of experiment, and a more versatile revision method, and that these improvements translate into an increased ability to correct models. We also conclude that the presented experiment design solution can not only design useful experiments, but also handle complicated tasks that require multiple interventions. More extensive *in silico* tests are still needed to test Huginn's performance in different configurations and under different circumstances. For example, we have not yet tested Huginn's ability to handle inconsistencies in results (*e.g.* introduced by experimental errors).

4   Related Work

Thagard demonstrated the use of various types of abduction in hypothesis formation using an AI system called PI. [18] Here, we used "simple abduction" to revise refuted models.

Substantial advancements have been done in the field of computational discovery. Langley *et al.* [12] describes BACON, DALTON, GLAUBER, and

---

[3] Tested using paired, one-tailed t-test.

STAHL – seminal systems designed to model historical discoveries of quantitative and qualitative laws.

Džeroski and Todorovski [7] described QMN and LAGRANGE – systems for discovering quantitative and qualitative laws governing dynamical systems. Schmidt and Lipson [17] developed a system for discovering non-trivial conservation laws from experimental data. Todorovski *et al.* [19] developed HIPM, a system for developing complex hierarchical models of dynamical systems using induction, while taking advantage of expert knowledge. Compared to these studies we have focussed on qualitative aspects of scientific discovery, which can provide necessary insight into functioning of biological systems in terms of mechanistic explanations. However, methods for developing quantitative models are likely to be useful in further steps of building biological models.

Valdés-Pérez created MECHEM, a system for proposing possible intermediate steps of chemical transformations. The system uses information about chemicals' composition and structure to constrain the search-space as well as divide-and-conquer and Ockham's razor heuristics to make the search more efficient. An interesting feature of the system is its ability to propose new reactants. [20] Compared to MECHEM, Huginn focuses on developing larger models of metabolism from pre-defined reactions and on using biological experiments to gradually constraint the search-space.

Langley [13] summarises the lessons learned from their experience with developing computational tools for scientific discovery. They advise one to use the scientists' representations and their knowledge; tools should not just summarise, but provide explanations. Our approach follows these lessons. Representation of metabolism used by Huginn is taken from biochemistry, ensuring that it is easily understandable by biologists. Huginn records all produced models and results so checking why particular models were produced is possible.

Acknowledgment. This work is supported by an EPSRC-EU Doctoral Training Award and the Faculty Engineering and Physical Sciences of the University of Manchester.

References

1. Aung, H.W., Henry, S.A., Walker, L.P.: Revising the representation of fatty acid, glycerolipid, and glycerophospholipid metabolism in the consensus model of yeast metabolism. Ind. Biotechnol. **9**(4), 215–228 (2013)
2. Bechtel, W., Richardson, R.C.: Discovering Complexity: Decomposition and Localization as Strategies in Scientific Research. MIT Press, Cambridge (2010)
3. Collet, G., Eveillard, D., Gebser, M., Prigent, S., Schaub, T., Siegel, A., Thiele, S.: Extending the metabolic network of *ectocarpus siliculosus* using answer set programming. In: Cabalar, P., Son, T.C. (eds.) LPNMR 2013. LNCS, vol. 8148, pp. 245–256. Springer, Heidelberg (2013)
4. Craver, C., Darden, L.: Discovering mechanisms in neurobiology. In: Machamer, P.K., et al. (eds.) Theory and Method in the Neurosciences, pp. 112–137. University of Pitt Press, Pittsburgh (2001)

5. Craver, C.F., Darden, L.: In Search of Mechanisms: Discoveries Across the Life Sciences. University of Chicago Press, Chicago (2013)
6. Darden, L.: Reasoning in Biological Discoveries. Cambridge University Press, Cambridge (2006)
7. Džeroski, S., Todorovski, L.: Discovering dynamics: from inductive logic programming to machine discovery. J. Intell. Inf. Syst. **4**(1), 89–108 (1995)
8. Gebser, M., Kaufmann, B., Neumann, A., Schaub, T.: *clasp*: A conflict-driven answer set solver. In: Baral, C., Brewka, G., Schlipf, J. (eds.) LPNMR 2007. LNCS (LNAI), vol. 4483, pp. 260–265. Springer, Heidelberg (2007)
9. Gebser, M., Schaub, T., Thiele, S.: Gringo: a new grounder for answer set programming. In: Baral, C., Brewka, G., Schlipf, J. (eds.) LPNMR 2007. LNCS (LNAI), vol. 4483, pp. 266–271. Springer, Heidelberg (2007)
10. King, R., Rowland, J., Oliver, S., Young, M., Aubrey, W., Byrne, E., Liakata, M., Markham, M., Pir, P., Soldatova, L., et al.: The automation of science. Science **324**(5923), 85–89 (2009)
11. King, R.D., Whelan, K.E., Jones, F.M., Reiser, P.G., Bryant, C.H., Muggleton, S.H., Kell, D.B., Oliver, S.G.: Functional genomic hypothesis generation and experimentation by a robot scientist. Nature **427**(6971), 247–252 (2004)
12. Langley, P.: Scientific Discovery: Computational Explorations of the Creative Processes. MIT Press, Cambridge (1987)
13. Langley, P.: Lessons for the computational discovery of scientific knowledge. In: Proceedings of First International Workshop on Data Mining Lessons Learned, pp. 9–12. University of New South Wales (2002)
14. Machamer, P., Darden, L., Craver, C.F.: Thinking about mechanisms. Philos. Sci. **67**, 1–25 (2000)
15. Ray, O.: Nonmonotonic abductive inductive learning. J. Appl. Logic **7**(3), 329–340 (2009)
16. Ray, O., Whelan, K., King, R.: Automatic revision of metabolic networks through logical analysis of experimental data. In: De Raedt, L. (ed.) ILP 2009. LNCS, vol. 5989, pp. 194–201. Springer, Heidelberg (2010)
17. Schmidt, M., Lipson, H.: Distilling free-form natural laws from experimental data. Science **324**(5923), 81–85 (2009)
18. Thagard, P.: Computational Philosophy of Science. MIT Press, Cambridge (1993)
19. Todorovski, L., Bridewell, W., Shiran, O., Langley, P.: Inducing hierarchical process models in dynamic domains. In: Proceedings of the National Conference on Artificial Intelligence, vol. 20, p. 892. AAAI Press, MIT Press, Menlo Park, Cambridge (1999, 2005)
20. Valdés-Pérez, R.E.: Machine discovery in chemistry: new results. Artif. Intell. **74**(1), 191–201 (1995)

Qualitative Reasoning for Reaction Networks with Partial Kinetic Information

Joachim Niehren[2,3], Mathias John[1,2], Cristian Versari[1,2]([✉]),
François Coutte[1,4], and Philippe Jacques[1,4]

[1] Université de Lille, Lille, France
[2] Cristal, Lille, France
`joachim.niehren@inria.fr`, `cristian.versari@univ-lille1.fr`
[3] Inria Lille, Lille, France
[4] Research Institute Charles Viollette, Lille, France

Abstract. We propose a formal modeling language for reaction networks with partial kinetic information. The language has a graphical syntax reminiscent to Petri nets. The kinetics of reactions need to be described only partially, so that the language can be used to model the regulation of metabolic networks. We present a qualitative reasoning method based on abstract interpretation of the steady state semantics of reaction networks modeled in our language. In particular, we can predict changes of influxes that lead to expected changes of outfluxes.

1 Introduction

Models of reaction networks in systems biology often require full kinetic information, while only partial information on activators and inhibitors is available in practice. In order to become applicable nevertheless, the existing model-based reasoning methods often ignore any kinetic information. Most typically, this holds for flux balance analysis [10,12] when applied to metabolic networks [11,15]. The missing information is then compensated heuristically by the adoption of ad hoc optimization criteria. Alternatively, pathway analysis approaches [12] rely on the structure of reactions networks, but the combinatorial nature of the problem makes difficult their application to densely interconnected networks. To both methods boolean constraints can be added in order to account for inhibitors that block reactions completely [6]. But blocking inhibitors is not appropriate in deterministic semantics, where the average over blocked and unblocked situations is to be considered. The problem therefore is how to model reaction networks with partial kinetic information and how to reason with such models.

In this paper, we propose a modeling language for reaction networks with partial kinetic information. Our language is parameterized by a similarity relation on kinetic functions, so that the rate laws of chemical reactions need only to be specified up to similarity. For instance, we could define two kinetic functions to be similar if they have the same monotonicity behavior. For instance, $2A$ is similar to $5A/(7 + A)$ since whenever A increases then both terms increase, and whenever A decreases than both terms decrease.

© Springer International Publishing Switzerland 2015
O. Roux and J. Bourdon (Eds.): CMSB 2015, LNBI 9308, pp. 157–169, 2015.
DOI: 10.1007/978-3-319-23401-4_14

The models of reaction networks in our language have a graphical syntax that is reminiscent of Petri nets, and also an equivalent XML syntax. To any model a standard steady state semantics can be assigned, which provides the usual flux balance equations and additional equations with variables for kinetic functions, that are subject to similarity constraints. The steady state semantics ensures that inhibitors slow down reactions, while activators speed them up. In particular, our language can be used to model metabolic networks with complex regulation such as for *B. subtilis* in the Subtiwiki [8]. As an example, we present in Fig. 3 the graphical model of the regulation network of the PIlv-Leu promoter of *B. subtilis*, which regulates the metabolism of the branched-chain amino acids Valine, Leucine, and Isoleucine. Previous models of these metabolic networks as in the Subtiwiki were not given any formal semantics, so that they could not be used for directly for qualitative prediction algorithms.

We then show how to lift the abstract interpretation method from [5] for qualitative reasoning [3] to models of reaction networks in our language. The main technical contribution is to overcome the previous limitation to mass action laws with unknown parameters. By applying abstract interpretation to the steady state semantics, we can abstract away the variables for kinetic functions, and discretize the available partial kinetic information. This yields so called difference constraints [5], which are finite domain constraints that can be solved by finite-domain constraint programming.

As an application of our qualitative reasoning method, we show how to predict changes of influxes when given the expected changes of the outfluxes. This can be done based on the difference constraints obtained from abstract interpretation, either by constraint simplification, by rules that we present in this paper, or else by constraint solving based on the solver from [5]. In particular, constraint simplification can be used for the PIlv-Leu network to predict that any increase of leucine outflux is due to a decrease of either the CodY influx or TnrA influx. For this simple example, a similar reasoning can be done by humans based on the graphical model. This illustrates that our algorithm formalizes a natural kind of qualitative reasoning. In a follow up work [2], the same method is extended to the prediction of gene knockouts leading to the overproduction of some target metabolites [13]. The arguments used there are by far too complicated to be performed manually without any computational support for qualitative reasoning.

2   Reaction Networks

We define reaction networks with complete kinetic information, and show how to compute their steady state semantics. This is basically standard, except for the treatment of inflows and outflows of reaction networks, by which we can model the interaction of the reaction network with its context. In this way, any reaction network can be considered as "module" of a larger biological system, or as part of a chemical experiment that interacts with the network.

Let $\mathbb{R}_+$ be the set of non-negative real numbers, S a finite set of species, and $\prec$ an arbitrary total order on S. A kinetic function of arity $k \geq 0$ is a function of type $\kappa : \mathbb{R}_+^k \to \mathbb{R}_+$. Kinetic functions will be used to define the rate laws of chemical reactions. A chemical reaction r is a tuple of the form: $s_1, \ldots, s_k \xrightarrow{\kappa} s_{k+1}, \ldots, s_l$ where $0 \le k \le l$, $s_1, \ldots, s_l \in S$, and $\kappa : \mathbb{R}_+^k \to \mathbb{R}_+$ is a kinetic function. Any reaction has a tuple of reactants $s_1, \ldots, s_k$ and a tuple of products $s_{k+1}, \ldots, s_l$. In order to account for the stoichiometry of a reaction, we write $\mathrm{rct}_r(s)$ for number of occurrences of s in the tuple of reactants of r, and $\mathrm{prd}_r(s)$ for number of occurrences of s in the tuple of products of r. A modifier of a reaction is a species s with $\mathrm{rct}_r(s) = 1 = \mathrm{prd}_r(s)$. Whether a modifier behaves as an activator or as an inhibitor depends on the choice of the rate law κ.

Definition 1. *A reaction network over a species set S is a triple $N = (S, R, I, O)$ where R is a finite set of chemical reactions over S, a set of inflow species $I \subseteq S$, and an outflow function $O : S \to \mathbb{R}_+$.*

A reaction network defines the evolution of a chemical solution in a context. Each inflow species $s \in I$ specifies an inflow that adds s to the chemical solution, and is controlled by the context. An *outflow species* is an element $s \in S$ with $O(s) \ne 0$; for any outflow species there is outflow into the context that consumes s from the chemical solution. The outflow kinetics for s follows the mass-action law with constant $O(s)$. Note that any species may have an inflow and an outflow at the same time.

Under the assumption of deterministic network behavior, for any initial chemical solution a unique limit will be reached that is called a steady state. Since we do not fix any initial chemical solution, many steady states may exist for the same reaction network. The rates of all inflows and outflows are also assumed to be constant in any steady state, as well as the rates of all reactions and the concentrations of all species of the network.

$$(\text{INFLOWS}) \; \frac{s \in S \setminus I}{x_s = 0} \qquad (\text{OUTFLOWS}) \; \frac{s \in S}{y_s = O(s) \cdot z_s}$$

$$(\text{CONS}) \; \frac{s \in S}{c_s = y_s + \sum_{r \in R} \mathrm{rct}_r(s) \cdot v_r} \qquad (\text{PROD}) \; \frac{s \in S}{p_s = x_s + \sum_{r \in R} \mathrm{prd}_r(s) \cdot v_r}$$

$$(\text{RATE}) \; \frac{r \text{ is } s_1, \ldots, s_k \xrightarrow{\kappa} s_{k+1}, \ldots, s_l}{v_r = \kappa(z_{s_1}, \ldots, z_{s_k})} \qquad (\text{STEADY STATE}) \; \frac{s \in S}{c_s = p_s}$$

Fig. 1. Steady state equations of a reaction network $N = (S, R, I, O)$.

The steady state semantics of a reaction network is given by a system of arithmetic equations. These use the following variables taking values in $\mathbb{R}_+$. For any species $s \in S$, there is a variable z_s that denotes the concentration of s in a steady state, a variable x_s that denotes the rate of the inflow, which also called the *influx*, and a variable y_s that stands for the rate of the outflow, which also called the *outflux*. For any reaction $r \in R$, variable v_r stands for the rate of reaction r in a steady state.

The steady state equations are inferred from the network by the inference rules in Fig. 1 which are mainly standard. Each inference rule can be seen as an implication, whose condition is written above the line and whose conclusions is written below the line. Rules (INFLOW) states that the influx for any non-inflow species $s \notin I$ is zero. Rule (OUTFLOW) requires for any species $s \in S$ that its outflux is equal to $O(s) \cdot z_s$ according to the mass-action law. The production rate p_s of a species s is defined by rule (PROD) and its consumption rate c_s by rules (CONS). Rule (RATE) provides the rate of reaction r by applying its kinetic function to the concentrations of all its reactants. The (STEADY STATE) states that consumption and production rates are balanced for all species.

Definition 2. *Any reaction network N with n inflow and m outflow species defines a exchange relation $R_N \subseteq \mathbb{R}^n_+ \times \mathbb{R}^m_+$, determined by the solutions of the steady state equations for N, when projected to the n-tuple of variables x_s for the inflow species s of N and the m-tuple of variables $y_{s'}$ for the outflow species s' of N. The order of both tuples is given by the order $\prec$ on S.*

3 Modeling Language

We now present a modeling language for reaction networks with partial kinetic information. As first parameter of our language, we assume a similarity relation $\sim$ on kinetic functions. Rather than specifying rate laws of chemical reactions by kinetic functions, we will describe them only up to similarity: a rate law belongs to $\sim\kappa$ if it is similar to the kinetic function κ.

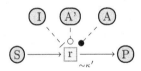

Fig. 2. An enriched reaction with a partially known rate law $\sim\kappa'$. It has substrate S, inhibitor I, accelerator A', activator A, and one product P beside of the modifiers I, A, and A'.

Enriched chemical reactions will be used to describe the chemical reactions of a reaction network. An example is given in Fig. 2. The graph there represents an enriched chemical reaction r with substrate S, activator A, an accelerator A', and inhibitor I and a product P. Please note that the same species may play different roles even in the same reaction, and several times. Both, activators and accelerators speed up a reaction. Activators are like enzymes. The difference is that all activators of a reaction must be present for its application, while the accelerators need not to be there.

For graphical representation, we use conventions similarly to Petri nets. Species are represented by rounded nodes Ⓢ containing the name s of the species, and enriched reactions are graphically represented by boxed nodes $\boxed{\text{r}}$ containing the name r of the reaction. More generally, enriched chemical reactions have different kinds of reactants, that are fixed by a finite set of roles Rol, which is the second parameter of our language. In our example, there will be substrates – that are consumed – and three kinds of modifiers: inhibitors, activators, and accelerators, so we set $Rol = \{inh, subs, act, acc\}$. For our graphical syntax, we assign to each role an edge type, for edges pointing from the reactant to the reaction. We will use $\longrightarrow$ for $subs$, $----\!\shortmid$ for inh, $---\!\bullet$ for act, and $---\!\circ$ for acc. The products of a reaction – beside of the above modifiers – will be linked by arrows $\longrightarrow$ pointing from the reaction to the product.

Reactant roles serve to order the arguments of the rate law of a enriched chemical reaction. Such a rate law is given by an enriched kinetic function:

$$\kappa' : (Rol \times \mathbb{R}_+)^k \to \mathbb{R}_+$$

We assume that any enriched kinetic function is well-behaved, in that any permutation of arguments with the same role does not change its value. When fixing the order of the arguments, any enriched kinetic function κ' can be replaced by a standard kinetic function κ, for instance such that $\kappa(z_S, z_I, z_{A'}, z_A) = \kappa'(subs: z_S, inh: z_I, act: z_{A'}, acc: z_A)$. An enriched chemical reaction can then be replaced by a chemical reaction, in which the kinetic function is replaced by a variable. With the same ordering as for obtaining κ from κ', we obtain for the example from Fig. 2:

$$S, I, A', A \xrightarrow{\sim\kappa} P, I, A', A$$

Here, $\sim\kappa$ stands for a fresh variable for a standard kinetic function that is similar to κ. A model in our language is a tuple (S, R, I, O) where R is a set of enriched reactions and $I, O \subseteq S$. Note that we do not require to specify rate constants for outflows. Graphically, inflow species in I and outflow species in O are indicated respectively by ingoing and outgoing arrows $\cdots\cdots\!\!\rightarrow$. An example model in graphical syntax is given in Fig. 3.

For any model in our language, we can generate a reaction network with variables for kinetic functions that are subject to similarity constraints. Therefore, we can define the steady state equations of any model in the language as before, except that kinetic functions will be represented by variables, as well as rate constants of outflows. An example is worked out in the next section.

Besides the graphical syntax, our language supports an XML syntax, which serves for writing the models, so that the graphs can be generated. We implemented tools for doing this in XSLT. These tools can also compute the steady state equations, and perform abstract interpretation.

4   Example: Regulation of Metabolism of *B. subtilis*

As an example, we model the leucine biosynthesis pathway of *B. subtilis* in our language. This is one of the complex regulation mechanisms of the metabolism of

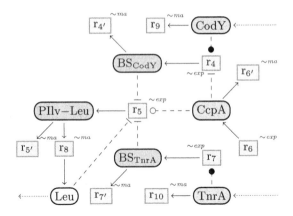

Fig. 3. Reaction network of the regulation of promoter PIlv-Leu in *B. subtilis* (Color figure online).

B. subtilis, for which informal models are given in the Subtiwiki [8]. The precise similarity relation of the model will be defined in Sect. 5.

The resulting model in graphical syntax is given in Fig. 3. For clearer visualization, nodes have different colors depending on the type of the species: in this paper we will use proteins ⓟ, metabolites Ⓜ and promoters or binding sites Ⓑ. The variable z_B stands for the activity of the promoter or binding site B, while z_P and z_M stand for the concentrations of P and M.

We consider an acceleration function with $Acc(d) = 1 + d$ and an inhibition function with $Inh(d) = 1/Acc(d)$. We define the enriched kinetic functions exp such that for all tuples $t = (r_1 : d_1, \ldots, r_k : d_k) \in (Rol \times \mathbb{R}_+)^k$:

$$exp(t) = \prod_{r_i \in \{subs, act\}} d_i \cdot \prod_{r_j = acc} Acc(d_j) \cdot Inh(\textstyle\sum_{r_l = inh} d_l)$$

Note that the order of arguments with the same role is not important, so that function exp is well-behaved. When a reaction has the exp kinetics, then its inhibitors slow down the reaction but do not block it. Accelerators and activators both speed up the reaction. Furthermore, if one of the activators is missing then the reaction is blocked. One might want to generalize exp with parameters defining the strenght of respective accelerations and inhibitions. We do not do so, since these parameters are typically unknown, and since all such generalized expression kinetics will turn out to be similar. Generally, we are only interested in $\sim exp$, so similar definitions would to the job as well. The enriched mass-action kinetics is the special case $ma(t) = exp(t)$ for all $t \in (\{subs\} \times \mathbb{R}_+)^*$.

Leucine biosynthesis is realized by enzymes which are coded by the genes of the *ilv-leu* operon. This operon is under the regulation of the promoter PIlv–Leu. For simplicity, we group the whole reaction network leading to the leucine biosynthesis into reaction r_8. The activation of PIlv–Leu is done by reaction r_5, under regulation by TnrA, CcpA, and CodY. Proteins TnrA and CodY are influx species added by the context and degraded by reactions r_{10} and r_9 respectively. Protein CcpA is expressed by reaction r_6 and degraded by reaction $r_{6'}$. Transcription at the *ilv-leu* promoter is well known to be inhibited by CodY through a binding of this latter on the promoter [7,9,14,16,17]. To model this action of CodY on the promoter PIlv–Leu, we introduce the reaction r_4 which activates the binding side BS_{CodY} of CodY at the promoter, which in turn slows down reaction r_5 and thus reduces the promoter's activity. The binding of CodY to the promoter's binding site BS_{CodY} can be prohibited when CcpA is bound to the promoter. Therefore the presence of CcpA slows down reaction r_4 [1,16] but it does not block it on average in a steady state. The promoter PIlv–Leu is also down-regulated by Leu in terms of a T-box [1,4], which is captured by the negative control of the reaction r_5 by Leu. Protein TnrA forms a further inhibitor whose impact on the PIlv–Leu promoter is represented by the binding side BS_{TnrA} through the reaction r_7. Protein CcpA is also independently up-regulating the *ilv-leu* operon transcription, and thus activating reaction r_5.

Flux balance equations:

(Leu) $\quad v_{r_8} = y_{Leu}$

(CcpA) $\quad v_{r_6} = v_{r_{6'}}$

(CodY) $\quad x_{CodY} = v_{r_9}$

(TnrA) $\quad x_{TnrA} = v_{r_{10}}$

(BS$_{CodY}$) $\quad v_{r_4} = v_{r_{4'}}$

(PIlv–Leu) $v_{r_5} = v_{r_{5'}} + v_{r_8}$

(BS$_{TnrA}$) $\quad v_{r_7} = v_{r_{7'}}$

Outfluxes:

$y_{Leu} = ma^{(8)}(subs\colon z_{Leu})$

Reaction rates:

$v_{r_4} = exp^{(1)}(inh\colon z_{CcpA}, act\colon z_{CodY})$

$v_{r_{4'}} = ma^{(1)}(subs\colon z_{BS_{CodY}})$

$v_{r_5} = exp^{(2)}(inh\colon z_{BS_{CodY}}, acc\colon z_{CcpA},$
$\qquad\qquad inh\colon z_{Leu}, inh\colon z_{BS_{TnrA}})$

$v_{r_{5'}} = ma^{(2)}(subs\colon z_{PIlv-Leu})$

$v_{r_6} = exp^{(3)}()$

$v_{r_{6'}} = ma^{(3)}(subs\colon z_{CcpA})$

$v_{r_7} = exp^{(4)}(act\colon z_{TnrA})$

$v_{r_{7'}} = ma^{(4)}(subs\colon z_{BS_{TnrA}})$

$v_{r_8} = ma^{(5)}(subs\colon z_{PIlv-Leu})$

$v_{r_9} = ma^{(6)}(subs\colon z_{CodY})$

$v_{r_{10}} = ma^{(7)}(subs\colon z_{TnrA})$

Fig. 4. Steady state equations for the PIlv–Leu network.

From the model, the steady state equations in Fig. 4 were inferred. These contain variables $exp^{(i)}$ for enriched kinetic functions similar to exp, and variables $ma^{(i)}$ for enriched kinetic functions similar to the mass-action law ma for any i. The equations can be simplified by replacing local variables by equal terms, yielding the equations in Fig. 5.

$v_{r_5} = v_{r_{5'}} + y_{Leu}$

$y_{Leu} = ma^{(8)}(subs\colon z_{Leu})$

$v_{r_4} = exp^{(1)}(inh\colon z_{CcpA}, act\colon z_{CodY})$

$v_{r_4} = ma^{(1)}(z_{BS_{CodY}})$

$v_{r_5} = exp^{(2)}(inh\colon z_{BS_{CodY}}, acc\colon z_{CcpA},$
$\qquad\qquad inh\colon z_{Leu}, inh\colon z_{BS_{TnrA}})$

$v_{r_{5'}} = ma^{(2)}(subs\colon z_{PIlv-Leu})$

$v_{r_6} = exp^{(3)}()$

$v_{r_6} = ma^{(3)}(subs\colon z_{CcpA})$

$v_{r_7} = exp^{(4)}(act\colon z_{TnrA})$

$v_{r_7} = ma^{(4)}(subs\colon z_{BS_{TnrA}})$

$y_{Leu} = ma^{(5)}(subs\colon z_{PIlv-Leu})$

$x_{CodY} = ma^{(6)}(subs\colon z_{CodY})$

$x_{TnrA} = ma^{(7)}(subs\colon z_{TnrA})$

Fig. 5. Simplified steady state equations for the PIlv–Leu network.

In order to illustrate the qualitative reasoning methods that we will develop, we consider the overproduction problem of Leu for the PIlv–Leu network. The question is which changes of the influxes may lead to an increase of the Leu outflux? Informally, the problem can be solved as follows. Leu is produced only from PIlv–Leu, which is solely produced by reaction r_5, so the speed of r_5 must be increased. This can be done by either decreasing one of its three inhibitors Leu, BS_{CodY}, or BS_{TnrA}, or by increasing its accelerator CcpA. But CcpA is not connected to any inflow, so it cannot be increased by changing the influxes. And inhibitor Leu cannot be decreased, when we want to increase its outflux. Hence, either BS_{CodY} or BS_{TnrA} must be decreased. This is possible only by decreasing the influxes of CodY or TnrA.

5 Similarity by Difference Abstraction

We now recall the similarity relation $\sim$ on kinetic functions from [5], which is obtained by abstracting from changes between real numbers.

We are interested in changes of the network raised for example by modification of inflows or outflows. A change of a concentration or a flow rate from one steady state to another is given by a pair of positive real numbers. We now want to abstract the space of all changes in $\mathbb{R}_+ \times \mathbb{R}_+$ into a finite set of *difference relations*. For this, we partition the set $\mathbb{R}_+ \times \mathbb{R}_+$ into a finite collection of subsets $\Delta \subseteq 2^{\mathbb{R}_+ \times \mathbb{R}_+}$, so that we can abstract any change in $\mathbb{R}_+ \times \mathbb{R}_+$ into a *difference relation* of Δ. In the examples that follow, we will use the partition $\Delta = \{<, >, \doteq\}$ where the symbols represent "increase" $< = \{(x,y) \in \mathbb{R}_+^2 \mid x < y\}$, "decrease" $> = \{(x,y) \in \mathbb{R}_+^2 \mid x > y\}$ and "no change" $\doteq = \{(x,x) \mid x \in \mathbb{R}_+\}$.

In general, we assume a relation $R \subseteq \mathbb{R}_+^p$, that may be either a kinetic function κ of arity $p-1$ or the relation R_N of a reaction network with p in- and outflows. We define the set of Δ-differences of the p-ary relation R as follows:

$$R^\Delta = \{(\delta_1, \ldots, \delta_p) \in \Delta^p \mid \forall i.\ (d_i, d_i') \in \delta_i,\ (d_1, \ldots, d_p) \in R,\ (d_1', \ldots, d_p') \in R\}$$

So for instance, consider the exchange relation R_N for some reaction network N. Its difference abstraction $R_N{}^\Delta$ then expresses how the tuples in R_N may change when moving from one steady state of N to another.

Definition 3. *Two kinetic functions $\kappa_1, \kappa_2 : (\mathbb{R}_+)^{p-1} \to \mathbb{R}_+$ are similar, written $\kappa_1 \sim \kappa_2$, iff $\kappa_1^\Delta = \kappa_2^\Delta$.*

Example 1. Let $ma_k(subs{:}d_1, subs{:}d_2) = k \cdot d_1 \cdot d_2$ be the mass action law with constant k. As usual, we identify binary functions with ternary relations. The differences abstraction $ma_k{}^\Delta$ is then equal for all choices of parameter k; it contains all triples of difference relations $(\delta_1, \delta_2, \delta_3) \in \Delta^3$ given in the table on the right.

δ_1	δ_2	δ_3
$<$	$<$	$<$
$<$	$>$	$<, \doteq, >$
$<$	$\doteq$	$<$
$>$	$<$	$<, \doteq, >$
$>$	$>$	$>$
$>$	$\doteq$	$>$

Example 2. We consider an enhanced Michaelis-Menten law with an additional activator: $mm_{k_1,k_2}(subs{:}d_1, act{:}d_2) = d_2 \frac{k_1 \cdot d_1}{k_2 + d_1}$. Again, it can be shown that the abstraction $(mm_{k_1,k_2})^\Delta$ is independent of the choice of $k_1, k_2 \in \mathbb{R}_+$. Indeed, it is equal to $ma_k{}^\Delta$ for all parameters k, i.e.: mass-action and the enhanced Michaelis-Menten kinetics are similar with respect to Δ-abstraction, where $\Delta = \{<, >, \doteq\}$. Of course, there exist more precise difference sets Δ for which the two families of kinetics can be distinguished.

It should be noticed that R^Δ is always a finite relation, since Δ is chosen to be finite. The relation R in contrast, may contain infinitely many tuples. As a consequence, infinitely much information may be abstracted away, in particular the details about the parameters of kinetic functions. This is why the relations $ma_k{}^\Delta$ and $(mm_{k_1,k_2})^\Delta$ in the above examples could be computed independently of the parameters. The information that is preserved, however, is still able to distinguish inhibitors and activators.

6   Abstract Interpretation to Difference Constraints

We next show how to interpret steady state equations abstractly as difference constraints, which will then be used for qualitative reasoning about reaction networks in our language in the next section.

The idea is to lift the difference abstraction $.^\Delta$ from relations over $\mathbb{R}_+$ to relations over Δ to the level of constraints defining such relations. For instance, the arithmetic equation $x_A = ma_k(z_A, z_B)$ can be abstracted to a difference constraint that defines the relation $ma_k{}^\Delta$. We write this difference constraint as $x_A \in ma_k(z_A, z_B)$, since now the variables are interpreted by values of Δ and ma_k is interpreted as the set valued function $ma_k{}^\Delta$. It should be noticed that the relation $ma_k{}^\Delta$ is finite and independent of the unknown parameter k, i.e., the unknown parameter has been abstracted away successfully.

Arithmetic constraints were used to define the steady state semantics of reaction networks. These are built from a totally ordered set of variables including those from the steady state equations. More formally, an arithmetic constraint is a conjunctive logic formula with existential quantifiers with the following abstract syntax:

$$\phi ::= x{=}\kappa^{(i)}(x_1, \ldots, x_k) \mid x = x_1 + x_2 \mid x_1 = x_2 \mid \phi \wedge \phi' \mid \exists x.\phi$$

where $i \in \mathbb{N}$, $\kappa : \mathbb{R}_+^k \to \mathbb{R}_+$, and all x'es are variables. The expression $\kappa^{(i)}$ is a variable for a kinetic function that is similar to κ, i.e., an implicitly existentially quantified variable that is subject to the similarity constraint $\kappa^{(i)} \sim \kappa$.

A solution of an arithmetic constraint ϕ with n variables can be identified with a tuple in $\mathbb{R}_+^n$ since we assumed a total order on the variables. The solution set $sol(\phi)$ of a formula ϕ satisfies $sol(\phi) \subseteq \mathbb{R}_+^n$. For a reaction network N, the steady state equations are an arithmetic constraint ϕ_N such that $sol(\phi_N) = R_N$.

A difference constraint is a conjunctive logic formula with existential quantifiers with the following abstract syntax, where x'es are variables and $\delta \in \Delta$:

$$
\begin{array}{lll}
\text{difference relation} & t ::= x \mid \delta \\
\text{set of difference relations} & s ::= \{t_1, \ldots, t_n\} \mid s + s' \mid s \cdot s' \\
& \quad \mid Inh(s) \mid Acc(s) \mid \kappa(s_1, \ldots, s_k) \\
\text{difference constraints} & \psi ::= t \in s \mid t{=}t' \mid \psi \wedge \psi' \mid \exists x.\psi
\end{array}
$$

In contrast to before, all arithmetic operations now return sets of values in difference constraints. Difference constraints are interpreted over Δ, so that variables x are assigned to elements of Δ (rather than elements of $\mathbb{R}_+$). Arithmetic functions such as $+$ are interpreted as set-valued functions on Δ such as $+^{\Delta}$, and similarly a kinetic function κ is interpreted as the set valued function κ^{Δ}.

Since variables are totally ordered, a solution of a difference constraint can be identified with a tuple in Δ^n, so that the solution set $sol(\psi)$ of any difference constraint ψ satisfied $sol(\psi) \subseteq \Delta^n$.

We can now abstract from arithmetic constraints by interpreting them as difference constraints:

$$
\begin{array}{ll}
[\![x{=}\kappa^{(i)}(x_1, \ldots, x_k)]\!] = x \in \kappa(x_1, \ldots, x_k) \\
[\![x = x_1 + x_2]\!] = x \in x_1 + x_2 & [\![x_1{=}x_2]\!] = (x_1{=}x_2) \\
[\![\phi \wedge \phi']\!] = [\![\phi]\!] \wedge [\![\phi']\!] & [\![\exists x.\phi]\!] = \exists x.[\![\phi]\!]
\end{array}
$$

An important point here is that the variables $\kappa^{(i)}$ for the partially known kinetic functions are replaced by well-known kinetic functions κ. For instance, we can abstract $x = ma^{(i)}(subs{:}x_1)$ to $x = x_1$, $x = ma^{(i)}(subs{:}x_1, subs{:}x_2)$ to $x \in x_1 \cdot x_2$, $x = exp^{(i)}()$ to $x = \dot{=}$, and $x = exp^{(i)}(subs{:}x_1, inh{:}x_2, inh{:}x_3)$ to $x \in x_1 \cdot Inh(x_2 + x_3)$. This way, the simplified steady state equations for the PIlv–Leu network are abstracted to the difference constraints in Fig. 6.

Theorem 1 (Soundness of Abstract Interpretation). $sol(\phi)^{\Delta} \subseteq sol([\![\phi]\!])$.

This theorem shows for any reaction network N that the solution set of the abstract interpretation $[\![\phi_N]\!]$ is a correct over-approximation of the abstraction of the exchange relation $R_N{}^{\Delta}$:

Corollary 1. $(R_N)^{\Delta} \subseteq sol([\![\phi_N]\!])$.

Proof. This follows immediately from Theorem 1, since $R_N = sol(\phi_N)$ by construction of ϕ_N.

$$
\begin{array}{lll}
v_{r_5} \in v_{r_{5'}} + y_{\text{Leu}} & v_{r_{5'}} = z_{\text{PIlv}-\text{Leu}} \\
y_{\text{Leu}} = z_{\text{Leu}} & v_{r_6} = \dot{=} & y_{\text{Leu}} = z_{\text{PIlv}-\text{Leu}} \\
v_{r_4} \in z_{\text{CodY}} \cdot Inh(z_{\text{CcpA}}) & v_{r_6} = z_{\text{CcpA}} & x_{\text{CodY}} = z_{\text{CodY}} \\
v_{r_4} = z_{\text{BS}_{\text{CodY}}} & v_{r_7} = z_{\text{TnrA}} & x_{\text{TnrA}} = z_{\text{TnrA}} \\
v_{r_5} \in z_{\text{CcpA}} \cdot Inh(z_{\text{BS}_{\text{CodY}}} + z_{\text{Leu}} + z_{\text{BS}_{\text{TnrA}}}) & v_{r_7} = z_{\text{BS}_{\text{TnrA}}}
\end{array}
$$

Fig. 6. Difference constraints for the PIlv-Leu network.

7   Qualitative Reasoning with Difference Constraints

Since difference constraints have finite domains, we can compute all solutions of difference constraints by using finite domain constraint programming. Or else we can use constraint simplification for qualitative reasoning.

For instance, we can reconsider the question, which changes of the influxes of the PIlv–Leu network may increase the outflux of Leu. To find the answers, we can ask for the top-n solutions of the difference constraint $y_{\text{Leu}} \; = \; <$ in conjunction with difference constraint inferred for the PIlv–Leu network in Fig. 6. These solutions can be computed by the solver for difference constraints from [5], but extended with functions *Inh* and *Acc* in difference constraints.

There are only 5 solutions for this difference constraint after projection to in- and outflux variables. These solutions are given to the right. The top-2 solutions with the fewest changes (1. and 2.) show that one can either decrease the influx of CodY or TnrA. The next three solutions show that one more change does not change the matter.

	x_{CodY}	x_{TnrA}	y_{Leu}
1.	$>$	$\doteq$	$<$
2.	$\doteq$	$>$	$<$
3.	$>$	$>$	$<$
4.	$>$	$<$	$<$
5.	$<$	$>$	$<$

Since the PIlv–Leu network is quite simple, one can obtain the same predictions based on constraint simplification. The simplification of the difference constraints in Fig. 6 based on the rewrite rules in Fig. 7 yields:

$$y_{\text{Leu}} \in Inh(x_{\text{CodY}} + x_{\text{TnrA}}).$$

When assuming $y_{\text{Leu}} \; = \; <$ in addition we can simplify the constraint further to: $< \in Inh(x_{\text{CodY}} + x_{\text{TnrA}})$ which is equivalent to $x_{\text{CodY}} \; = \; > \lor x_{\text{TnrA}} \; = \; >$. This can be satisfied by decreasing the influx of either *CodY* or *TnrA*. Thus, we obtain the same result as before.

In Fig. 7 we present simplification rules for difference constraints over the specific domain $\Delta = \{<, >, \doteq\}$. Rule (bv) replaces equal by equal while eliminating existentially bound variables (all variables z_A and v_{r_i} are implicitly existentially quantified). The simplification rules (no_i) remove the nochange value $\doteq$. The third rule (si) simplifies membership in singletons to equality. Rule (ip) expresses the idempotence of addition.

(no_1) $Inh(\doteq) \Rightarrow \doteq$	(no_3) $t \cdot \doteq \Rightarrow t$	(ip) $x + x \Rightarrow x$
(no_2) $Acc(\doteq) \Rightarrow \doteq$	(no_4) $\doteq \cdot t \Rightarrow t$	(si) $t \in t' \Rightarrow t = t'$
(bv) $\exists x. \, (x = t \land \psi) \Rightarrow \psi[t/x]$		(inh) $t \in Inh(t + s) \Rightarrow t \in Inh(s)$

Fig. 7. Simplification rules.

8   Conclusion

We have presented a formal modeling language for chemical reaction networks with partial kinetic information, and shown how to abstract away from the unknowns thanks abstract interpretation. We have illustrated that this allows us to reason qualitatively about such networks at the example of influx-change prediction. The same reasoning techniques are lifted to predict gene knockout strategies in follow-up work [2]. An important question for future work is how to develop finer abstractions for quantitative predictions.

References

1. Brinsmade, S.R., Kleijn, R.J., Sauer, U., Sonenshein, A.L.: Regulation of CodY Activity through Modulation of Intracellular Branched-Chain Amino Acid Pools. J. Bacteriol. **192**(24), 6357–6368 (2010)
2. Coutte, F., Niehren, J., Dhali, D., John, M., Versari, C., Jacques, P.: Knockout prediction in surfactin precursors biosynthetic pathway for its overproduction. J. Biotechnology. (to appear)
3. Forbus, K.D.: Qualitative reasoning. In: Tucker, A.B. (ed.) The Computer Science and Engineering Handbook, pp. 715–733. CRC Press (1997)
4. Grandoni, J.A., Zahler, S.A., Calvo, J.M.: Transcriptional regulation of the ilv-leu operon of Bacillus subtilis. J. Bacteriol. **174**(10), 3212–3219 (1992)
5. John, M., Nebut, M., Niehren, J.: Knockout prediction for reaction networks with partial kinetic information. In: Giacobazzi, R., Berdine, J., Mastroeni, I. (eds.) VMCAI 2013. LNCS, vol. 7737, pp. 355–374. Springer, Heidelberg (2013)
6. Jungreuthmayer, C., Zanghellini, J.: Designing optimal cell factories: integer programming couples elementary mode analysis with regulation. BMC Syst. Biol. **6**(1), 103 (2012)
7. Mäder, U., Hennig, S., Hecker, M., Homuth, G.: Transcriptional organization and posttranscriptional regulation of the Bacillus subtilis branched-chain amino acid biosynthesis genes. J. bacteriol. **186**(8), 2240–2252 (2004)
8. Mäder, U., Schmeisky, A.G., Flórez, L.A., Stülke, J.: Subtiwiki – a comprehensive community resource for the model organism bacillus subtilis. Nucleic Acids Res. **40**(D1), 278–287 (2012)
9. Molle, V., Nakaura, Y., Shivers, R.P., Yamaguchi, H., Losick, R., Fujita, Y., Sonenshein, A.L.: Additional targets of the Bacillus subtilis global regulator CodY identified by chromatin immunoprecipitation and genome-wide transcript analysis. J. bacteriol. **185**(6), 1911–1922 (2003)
10. Orth, J.D., Thiele, I., Palsson, B.O.: What is flux balance analysis? Nat. Biotechnol. **28**(3), 245–248 (2010)
11. Otero, J.M., Nielsen, J.: Industrial systems biology. Industrial Biotechnology: Sustainable Growth and Economic Success (2010)
12. Papin, J.A., Stelling, J., Price, N.D., Klamt, S., Schuster, S., Palsson, B.O.: Comparison of network-based pathway analysis methods. Trends Biotechnol. **22**(8), 400–405 (2004)
13. Price, N.D., Reed, J.L., Palsson, B.O.: Genome-scale models of microbial cells: evaluating the consequences of constraints. Nat. Rev. Microbiol. **2**(11), 886–897 (2004)
14. Shivers, R.P., Sonenshein, A.L.: Activation of the Bacillus subtilis global regulator CodY by direct interaction with branched-chain amino acids. Mol. Microbiol. **53**(2), 599–611 (2004)
15. Sohn, S.B., Kim, T.Y., Park, J.M., Lee, S.Y.: In silico genome-scale metabolic analysis of Pseudomonas putida kt2440 for polyhydroxyalkanoate synthesis, degradation of aromatics and anaerobic survival. Biotechnol. J. **5**(7), 739–750 (2010)

16. Tojo, S., Satomura, T., Morisaki, K., Deutscher, J., Hirooka, K., Fujita, Y.: Elaborate transcription regulation of the Bacillus subtilis ilv-leu operon involved in the biosynthesis of branched-chain amino acids through global regulators of CcpA, CodY and TnrA. Mol. Microbiol. **56**(6), 1560–1573 (2005)

17. Villapakkam, A.C., Handke, L.D., Belitsky, B.R., Levdikov, V.M., Wilkinson, A.J., Sonenshein, A.L.: Genetic and Biochemical Analysis of the Interaction of B. subtilis CodY with Branched-Chain Amino Acids. J. Bacteriol. **191**(22), 6865–6876 (2009)

Boolean Network Identification from Multiplex Time Series Data

Max Ostrowski[1], Loïc Paulevé[2]([✉]), Torsten Schaub[1],
Anne Siegel [3], and Carito Guziolowski[4]

[1] Computer Science Department, Potsdam University,
Postdam, Germany
loic.pauleve@lri.fr
[2] CNRS, Université Paris-Sud LRI-UMR 8623, Orsay, France
[3] CNRS, Université de Rennes 1, IRISA-UMR 6074, Rennes, France
[4] École Centrale de Nantes, IRCCyN-UMR CNRS 6597, Nantes, France

Abstract. Boolean networks (and more general logic models) are useful frameworks to study signal transduction across multiple pathways. Logical models can be learned from a prior knowledge network structure and multiplex phosphoproteomics data. However, most efficient and scalable training methods focus on the comparison of two time-points and assume that the system has reached an early steady state. In this paper, we generalize such a learning procedure to take into account the time series traces of phosphoproteomics data in order to discriminate Boolean networks according to their transient dynamics. To that goal, we exhibit a necessary condition that must be satisfied by a Boolean network dynamics to be consistent with a discretized time series trace. Based on this condition, we use a declarative programming approach (Answer Set Programming) to compute an over-approximation of the set of Boolean networks which fit best with experimental data. Combined with model-checking approaches, we end up with a global learning algorithm and compare it to learning approaches based on static data.

1 Introduction

Generic prior knowledge about canonical cell signaling networks can be retrieved from database sources. They provide a first insight on how cells respond to their environment by triggering processes such as growth, survival, apoptosis (cell death), and migration. However, little is known about the exact chaining and composition of signaling events within these networks in specific cells and specific conditions, as provided by the simulations of predictive mathematical models (e.g. a set of differential equations or a set of logic rules). When building predictive models, the parameters of a model (built accordingly to generic prior knowledge) can be fitted to the data to obtain the most plausible model for a specific cell type, if enough experimental data is available. This is normally achieved by defining an objective fitness function to be optimized. In this context,

---

M. Ostrowski and L. Paulevé—Co-first authors

© Springer International Publishing Switzerland 2015
O. Roux and J. Bourdon (Eds.): CMSB 2015, LNBI 9308, pp. 170–181, 2015.
DOI: 10.1007/978-3-319-23401-4_15

post-translational modifications, notably protein phosphorylation, play a key role in signaling. They are very useful for the training of model parameters through the use of multiplex phosphorylation assays, a recent form of high-throughput data providing information about protein-activity modifications in a specific cell type upon various perturbations (clamping) [1].

Boolean logical networks [12] provide a simple yet powerful qualitative framework which has become very popular during the last decade to model signaling or regulatory networks [16]. In contrast to quantitative methods which permit fine-grained kinetic analysis, qualitative approaches allow for addressing large-scale biological networks. In this context, the manual identification of logic rules underlying the system has been addressed under different hypotheses and methods [4]. Although, scalable methods restrain themselves to learning models from two time points (start; end), assuming the system has reached an early steady-state when the measurements are performed. As shown in [14], this assumption prevents capturing important characteristics of signaling networks such as loops.

The goal of this paper is to introduce a new method to infer Boolean networks (BNs) from time series datasets which scales to the size of currently studied BNs. Given multiplex time series data from the measurement of a partial set of biological entities under different experimental conditions, we want to identify all the BNs that have a structure compatible with a given prior knowledge interaction graph and that can reproduce all the (experimentally) observed time series. Time series data are assumed incomplete, i.e., only a subset of network components are observed, with measurements made at discrete time points and with normalized continuous values. It is possible that no BN, constrained by the prior interaction graph, reproduces all the input time series. In such a case we introduce a fitness function to measure the distance between a trace of a BN simulation and a measured time series. Therefore, we aim to infer the BNs whose dynamics contains traces with the best fitness to all measurements.

Our approach relies on the combination of several techniques. First, we introduce a necessary condition for a discretized time series data to be the trace of a BN. This provides an over-approximation of the successive reachability properties, leading to reject BNs that cannot reproduce the time series without a costly exhaustive analysis of the dynamics. Then, we use efficient declarative programming approaches (Answer Set Programming; ASP) to enumerate BNs which approximate the best experimental data while satisfying the necessary condition on the dynamics. At the end, we obtain a set of BNs associated with traces which both satisfy the necessary condition and optimally fit with experimental data. Because of the reachability over-approximation, part of the returned BNs cannot reproduce the associated Boolean traces. Such false positives can be detected *a posteriori* using a model-checking approach on the returned results.

We evaluated our inference method on synthetic data generated from BNs between 13 and 17 nodes. On those BNs, six nodes have been selected as observable, and several experimental conditions have been simulated. Our prototype implementation has been able to identify efficiently all BNs satisfying the necessary condition with a very low rate of false positives. Finally, we estimated the added-value of models identified with our method on the full time series with models learned from two time points, considered as a steady state.

2   Boolean Network Identification

2.1   Admissible Boolean Networks and Multiplex Time Series Data

Boolean Networks (BNs). A BN with n components $\{1, \ldots, n\}$ consists of a tuple of n functions $F = (f_1, \ldots, f_n)$ where each function $f_i : \mathbb{B}^n \to \mathbb{B}$, $\mathbb{B} \overset{\Delta}{=} \{0,1\}$, $i \in \{1, \ldots, n\}$, associates to each global state $x \in \mathbb{B}^n$ of the network with the next value of the i-th component. The value of the i-th component in x is noted x_i. The transitions between global states of the network are specified with a reflexive transition relation $\to \subseteq \mathbb{B}^n \times \mathbb{B}^n$. The transitive closure of $\to$ is denoted by $\to^*$. Given $x, x' \in \mathbb{B}^n$, $x \to^* x'$ if and only if, either $x = x'$, or $x \to \cdots \to x'$.

Concrete Semantics for the Transition Relation. Several definitions of the transition relation $\to$ can be used depending on the update schedule of the components [2], ranging from so-called parallel (or synchronous) updates where each transition updates the value of all the components, to the asynchronous update where each transition updates the value of only one component chosen non-deterministically. As the over-approximation results presented in this article are independent from the update schedule, we use the general definition, where any number of components can be updated during a transition: for any $x, x' \in \mathbb{B}^n$,

$$x \to x' \overset{\Delta}{\Leftrightarrow} \forall i \in \{1, \ldots, n\}, x'_i \neq x_i \Rightarrow x'_i = f_i(x). \tag{1}$$

Prior Knowledge Network and Admissible BNs. An *interaction graph* between n components is a digraph between nodes $\{1, \ldots, n\}$ where each edge is signed, i.e., either positive or negative. The interaction graph of a BN F, noted $\mathsf{IG}(F)$, has a positive (resp. negative) edge from node j to node i if and only if there exists $x, x' \in \mathbb{B}^n$ which are identical except on the j-th coordinate where $x_j = 0$ and $x'_j = 1$ and such that $f_i(x) < f_i(x')$ (resp. $f_i(x) > f_i(x')$).

In the rest of the paper, the *Prior Knowledge Network* (PKN) is an interaction graph which delimits the set of *admissible BNs*: a BN F is admissible with respect to a PKN $\mathcal{G}$ if and only if $\mathsf{IG}(F)$ is a sub-graph of $\mathcal{G}$ and $\mathsf{IG}(F)$ has at one most (signed) edge between two nodes.

Multiplex Time Series Data. We consider classical biology experimental settings where the activity of a subset of biological species is observed over time, at discrete time points, in different experimental conditions, ranging over various input signals and *clamping* operations. Clampings consist of a subset A of components with a forced activation, and a subset I of components with a forced inhibition. Given a BN $F = (f_1, \ldots, f_n)$, the corresponding *clamped BN* $F_{[A,I]} = (f'_1, \ldots, f'_n)$ is defined for all $i \in \{1, \ldots, n\}$ as:

$$f'_i \overset{\Delta}{=} \begin{cases} x \mapsto 1 & \text{if } i \in A \\ x \mapsto 0 & \text{if } i \in I \\ f_i & \text{otherwise.} \end{cases}$$

Without loss of generality, we assume that the time series data relate to the observation of $m \leq n$ nodes that match the nodes $\{1, \ldots, m\}$ of the BN (so the nodes $\{m + 1, \ldots, n\}$ are not observed). The observations consist of normalized continuous values: a time series of k data points is denoted by $T = (t^1, \ldots, t^k)$, with $\forall j \in \{1, \ldots, k\}, t^j \in [0; 1]^m$.

Hereafter, we consider a simple binarization of observations using a 0.5 threshold: given a continuous observation $t_i^j \in [0; 1]$ of a component, its Boolean value is noted $\eta(t_i^j)$ where $\eta(t_i^j) \triangleq 1$ when $t_i^j \geq 0.5$, and $\eta(t_i^j) \triangleq 0$ otherwise. The distance between a binary sequence $X = (x^1, \ldots, x^k)$, where $\forall i \in \{1, \ldots, k\}, x^i \in \mathbb{B}^m$, and a time series T is evaluated with the standard *Mean Squared Error*:

$$\mathsf{mse}(X, T) \triangleq \sqrt{\sum_{j=1}^{k} \sum_{i=1}^{m} \left(x_i^j - t_i^j \right)^2}.$$

2.2 Over-Approximation of Boolean Network Verification

Given a BN F and a pair of states $x, y \in \mathbb{B}^n$, checking the reachability of y from x ($x \rightarrow^* y$) is a standard model-checking task, known to have a limited scalability due to its theoretical complexity (NP-complete [11]). In this section, we introduce a so-called *meta-state semantics* ($\rightrightarrows$) for BNs. From such semantics, we express a necessary condition for reachability in the concrete semantics ($\rightarrow$), referred to as *support consistency* ($\rightsquigarrow^*$). Meta-state semantics offers properties (notably monotonicity) that make support consistency efficient to verify, in particular with ASP. However, support consistency is not a sufficient condition for reachability, so this approach may lead to false positives but guarantees the absence of false negatives. Therefore, we will apply exact model-checking approaches on the inferred BNs in order to rule out false positives. Thanks to the over-approximation criteria, one can expect that the set of BNs satisfying the necessary condition is small compared to the full domain of BNs delimited by the PKN, leading to a global gain in terms of performance.

Meta-state Semantics. A *meta-state* u of dimension n is a vector of n non-empty subsets of $\mathbb{B}$, noted $\mathbb{M} \triangleq \{\{0\}, \{1\}, \{0, 1\}\}$; the set of meta-states is $\mathbb{M}^n$. In the following, meta-states characterize a set of Boolean states: a state $x \in \mathbb{B}^n$ belongs to a meta-state $u \in \mathbb{M}^n$, noted $x \in u$, iff each Boolean component x_i belongs to the set u_i, i.e., $\forall i \in \{1, \ldots, n\}, x_i \in u_i$. Given a state $x \in \mathbb{B}^n$, $\overline{x}$ is the meta-state such that $\forall i \in \{1, \ldots, n\}, \overline{x}_i = \{x_i\}$. In the scope of a BN $F = (f_1, \ldots, f_n)$, we define a reflexive transition relation between meta-states $\rightrightarrows \subseteq \mathbb{M}^n \times \mathbb{M}^n$ as follows: from a meta-state u, there is one transition for each $i \in \{1, \ldots, n\}$ which adds to u_i all the possible values of the function f_i applied to every $x \in u$:

$$u \rightrightarrows v \triangleq \exists i \in \{1, \ldots, n\}, v = \langle u_1, \ldots, u_i \cup \{f_i(x) \mid x \in u\}, \ldots u_n \rangle. \quad (2)$$

Several properties arise from this definition, in particular $u \rightrightarrows v$ implies that $\forall i \in \{1, \ldots, n\}, u_i \subseteq v_i$; therefore $x \in u \Rightarrow x \in v$ (monotonicity). Moreover, $u_i \neq v_i$ if and only if $v_i = \{0, 1\}$ and $\exists x \in u$ such that $f_i(x) \notin u_i$.

Lemma 1 establishes the consistency of the meta-semantics ($\Rightarrow$) with the concrete semantics ($\rightarrow$): given $x, y \in \mathbb{B}^n$, $x \rightarrow y$ requires that there exists a meta-state u such that $y \in u$ and $\overline{x} \Rightarrow^* u$, where $\Rightarrow^*$ is the transitive closure of $\Rightarrow$.

Lemma 1. $\forall x, y \in \mathbb{B}^n$, $x \rightarrow y \implies \exists u \in \mathbb{M}^n, y \in u : \overline{x} \Rightarrow^* u$.

Proof. Assuming $x \rightarrow y$, let us define the set $I \overset{\Delta}{=} \{i \in \{1, \ldots, n\} \mid y_i \neq x_i\}$. From Eq. (1), $\forall i \in I, y_i = f_i(x)$. Let us assume that for some strict subset $J \subsetneq I$, $\exists v \in \mathbb{M}^n, \overline{x} \Rightarrow^* v$ with $\forall i \in J, y_i \in v_i$. It is notably the case with $J = \emptyset$. By induction, we show that, for any $k \in I \setminus J$, $\exists u \in \mathbb{M}^n$ such that $\overline{x} \Rightarrow^* v \Rightarrow u$ with $\forall i \in J \cup \{k\}, y_i \in u_i$. Remarking that $x \in v$ and defining $u \in \mathbb{M}^n$ such as $u_i = v_k \cup \{f_k(z) \mid z \in v\}$ if $i = k$ and $u_i = v_i$ if $i \neq k$, we obtain that $v \Rightarrow u$, with $y_k = f_k(x) \in u_k$. □

Such a necessary condition for reachability can be furthermore refined by ensuring that for each component $i \in \{1, \ldots, n\}$ that is equal in x and y, if all meta-states u containing y with $\overline{x} \Rightarrow^* u$ are such that $u_i = \{0, 1\}$, then u contains a state z with $f_i(z) = y_i = x_i$. Intuitively, this refinement ensures that if the i-th component has to temporarily change its value for reaching y, a state from which it can recover its initial (and final) value has to be reached in between. Such a condition is referred to as *support consistency* (Definition 1). Theorem 1 states that support consistency is a necessary condition for reachability.

Definition 1 (Support Consistency ($\rightsquigarrow^*$)). *A state $x \in \mathbb{B}^n$ is support-consistent with $y \in \mathbb{B}^n$, denoted by $x \rightsquigarrow^* y$, if and only if there exists $u \in \mathbb{M}^n$ with $x \Rightarrow^* u$ such that $y \in u$ and for all $i \in \{1, \ldots, n\}$ where $y_i = x_i$, $u_i = \{0, 1\} \implies \exists z \in u : f_i(z) = y_i$.*

Theorem 1. $\forall x, y \in \mathbb{B}^n$, $x \rightarrow^* y \implies x \rightsquigarrow^* y$.

Proof. Let us consider any tuple of states $(x^1, \ldots, x^k)$ with $x^1 = x$, $x^k = y$, and $\forall j \in \{1, \ldots, k-1\}, x^j \rightarrow x^{j+1}$. From Lemma 1, $\exists u \in \mathbb{M}^n$ such that $x \Rightarrow^* u$ and $\forall j \in \{1, \ldots, k\}, x^k \in u$. If for all such u, for any $i \in \{1, \ldots, n\}$, $u_i = \{0, 1\}$ implies that there exists $l \in \{1, \ldots, k\}$ with $x_i^l \neq x_i$. If $y_i = x_i$, there necessarily exists $m \in \{l, \ldots, k-1\}$ such that $f_i(x^m) = y_i$. Therefore $x^m \in u$. □

2.3 Optimization with Respect to Time Series Data

Our objective is to infer BNs that are admissible with a given PKN and that verify the sequential reachability of binary states in $\mathbb{B}^m$ that are as close as possible to a given time series data and its associated experimental settings.

Distance Between a Time Series Data and a BN. Given a time series T with associated clamping A, I, the distance between a BN F and (T, A, I), noted $\mathsf{mse}(F_{[A,I]}, T)$, is the minimal MSE between T and a sequence of binary states $X = (x^1, \ldots, x^k)$, with $\forall j \in \{1, \ldots, k\}, x^j \in \mathbb{B}^n$, that are successively reachable in $F_{[A,I]}$: $x^1 \rightarrow^* x^2 \ldots \rightarrow^* x^k$. We notice that the lowest possible

$\mathsf{mse}(X, T)$ among all Boolean traces is the MSE between T and its binarization $\eta(T) = ((\eta(t_i^1))_{i=1...m}, \cdots, (\eta(t_i^k))_{i=1...m})$. Let us call $\mathrm{MSE}_T \overset{\Delta}{=} \mathsf{mse}(\eta(T), T)$ this *minimum MSE* which is intrinsic to the time series T and to the threshold for binarization (0.5); $\mathsf{mse}(F_{[A,I]}, T) \geq \mathrm{MSE}_T$. Whenever $\mathsf{mse}(F_{[A,I]}, T) = \mathrm{MSE}_T$, we say the BN F *reproduces* the time series data T.

Relaxing the Semantics Sonstraint. In order to prevent an exhaustive exploration of the BN dynamics for characterizing the sequences of reachable ($\rightarrow^*$) Boolean states, we consider any sequence $X = (x^1, \ldots, x^k)$, with $\forall j \in \{1, \ldots, k\}, x^j \in \mathbb{B}^n$, that are *support-consistent* ($\rightsquigarrow^*$), i.e., $x^1 \rightsquigarrow^* x^2 \cdots \rightsquigarrow^* x^k$ in the scope of the BN $F_{[A,I]}$. The MSE of such a support-consistent Boolean state sequence X w.r.t. the time series T is noted $\widehat{\mathsf{mse}}(X, T)$; and the minimal distance among all support-consistent sequences in $F_{[A,I]}$ with T is referred to as $\widehat{\mathsf{mse}}(F_{[A,I]}, T)$. Because any reachable sequence is support-consistent (Theorem 1), we obtain that $\mathsf{mse}(F_{[A,I]}, T) \geq \widehat{\mathsf{mse}}(F_{[A,I]}, T) \geq \mathrm{MSE}_T$; and in particular $\mathsf{mse}(F_{[A,I]}, T) \neq \widehat{\mathsf{mse}}(F_{[A,I]}, T)$ only if none of the support-consistent sequences X with minimal $\widehat{\mathsf{mse}}(X, T)$ are actually sequences of reachable Boolean states. In such cases, F is a *false positive*. Determining if F is a true positive can be done *a posteriori* with a model-checking approach: if $\widehat{\mathsf{mse}}(F_{[A,I]}, T) = \mathrm{MSE}_T$, we check that $\eta(T)$ is a valid sequence of reachable states in $F_{[A,I]}$; otherwise, we check the validity with respect to reachability of at least one sequence X with minimal $\widehat{\mathsf{mse}}(X, T)$.

Optimization Problem. We consider a PKN $\mathcal{G}$ and a set of r multiplex time series $D = (T^1, A^1, I^1), \ldots, (T^r, A^r, I^r)$. The distance between a BN F and the dataset is the sum of distances $\widehat{\mathsf{mse}}(F, D) \overset{\Delta}{=} \sum_{l=1}^r \widehat{\mathsf{mse}}(F_{[A^l, I^l]}, T^l)$. The optimization procedure identifies the BNs compatible with the PKN $\mathcal{G}$ that have the minimal distance $\widehat{\mathsf{mse}}(F, D)$. In the scope of this paper, we enforce that each non-observed node starts with the same initial value in all the time series: for each $l \in \{1, \ldots, r\}$, if X^l is a sequence of support-consistent Boolean states in $\mathbb{B}^n$ such that $\widehat{\mathsf{mse}}(F_{[A^l, I^l]}, T^l) = \mathsf{mse}(X^l, T^l)$, for all $i \in \{m+1, \ldots, n\}$, $X_{1,i}^l = X_{1,i}^1$. Whereas this constraint reduces the space of sequences to explore, it also ensures consistency between the different experimental settings.

Depending on the number of nodes in the PKN, and on the discriminative power of the time series dataset, a rather large number of BNs may be expected to be inferred. As an alternative, we can output only the BNs having the smallest Disjunctive Normal Form (DNF) representation with respect to clause inclusion, i.e., no literal nor clause can be removed. This means that no unnecessary edges occur in the BNs, thus providing only the simplest BNs. In the following, we refer to such a set of solutions as *subset-minimal*.

2.4 Implementation

Answer Set Programming (ASP; [3,7]) is a declarative approach to solving knowledge-intense combinatorial (optimization) problems comprising up to tens of millions of variables. ASP's distinguishing combination of a high-level modeling language with high-performant solving tools allows for concentrating on an actual problem, rather than a smart way of implementing it. The basic idea of ASP is to express a problem in a logical format so that the (logical) models of its representation provide the solutions to the original problem. Problems are expressed as logic programs and the resulting models are referred to as answer sets. Although determining whether a program has an answer set is the fundamental decision problem in ASP, modern ASP solvers like *clasp* [9] support various combinations of reasoning modes, among them, regular and projective enumeration, intersection and union, multi-criteria optimization and subsets [8] and/or sum-based minimal (maximal, resp.) model enumeration.

Here we describe the general design of the encoding, while the complete version is available online (see Footnote 1). For the encoding we follow the general design approach in ASP in a way that we first guess all admissible BNs given a PKN. Guessing in this context does not mean choosing a BN by some heuristic, but exhaustively trying all possible combinations of edges and logical connectives. We also guess time series, a value $\{0, 1\}$ for every species in every experiment and every time point. In the case of non-observed nodes we add a constraint that fixes their initial value, at time point 0, across all the experiments. We then restrict this search space by posting constraints that the guessed time series shall be support-consistent with the guessed BN. In this way all enumerated BNs are consistent with the guessed time series. As an optimization function we minimize the distance between the guessed time series and the measured one. In the optimal case, this means that the guessed time series is equal to the measured one, and the BN is support-consistent with the measured data.

3   Evaluation

3.1   Case Study

As a proof of concept we used the PKN published in [14] (see the compressed PKN in Fig. 1A). From this PKN, the authors of [14] randomly generated an admissible golden-standard BN to simulate synthetic time series data (Fig. 1C). Afterwards, they removed the link from tnfa to ap1 from the PKN to represent incomplete regulatory knowledge. After confronting the incomplete PKN with the time series data, our method learned a family of BNs consisting of 3 subset-minimal BNs. All BNs were checked to be true positives, therefore they have an optimal MSE score of 0.07 with respect to the data. The family of optimal BNs was learned after 0.04 s of computation on a standard desktop computer. In Fig. 1B we plot the subset-minimal family of BNs learned for this case study. It recovers the complete logical behaviors of the golden-standard, except for the one regulation from tnfa to ap1 which was removed from the PKN. Only the logical function of the regulation over p38 is not consensually learned across all BNs in the family; the rest of logical functions learned are shared by all models. The quality of our results concerning the learned BNs is comparable to the one obtained in [14] for the same case study. The computation time of our method improves the one of published methods in a range of 2 to 4 orders of magnitude. Moreover, our method is exhaustive: all logical networks are learned. The full

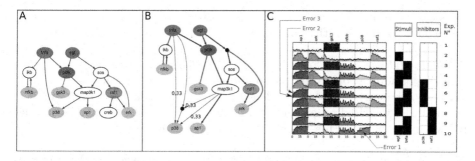

Fig. 1. (A) Compressed PKN from [14]. Green and red edges indicate activations and inhibitions respectively. Colors of the nodes represent the chosen experimental design: green refers to inputs/stimuli, red, to inhibited nodes, and blue, to measured species. **(B)** Boolean networks (BNs) learned from time series data which are subset-minimal. All BNs predictions have minimal ΔMSE with respect to the synthetic time series data. A black circle represents a logical AND gate. A number written over an edge represents the frequency of this logical gate or edge with respect to the family of BNs when the edge is not shared by all BNs. **(C)** Synthetic time series data used in [14] simulated using a BN admissible for the PKN in A. In total 10 experimental conditions were simulated. Red boxes indicate the minimal set of 3 error time-points detected (Color figure online).

set of solutions (not only the subset-minimal BNs) was also computed showing one more BN with an OR gate above p38 from tnfa and map3k1.

The method also automatically identified the list of minimal errors in the time series data, selecting time-points that cannot be explained by the learned BNs. For the case of all optimal BNs, we found the following 3 errors (see Fig. 1C) in all of them. For *experiment 10, time-point 10, species p38,* the error can be explained by the noise artificially introduced in the dataset. The predecessors of p38 are tnfa and egfr, both active in experimental condition 10 (see Fig. 1C). The signal of p38 can therefore only increase (or stay the same). However, the measure of p38 slightly decreases (due to noise) at time-point 10; this generates an error since the BNs cannot satisfy the data at this particular time-point. For *experiment 6, time-point 2 and 4, species ap1,* the errors can be explained by the fact that one edge (the link from tnfa to ap1) was deleted from the PKN, but was kept to generate the synthetic time series data. All BNs agree on a regulation of p38 and ap1 from map3k1. In experiment 6 tnfa is stimulated and pi3k is inhibited (see Fig. 1C). At time-point 2 the value of map3k1 has to be activated (transition $0 \rightarrow 1$) to justify the activation of ap1. However, since map3k1 is the only regulator of p38, which is all the experiment at value 0, this cannot be explained by the BN and generates an error.

3.2 Benchmarks

In this section we evaluate our method for BN identification on synthetic multiplex time series data. Given a PKN, a dataset and a set of inferred BNs, we focus on two evaluation criteria: the MSE distance of the BNs to the dataset, and the rate of false positives due to our reachability over-approximation.

Synthetic Multiplex Time Series Datasets. 10 PKNs were derived by randomly removing or adding edges from the compressed PKN published in [14]. For each PKN we randomly selected 3 golden-standard admissible BNs. Each golden-standard BN was used to generate synthetic time series data by simulating the BN with logic-based ODEs. In total we generated 30 datasets[1].

MSE Computation. Following Sect. 2.3, our method optimizes the MSE of the BNs F to the dataset D up to the reachability over-approximation criteria: if the BN is a true positive, the estimated MSE $\widehat{\mathsf{mse}}(F, D)$ is the exact MSE $\mathsf{mse}(F, D)$, otherwise the estimated MSE is an under-approximation - the exact MSE may be larger. Due to the optimization, all the BNs have the same estimated MSE. The value of the estimated MSE can be computed using the equation given in Sect. 2.1 by sampling one BN from the result set with one Boolean trace X for each time series T of dataset D such that $\widehat{\mathsf{mse}}(X, T)$ is minimal.

True-positive Rate Computation. Any BN inferred by our method satisfies the necessary condition depicted in Sect. 2.2 for producing Boolean traces as close as possible to a given time series dataset. Verifying that the BN can actually reproduce those Boolean traces requires an exhaustive analysis of the dynamics to ensure the successive reachability of the Boolean states. In the scope of this paper, we performed such a verification using a model-checking approach. The presented experiments have been conducted using the tool NuSMV [5] which allows an efficient encoding of the dynamics accounting for the range of clamping settings of the different time series in the dataset[2]. The true-positive rate evaluation proceeds by iteratively checking each inferred BN. In the case when the estimated MSE is MSE_T (Sect. 2.3), the model-checking is performed with respect to the binarized time series. Otherwise, we iterate over the closest Boolean traces computed in Sect. 2.3 until a sample is validated by model-checking; if no such a sample exists, the BN is a false positive.

Results. For each dataset, the model identification has been performed with respect to the PKNs from which the BNs used for data generation have been extracted; and with respect to the PKNs where some edges have been deleted so the BNs used to generate the data are not in the considered domain. With the exact PKNs, the estimated MSE is always the minimum MSE_T; moreover, the rate of true positive is 100 % in 28 benchmark datasets, and above 90 % in the 2 others[3]. With the PKNs with deleted edges, most of the cases show a very high true positive rate (often 100 %) and an estimated MSE close to (often equal to) MSE_T. Note that for some dataset, no true positive has been found. For the cases

---

[1] Details in http://loicpauleve.name/cmsb15-suppl-A.pdf.

[2] Scripts and data available at http://loicpauleve.name/cmsb15-suppl.tbz2.

[3] Detailed results are given in http://loicpauleve.name/cmsb15-suppl-B.pdf.

when the estimated MSE is different from MSE_T, the true positive rate can only be evaluated by sampling Boolean traces close to the time series data. Because of the very high combinatorics of such sampling space, the computation has been aborted after one hour, hence we cannot guarantee that no true positive exists. When no true positives have been identified, the MSE may be under-estimated.

The inference of the subset-minimal solutions for the 30 benchmarks with exact PKNs took less than 2 s on average. The performance is similar for the benchmarks with incomplete PKNs that contained a true positive BN in the result. The number of results varies between 12 and 2640 with the exact PKN and from 2 to 1188 with modified PKNs. Depending on the size and the complexity of its dynamics, the model-checking of one BN took between 1 s and 5 min. The full set of solutions (not only the subset-minimal BNs) have also been performed with the exact PKN, showing very similar results and running time, with subsequently more results (up to 54,000 BNs, data not shown).

Same experiments have been conducted on the time series generated with noise but show no difference in the results (data not shown). This may indicate that the noise influence may be tempered by the binarization.

3.3 Comparison with Inferences Using Pseudo Steady-States

In this section we compare our results with the previously developed approach *Caspo* [10]. Caspo, as well as other state-of-the-art approaches such as CellNopt [15], considers two time-points (an initial point and a pseudo-steady state) and a PKN. It computes a set of BNs with minimal size that can explain the best the transition between the two time-points. Due to its static nature and the minimal size condition, it is not possible to infer feedback loops or dynamic behaviour, because models with loops would not improve the fitting with the data assuming a steady state. With this comparison, we aim at emphasizing the importance of taking into account model dynamics to obtain accurate model predictions.

Applied on the 30 synthetic datasets of Sect. 3.2 with the PKNs used for the data generation, we compared the best MSE obtained applying our optimization procedure on the BNs returned by *Caspo* on one time point (assumed steady by *Caspo*) and on all BNs delimited by the PKNs. Therefore, we compare the best estimated MSE with respect to the multiplex time series for both the *Caspo* approach and the method introduced by this paper. As explained in Sect. 2.3, and as in Sect. 3.2, the computed MSE may be under-estimated so we used model-checking *a posteriori* to verify the presence of true positive BNs.

Figure 2 plots the estimated MSEs, where the 6th time point of the time series has been selected for the learning with *Caspo*; other time points give very similar results (data not shown). On the contrary to our approach where it was always possible to find a BN which was fully consistent with the data, having the minimum $MSE = MSE_T$, *Caspo* failed to identify a consistent model with the data in 25 over the 30 experiments. Among those 25 experiments, the estimated MSE on *Caspo* results may be under-estimated in 5 experiments where the returned BNs are actually false positives (streaked bars). This evidences the role of feedback loops which cannot be captured with a two-timepoints learning procedure and the information gain brought by time series data.

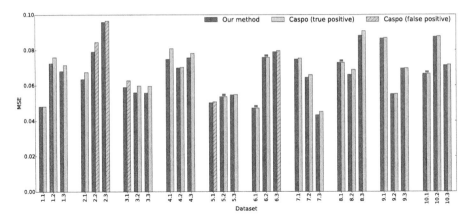

Fig. 2. Comparing MSE with *Caspo* for 10 different PKNs with 3 datasets each. "=" indicates equal MSE.

4   Conclusion

We have introduced a procedure based on combinatorial optimization with declarative programming approaches and model checking to identify BNs from multiplex time series data given a prior network structure. To cope with the complexity of an exhaustive analysis of BNs dynamics, we defined an abstract semantics of BNs from which we derived a necessary condition for the satisfaction of successive reachability properties, induced by the time series data. Our procedure identifies all the BNs that satisfy this necessary condition with the shortest distance (in terms of MSE) to the observed experimental data. Because the satisfaction criteria for the dynamics is over-approximated, our method may lead to BNs that are false positive, and have an under-estimated MSE. Applied to synthetic multiplex time series datasets on networks composed of 13 to 17 nodes, the identification of BNs takes only a few seconds and exhibits a very low rate of false positives, showing a remarkable efficiency.

In the present form, we assume that the experimental data is normalized between 0 and 1 and use a discretization threshold at 0.5. Whereas such a setting is relevant for phosphoproteomics data, future work may generalize our optimization framework to account for adaptive and multiple discretization levels. Moreover, application to larger networks should be considered, although few of such data are currently available, and generating synthetic data with sufficient discriminant power may be challenging.

Because our identification method can be exhaustive, the framework we propose is suited for the complete *Thomas parameters identification* for BNs from incomplete time series data [6,13]. Thanks to our abstract semantics, our method is able to filter out very efficiently a large number of candidate BNs without a costly exact model-checking, which is postponed to the validation of the results. In that way, future work may further explore the combination of dynamics over-approximations with model-checking approaches to provide scalable and exact inference of BNs from time series data.

References

1. Alexopoulos, L.G., Saez-Rodriguez, J., Cosgrove, B., Lauffenburger, D.A., Sorger, P.: Networks inferred from biochemical data reveal profound differences in toll-like receptor and inflammatory signaling between normal and transformed hepatocytes. Mol. Cell. Proteomics **9**(9), 1849–1865 (2010)
2. Aracena, J., Goles, E., Moreira, A., Salinas, L.: On the robustness of update schedules in boolean networks. Biosystems **97**(1), 1–8 (2009)
3. Baral, C.: Knowledge Representation. Reasoning and Declarative Problem Solving. Cambridge University Press, Cambridge (2003)
4. Berestovsky, N., Nakhleh, L.: An evaluation of methods for inferring boolean networks from time-series data. PLoS ONE **8**(6), e66031 (2013)
5. Cimatti, A., Clarke, E., Giunchiglia, E., Giunchiglia, F., Pistore, M., Roveri, M., Sebastiani, R., Tacchella, A.: NuSMV 2: an opensource tool for symbolic model checking. In: Brinksma, E., Larsen, K.G. (eds.) CAV 2002. LNCS, vol. 2404, pp. 359–364. Springer, Heidelberg (2002)
6. Gallet, E., Manceny, M., Le Gall, P., Ballarini, P.: An LTL model checking approach for biological parameter inference. In: Merz, S., Pang, J. (eds.) ICFEM 2014. LNCS, vol. 8829, pp. 155–170. Springer, Heidelberg (2014)
7. Gebser, M., Kaminski, R., Kaufmann, B., Schaub, T.: Answer set solving in practice. In: Synthesis Lectures on Artificial Intelligence and Machine Learning. Morgan and Claypool Publishers (2012)
8. Gebser, M., Kaufmann, B., Otero, R., Romero, J., Schaub, T., Wanko, P.: Domain-specific heuristics in answer set programming. In: Proceedings of the 27th National Conference on Artificial Intelligence (AAAI 2013), pp. 350–356. AAAI Press (2013)
9. Gebser, M., Kaufmann, B., Schaub, T.: Multi-threaded ASP solving with clasp. Theory and Pract. Log. Program. **12**(4–5), 525–545 (2012)
10. Guziolowski, C., Videla, S., Eduati, F., Thiele, S., Cokelaer, T., Siegel, A., Saez-Rodriguez, J.: Exhaustively characterizing feasible logic models of a signaling network using answer set programming. Bioinformatics **29**(18), 2320–2326 (2013)
11. Harel, D., Kupferman, O., Vardi, M.Y.: On the complexity of verifying concurrent transition systems. Inf. Comput. **173**(2), 143–161 (2002)
12. Kauffman, S.: Metabolic stability and epigenesis in randomly constructed genetic nets. J. Theor. Biol. **22**(3), 437–467 (1969)
13. Klarner, H., Streck, A., Šafránek, D., Kolčák, J., Siebert, H.: Parameter identification and model ranking of thomas networks. In: Gilbert, D., Heiner, M. (eds.) CMSB 2012. LNCS, vol. 7605, pp. 207–226. Springer, Heidelberg (2012)
14. MacNamara, A., Terfve, C., Henriques, D., Bernabe, B.P., Saez-Rodriguez, J.: State-time spectrum of signal transduction logic models. Phys. Biol. **9**(4), 045003 (2012)
15. Saez-Rodriguez, J., Alexopoulos, L.G., Epperlein, J., Samaga, R., Lauffenburger, D.A., Klamt, S., Sorger, P.K.: Discrete logic modelling as a means to link protein signalling networks with functional analysis of mammalian signal transduction. Molecular Systems Biology **5**, 331 (2009)
16. Wang, R., Saadatpour, A., Albert, R.: Boolean modeling in systems biology: an overview of methodology and applications. Phys. Biol. **9**(5), 055001 (2012)

BioPSy: An SMT-based Tool for Guaranteed Parameter Set Synthesis of Biological Models

Curtis Madsen$^{(\boxtimes)}$, Fedor Shmarov, and Paolo Zuliani

School of Computing Science, Newcastle University,
Newcastle upon Tyne, UK
{curtis.madsen,f.shmarov,paolo.zuliani}@ncl.ac.uk

Abstract. The *parameter set synthesis* problem consists of identifying *sets* of parameter values for which a given system model satisfies a desired behaviour. This paper presents BioPSy, a tool that performs *guaranteed* parameter set synthesis for *ordinary differential equation* (ODE) biological models expressed in the *Systems Biology Markup Language* (SBML) given a desired behaviour expressed by time-series data. Three key features of BioPSy are: (1) BioPSy computes parameter intervals, not just single values; (2) for the identified intervals the model is formally guaranteed to satisfy the desired behaviour; and (3) BioPSy can handle virtually any Lipschitz-continuous ODEs, including nonlinear ones. BioPSy is able to achieve guaranteed synthesis by utilising *Satisfiability Modulo Theory* (SMT) solvers to determine acceptable parameter intervals. We have successfully applied our tool to several biological models including a prostate cancer therapy model, a human starvation model, and a cell cycle model.

1 Introduction

Computational modelling is central to many scientific and engineering disciplines. For instance, the field of *systems biology* [17] uses modelling to gain a greater understanding of how biology works. Similarly, in *synthetic biology* [1], models are created in an attempt to engineer new, useful biological systems. This field typically develops models and analyses them *in silico* (on a computer) before synthesising the object of the model *in vitro* (in a test tube in the lab) or *in vivo* (within an organism). Biological systems in both systems biology and synthetic biology are often constructed with deterministic dynamics and can be readily translated into *ordinary differential equation* (ODE) models using *mass action kinetics*. There are many well known methods and tools for simulating ODEs that can be used to obtain results on the behaviour of the biological systems (*e.g.*, MATLAB). However, obtaining reliable results requires that all parts of a model are accurately defined. In particular, a key component to modelling biological systems is selecting the correct model parameters. Since quantitative parameters are often difficult or impossible to measure experimentally, a problem that often arises is how to select parameter values to achieve desired model behaviours. Indeed, small parameter variations can lead to vastly different results when simulating biological models.

© Springer International Publishing Switzerland 2015
O. Roux and J. Bourdon (Eds.): CMSB 2015, LNBI 9308, pp. 182–194, 2015.
DOI: 10.1007/978-3-319-23401-4_16

In order to determine acceptable values for the parameters of a system, modellers have employed methods that perform *parameter synthesis*. The parameter set synthesis problem consists of determining *ranges* (intervals) of parameters for which a model's temporal behaviour remains in satisfactory states, usually described by time-series data. Formally, parameter synthesis is categorised as a *reachability problem* [2] where the solution to a set of ODEs is known for a finite number of time points, but some of the parameter values that lead to that solution are missing. For instance, the parameter, k, can be synthesised in the ODE model given by $x'(t) = kt$. Given the time-series data in which $x = \{0, 1, 4, 9\}$ for $t = \{0, 1, 2, 3\}$, it is easy to see that k should be 2. However, if the system is noisy and the values of x can vary by, say, 0.1, solving the parameter synthesis problem for k will produce an interval such as $[1.978, 2.022]$.

This paper presents BioPSy, a tool that performs parameter set synthesis on biological models comprised not only of mass action kinetics, but also of general Lipschitz-continuous ODEs. Models are specified using the well-known *Systems Biology Markup Language* (SBML) [14]. BioPSy accomplishes parameter synthesis by extracting a collection of ODEs from an SBML model and formulating these ODEs along with time-series data into a *Satisfiability Modulo Theory* (SMT) problem. It then leverages the SMT solver dReal [12] to incrementally narrow down the parameter search space. Given a parameter domain, precision, and time-series data expressing desired behaviour, BioPSy returns

- a set of feasible (acceptable) parameter ranges - these are *formally* and *numerically* guaranteed to satisfy the synthesis problem;
- a set of infeasible (unsuitable) parameter ranges - these are *formally* and *numerically* guaranteed not to satisfy the synthesis problem; and
- a set of parameter ranges where, because of the given precision, BioPSy is unable to determine if they satisfy the synthesis problem.

Note that, depending on the problem at hand, any of the three sets above may be empty, although not all at the same time. We remark again that BioPSy can handle nonlinear ODEs, and its answers have mathematical proof strength.

Related Work. A simple way to perform parameter synthesis is first to discretise the parameter space (if necessary) and then use exhaustive simulation or Monte Carlo methods to determine which simulations satisfy a desired behaviour. Indeed, many tools utilise simulation-based approaches to find acceptable parameter values. For example, COPASI [13], a well known biochemical network simulator, uses methods such as genetic algorithms, particle swarm simulations, differential evolution, and simulated annealing among others to perform parameter estimation on SBML models. Tools like COPASI as well as others such as SBML-PET [28] can also give confidence intervals for the parameters that they estimate. Furthermore, there is a collection of applications that leverage the MATLAB framework to provide similar parameter estimation methods. These tools include AMIGO [3], a tool that uses a collection of initial value problem and non-linear optimization methods; PottersWheel [21], a tool that uses deterministic and stochastic optimisation techniques in concert to explore a logarithmic parameter space; and SBT [23], a tool that allows users to define their own cost-functions and use custom optimisation methods. These tools trade-off between how fine-grained the parameter search is and how much computation time is required to find acceptable parameter values.

Other approaches utilise numerical and formal methods to prove that a model meets certain criteria [4,5,27]. For example, Bernstein polynomials and linear programming [10], and probabilistic model checking [26] approaches can be applied to the parameter set synthesis problem. Model checking methods work by partitioning the parameter space into classes of equivalent behaviours for the various parameter values, which are then systematically validated. Simulation and model checking can be combined in a hybrid approach to efficiently search the parameter space. For example, the statistical model checking technique proposed in [16] enables parameter synthesis for stochastic biological models formulated as *continuous-time Markov chains* using temporal logic specifications (bounded LTL formulae) to express desired behaviours. Simulations can also be used to perform sensitivity analysis limiting how much of the state space the model checker will have to analyse [8]. Additionally, some methods formulate parameter synthesis as an SMT problem, but they usually handle restricted classes of models, *e.g.*, transition systems with linear dynamics [7] or with monotone dynamics [22], while we support very general dynamics such as nonlinear ODEs. Although some approaches can handle complex systems with a large number of parameters [9], their implementations are usually problem-specific.

Finally, a notion related to parameter synthesis is that of parameter identifiability, *i.e.*, whether parameters can be uniquely identified from data. This notion is usually explored in the context of specific classes of dynamics and error behaviours — see, *e.g.*, [19] and references therein.

2   Methods

We sketch the parameter synthesis technique and give implementation details of BioPSy. Full details of the theory will appear in a forthcoming paper.

Algorithm. BioPSy takes as input an SBML model file, a time-series data file, a list of model parameters to synthesise with their initial ranges, a noise value (η), a precision value (δ) for the SMT solver, and a precision value (ϵ) for the parameter synthesis algorithm. Time-series data is typically too constrictive as it contains an exact value for each variable at each time point. Also, measured data is often subject to noise. BioPSy utilises η to relax the time-series data and create an interval of acceptable states for each time point, and returns:

- a set of *feasible* (acceptable) parameter ranges: for all the points in this set, the model is *formally* and *numerically* guaranteed to satisfy the noisy time-series data;
- a set of *infeasible* (unsuitable) parameter ranges: for no point in this set, the model satisfies the noisy time-series data. Again, this is *formally* and *numerically* guaranteed; and
- a set of parameter ranges where, because of the given precision, BioPSy is unable to determine if they satisfy the noisy time-series data.

BioPSy works by extracting ODEs from the given SBML model along with the list of model parameters. The user can select which parameters ($\bar{p}$) to synthesise and provide initial parameter ranges to search through. BioPSy converts the ODEs, parameters, and the noisy time-series data into a collection of SMT problems. Each problem represents an *initial value problem* (IVP) constrained by the initial time point and one of the subsequent time points. Informally, the individual SMT problems contain assertions declaring that the values of each variable (*i.e.*, ODE solution) should be in the interval found in the noisy time-series data after integrating the ODEs for the amount of time between the initial time point and the time point being processed for the file. (Note that for every time point we solve an IVP, and therefore, the first value of the time-series data should not be noisy.) Assertions constraining the parameters being synthesised to be within the synthesised ranges from the previous time point are also added. These constraints help reduce the search space.

The initial boxes for the parameter set are passed one-by-one to the parameter synthesis algorithm, which generates appropriate SMT problems and calls the SMT solver dReal [12] to evaluate them. Basically, the synthesis algorithm iteratively splits each box until the minimum size, ϵ, is reached or the current box is either **unsat** or **sat**. (A box needing to be split is denoted **undet**.) An **unsat** outcome means that for no value in the box, the model reaches an acceptable state. A **sat** outcome means that all the values in the box lead the system to an acceptable state. An **undet** outcome means that the algorithm could not decide between **unsat** and **sat**. This indecision might be because the box contains both **sat** and **unsat** regions, or because of the precision, δ, used when solving the SMT problems. This process continues incrementally until all the points in the time series are processed. A high-level workflow for the BioPSy tool is presented in Fig. 1.

As mentioned, BioPSy returns three sets of synthesised parameter ranges, corresponding to **sat**, **unsat**, and **undet**. The **sat** and **unsat** parameter ranges are *formally* and *numerically* guaranteed to be correct. Essentially, these guarantees are made possible by dReal, which is based on validated ODE integration and rigorous constraint processing via interval arithmetics.

We note that the precision value, δ, for dReal can be arbitrarily small. However, it cannot be zero since solving first-order real formulae with general nonlinear functions is an undecidable problem. (For more information on the theory behind dReal, please refer to [11].) Additionally, the precision value, ϵ, can be arbitrarily small. This value determines the level of granularity that the parameter synthesis algorithm uses to search the parameter space. Smaller values mean that BioPSy will try to break parameter ranges into smaller segments when searching for acceptable values. The noise parameter, η, controls the size of the intervals produced from the time-series data. Choosing a small η makes it more difficult to identify acceptable ranges, but it produces parameter enclosures that result in the system having better compliance with the original time-series data.

Finally, we remark that the main advantage of solving the parameter synthesis problem in a *point-by-point* manner (as we do) is that it reduces the computational complexity, since fewer variables are passed to the SMT solver.

Implementation and Usage. The `BioPSy` *graphical user interface* (GUI) is implemented in Java. The parameter synthesis algorithm is implemented in C++, and it utilises the CAPD library[1] for interval arithmetics and `dReal`[12] as a standalone application. The algorithm is additionally parallelised using OpenMP. The GUI is launched using the `BioPSy` JAR file (Java JRE 1.6 or higher required). The user can browse for a model file and a time-series data file. Once selected, these files are shown in the SBML and Time-Series tabs, respectively. The files are also parsed, and the data is displayed in the Parameters and Variables tabs. Under the Parameters tab, the user can select which parameters are to be synthesised, and their precision (ϵ). For synthesised parameters, the user is also able to define a lower bound and upper bound that is used to constrain the parameter search space. Similarly, the Variables tab allows the user to specify bounds on the acceptable values and noise (η) for each variable in the model. Once the bounds are set, clicking the Run button will perform the synthesis. The Advanced Options button enables the user to specify the path

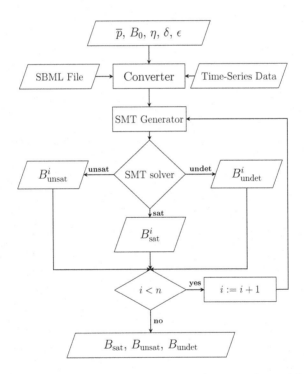

Fig. 1. Workflow diagram for `BioPSy`. *Legend:* $\bar{p} = \{p_1, \ldots p_m\}$: model parameters to synthesise, B_0 = initial set of parameter ranges, η = acceptable noise, δ = SMT solver precision, ϵ = precision of parameter synthesis, n = number of points in the time-series, B^i_{unsat}, B^i_{undet}, and B^i_{sat} = parameter sets for the i-th time point containing boxes for which synthesis is not feasible, undetermined, and feasible, respectively.

---

[1] http://capd.ii.uj.edu.pl.

to the `dReal` binary as well as the desired level of precision, δ, used by `dReal` ($\delta = 0.001$ is the default value). Once the synthesis has started, the Output tab displays the output file as it is being produced allowing a user to watch as the infeasible ranges, feasible ranges, and undetermined ranges are generated for each time point in the data. The Plot tab displays an updating in real-time graphical representation of the contents of the Output tab (for two parameters only). `BioPSy`'s source code, binary, and the models used in the experiments are available at https://github.com/dreal/biology.

3   Results

`BioPSy` has successfully been applied to several biological models including a model of prostate cancer treatment [15, 20], a model on human starvation [24], and a cell cycle model [25]. In each experiment, two parameters are selected for synthesis while the rest are fixed to the values found in the SBML file. Additionally, the experiments are performed on a 32-core (2.9 GHz) Ubuntu Linux machine. The models analysed and their parameters are available at https://github.com/dreal/biology/tree/master/models/CMSB2015.

3.1   Personalized Prostate Cancer Treatment

This model tracks the level of *prostate specific antigen* (PSA) (v) with comprises of two types of cancer cells: *hormone sensitive cells* (HSCs) (x) and *castration resistant cells* (CRCs) (y). In this treatment model, a patient is deprived of androgen (z) causing HRC survival rates to decline. However, lower androgen levels cause HRCs to convert to CRCs and increase the proliferation rate of CRCs. Administrators of this treatment must, therefore, alternate patients between phases of being 'on' and 'off' the treatment in order to prevent both the HSC and the CRC levels from getting out of hand. The ODEs [15, 20] describing the dynamics of a patient on the treatment are shown in Eq. (1).

$$\frac{dv}{dt} = \left(\frac{\alpha_x}{1 + e^{(k_1 - z)k_2}} - \frac{\beta_x}{1 + e^{(z - k_3)k_4}} - m_1 \left(1 - \frac{z}{z_0} \right) - c_1 \right) x + c_2 +$$

$$m_1 \left(1 - \frac{z}{z_0} \right) x + \left(\alpha_y \left(1 - d_0 \frac{z}{z_0} \right) - \beta_y \right) y$$

$$\frac{dx}{dt} = \left(\frac{\alpha_x}{1 + e^{(k_1 - z)k_2}} - \frac{\beta_x}{1 + e^{(z - k_3)k_4}} - m_1 \left(1 - \frac{z}{z_0} \right) - c_1 \right) x + c_2 \qquad (1)$$

$$\frac{dy}{dt} = m_1 \left(1 - \frac{z}{z_0} \right) x + \left(\alpha_y \left(1 - d_0 \frac{z}{z_0} \right) - \beta_y \right) y$$

$$\frac{dz}{dt} = -z\gamma - c_3$$

In this case study, we investigate two applications of `BioPSy`:

1. **Parameter Synthesis**: given an initial parameter domain and time-series data, we synthesise the parameter sets for which the model is guaranteed to satisfy the time-series; and
2. **Parameter Checking**: we check whether parameter estimates obtained by other methods actually satisfy the time-series.

The first application is, in general, very computationally intensive — its worst-case time complexity grows exponentially with the number of parameters to synthesise. The second application is lighter and gives the user the ability to check if a given parameter value satisfies a desired behaviour of the system.

Parameter Synthesis. We perform parameter synthesis using real clinical data [6][2] of a patient who was on treatment for 5 nonconsecutive times throughout 6 years (for about 9 months in each period). The patient was monitored every month and some of the observations (such as PSA and androgen levels) were documented. Overall, every period of time-series data contains around 4–5 time points. For each time-series, we synthesise the parameter set that satisfies the patient's clinical data with noise $\eta = 1.4$ ($\epsilon = 10^{-3}$ and $\delta = 10^{-3}$). The synthesised parameters, α_y and β_x, are explored on the set $[0.0, 0.05] \times [0.0, 0.05]$. The resulting parameter set satisfying all time-series is constructed as the intersection of parameter sets synthesised for each time-series. The feasible set including the ranges $[0.0225, 0.025] \times [0.0325, 0.0332031]$ and $[0.0210938, 0.0225] \times [0.0325, 0.0327344]$. Each time-series evaluation on the specified range took about 12 h of CPU time. The parameter sets synthesised for each time-series are presented in Fig. 2 and the resulting set intersection is shown in Fig. 3. We remark that the values for η and ϵ used in our experiments have been chosen purely for didactic reasons. The user can choose more appropriate values depending on the model being studied.

Parameter Checking. For this application, parameter values are obtained using the different parameter estimation methods available in COPASI, and the results are verified using BioPSy. These parameter estimation methods utilise simulation-based techniques to explore the parameter space and find a vector of parameters that cause the model to best approximate some time-series data. One downside to these methods is that they are not always capable of finding a satisfying vector of parameters due to the trial-and-error approaches they employ. For the prostate cancer treatment model, every parameter estimation method in COPASI is run using default parameters, and each result takes around 5 s to obtain. Some of the methods fail to produce results, but for those methods that are able to find parameters, the estimated values are checked and verified in BioPSy using the same time-series data and $\eta = 1.4$. Each verification took about 10–20 s of CPU time, depending on the time-series length. The verification results are presented in Table 1. Here, it can be seen that some of the methods produce results that only satisfy a few of the time-series data, and only one

---

[2] Data available at: http://www.nicholasbruchovsky.com/clinicalResearch.html.

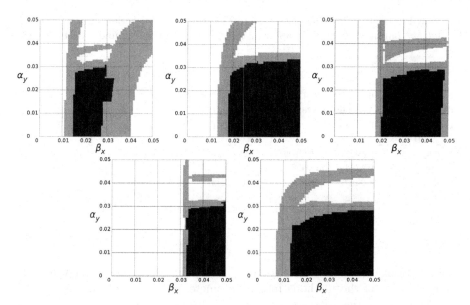

Fig. 2. Prostate cancer model: parameter synthesis results for β_x and α_y for five time-series (ordered *clockwise*) obtained for each 'on' treatment stage. *Legend*: ***white*** - infeasible boxes; ***black*** - feasible boxes; and ***gray*** - undetermined boxes.

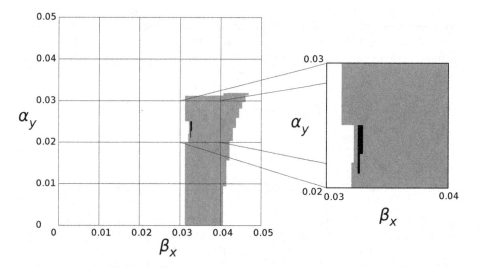

Fig. 3. Prostate cancer model: parameter synthesis results for β_x and α_y over five time-series obtained for each 'on' treatment stage. *Legend*: ***white*** - infeasible boxes; ***black*** - feasible boxes; and ***gray*** - undetermined boxes.

Table 1. Parameter Checking of COPASI results over five time-series on cancer model. *Legend*: n = parameters found invalid by BioPSy; y = parameters found valid by BioPSy

Method	α_x	α_y	β_x	β_y	BioPSy				
					S_1	S_2	S_3	S_4	S_5
Evolut. Prog.	-0.215799	-2.67586×10^{-6}	0.0271774	0.000135248	n	y	y	n	y
Hooke & Jeeves	-0.308608	-0.278566	0.029312	-0.24288	y	y	y	y	y
Levenberg-Marquardt	-0.17045	-31.9428	0.00661261	-10.5429	n	n	n	n	n
Praxis	-0.233483	-0.00697965	0.0240299	0.186801	y	y	y	n	y
Scatter Search	-0.17045	-31.9428	0.00661261	-10.5429	n	n	n	n	n
Simulated Annealing	-0.248778	6.3856×10^{149}	0.0226673	-2.27061×10^{148}	n	n	n	n	n
Truncated Newton	-0.236403	-0.00791949	0.0243545	0.0116282	y	y	y	n	y

method (Hooke &Jeeves) satisfies all of the data with its parameter values. In contrast, three of the methods return parameters that do not formally satisfy any of the time-series with a noise value equal to 1.4.

3.2   Human Starvation

The human starvation model [24] tracks the amount of fat (F), protein in muscle mass (M), and ketone bodies (K) in the human body after glucose reserves have been depleted from three to four days of fasting. These three variables are modelled using material and energy balances to ensure that the behaviour of the model tracks what is observed in actual experiments involving fasting. The ODEs for this model are presented in Eq. (2).

$$\frac{dF}{dt} = F\left(\frac{-a}{1+K} - \frac{1}{\lambda_F}\left(\frac{C+gL_0}{F+M} + \kappa\right)\right)$$
$$\frac{dM}{dt} = -\frac{M}{\lambda_M}\left(\frac{C+\kappa L_0}{F+M} + \kappa\right) \qquad (2)$$
$$\frac{dK}{dt} = \frac{VaF}{1+K} - b$$

Two parameters are synthesised, κ and b, on the ranges $[9, 11]$ and $[0.05, 0.08]$, respectively, using simulated time-series data that includes 25 time points, $\eta = 0.1$, $\delta = 0.001$, and $\epsilon = 0.1$. BioPSy took 5 min and 7 feasible ranges were obtained: $[9.88077, 9.8832] \times [0.0764844, 0.0771875]$, $[9.92213, 10] \times [0.0785938, 0.08]$, $[10, 10.0791] \times [0.0726172, 0.0744629]$, $[9.9416, 10] \times [0.0712109, 0.0761328]$, $[9.8832, 10] \times [0.0761328, 0.0785938]$, $[10, 10.1187] \times [0.0744629, 0.08]$, and $[9.88198, 9.8832] \times [0.0750781, 0.0757813]$. A graphical representation of the final result is shown in Fig. 4.

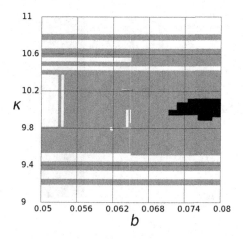

Fig. 4. Human starvation model: parameter synthesis results for κ and b. *Legend:* ***white*** - infeasible boxes; ***black*** - feasible boxes; and ***gray*** - undetermined boxes.

3.3 Cell Cycle

In the cell cycle model [25], two proteins, CDC2 (u) and Cyclin (v), combine to form a heterodimer that controls major events in a cell causing it to reach a steady state, act as a spontaneous oscillator, or act as an excitable switch. The ODEs for this model are presented in Eq. (3).

$$\frac{du}{dt} = k_4 \left(v - u\right) \left(\frac{k_4'}{k_4} + u^2\right) - k_6 u \tag{3}$$
$$\frac{dv}{dt} = \kappa - k_6 u$$

The cell cycle model used has reference BIOMD0000000006 in the BioModels Database [18]. In this example, two parameters are synthesised, k_4' and k_4, on the ranges $[0.01, 0.02]$ and $[175, 185]$, respectively, using $\eta = 0.001$, $\delta = 0.001$, $\epsilon = 0.1$, from 10 simulated data points. BioPSy took 10 min to find one feasible range, $[0.0166691, 0.0192934] \times [175, 185]$. This result is shown in Fig. 5.

4   Conclusions and Future Work

Here, we present BioPSy, an open-source tool for guaranteed parameter set synthesis on biological models from time-series data. BioPSy accepts SBML models, so it can be applied to a large number of existing biological models. Indeed, BioPSy is not only limited to biological models with mass action kinetics but can handle models involving general ODEs. An important feature about our tool is that models using parameters synthesised with BioPSy are formally guaranteed to behave as desired. Also, BioPSy can formally validate parameter estimates

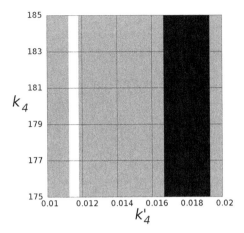

Fig. 5. Cell cycle model: parameter synthesis results for k_4 and k_4'. *Legend:* **white** - infeasible boxes; **black** - feasible boxes; and **gray** - undetermined boxes.

generated by other methods. We apply BioPSy to non-trivial biological models, including a highly nonlinear model of prostate cancer treatment. For this model in particular, BioPSy is able to synthesise parameters from real clinical data. Despite the complexity of parameter synthesis and of the models involved, BioPSy performs reasonably well, and it is usable in practice. We believe BioPSy can be useful for design space exploration in both synthetic and systems biology. In the future, we plan to extend BioPSy to handle biological models that contain both continuous and discrete dynamics — so called *hybrid* models.

Acknowledgements. C. M. has been supported by the Engineering and Physical Sciences Research Council (UK) grant EP/K039083/1; F. S. has been supported by award N00014-13-1-0090 of the US Office of Naval Research.

References

1. Arkin, A.: Setting the standard in synthetic biology. Nature Biotech. **26**, 771–774 (2008)
2. Asarin, E., Dang, T., Frehse, G., Girard, A., Guernic, C.L., Maler, O.: Recent progress in continuous and hybrid reachability analysis. In: IEEE Conference on Computer Aided Control System Design, pp. 1582–1587 (2006)
3. Balsa-Canto, E., Banga, J.R.: AMIGO, a toolbox for advanced model identification in systems biology using global optimization. Bioinformatics **27**(16), 2311–2313 (2011)
4. Barnat, J., Brim, L., Krejci, A., Streck, A., Safranek, D., Vejnar, M., Vejpustek, T.: On parameter synthesis by parallel model checking. IEEE/ACM Trans. Comput. Biol. Bioinform. **9**(3), 693–705 (2012)
5. Batt, G., Yordanov, B., Weiss, R., Belta, C.: Robustness analysis and tuning of synthetic gene networks. Bioinformatics **23**(18), 2415–2422 (2007)

6. Bruchovsky, N., Klotz, L., et al.: Final results of the Canadian prospective phase II trial of intermittent androgen suppression for men in biochemical recurrence after radiotherapy for locally advanced prostate cancer. Cancer **107**, 389–395 (2006)
7. Cimatti, A., Griggio, A., Mover, S., Tonetta, S.: Parameter synthesis with IC3. In: FMCAD, pp. 165–168. IEEE (2013)
8. Donzé, A., Clermont, G., Langmead, C.J.: Parameter synthesis in nonlinear dynamical systems: application to systems biology. J. Comput. Biol. **17**(3), 325–336 (2010)
9. Dötschel, T., Auer, E., Rauh, A., Aschemann, H.: Thermal behavior of high-temperature fuel cells: reliable parameter identification and interval-based sliding mode control. Soft Comput. **17**(8), 1329–1343 (2013)
10. Dreossi, T., Dang, T.: Parameter synthesis for polynomial biological models. In: HSCC 2014, pp. 233–242. ACM (2014)
11. Gao, S., Avigad, J., Clarke, E.M.: Delta-decidability over the reals. In: LICS, pp. 305–314 (2012)
12. Gao, S., Kong, S., Clarke, E.M.: dReal: an SMT solver for nonlinear theories over the reals. In: Bonacina, M.P. (ed.) CADE 2013. LNCS, vol. 7898, pp. 208–214. Springer, Heidelberg (2013)
13. Hoops, S., Sahle, S., Gauges, R., Lee, C., Pahle, J., Simus, N., Singhal, M., Xu, L., Mendes, P., Kummer, U.: COPASI - a Complex PAthway SImulator. Bioinformatics **22**(24), 3067–3074 (2006)
14. Hucka, M., Finney, A., Sauro, H.M., et al.: The Systems Biology Markup Language (SBML): a medium for representation and exchange of biochemical network models. Bioinformatics **19**(4), 524–531 (2003)
15. Ideta, A.M., Tanaka, G., Takeuchi, T., Aihara, K.: A mathematical model of intermittent androgen suppression for prostate cancer. J. Nonlinear Sci. **18**(6), 593–614 (2008)
16. Jha, S.K., Langmead, C.J.: Synthesis and infeasibility analysis for stochastic models of biochemical systems using statistical model checking and abstraction refinement. Theor. Comput. Sci. **412**(21), 2162–2187 (2011)
17. Kitano, H.: Foundations of Systems Biology. MIT Press, Cambridge (2001)
18. Li, C., et al.: BioModels Database: an enhanced, curated and annotated resource for published quantitative kinetic models. BMC Syst. Biol. **4**, 92 (2010)
19. Little, M.P., Heidenreich, W.F., Li, G.: Parameter identifiability and redundancy: theoretical considerations. PLoS ONE **5**(1), e8915 (2010)
20. Liu, B., Kong, S., Gao, S., Zuliani, P., Clarke, E.M.: Towards personalized cancer therapy using delta-reachability analysis. In: HSCC, pp. 227–232. ACM (2015)
21. Maiwald, T., Timmer, J.: Dynamical modeling and multi-experiment fitting with PottersWheel. Bioinformatics **24**(18), 2037–2043 (2008)
22. Meslem, N., Ramdani, N., Candau, Y.: Guaranteed parameter set estimation with monotone dynamical systems using hybrid automata. Reliable Comput. **14**, 88–104 (2010)
23. Schmidt, H., Jirstrand, M.: Systems Biology Toolbox for MATLAB: a computational platform for research in systems biology. Bioinformatics **22**(4), 514–515 (2006)
24. Song, B., Thomas, D.: Dynamics of starvation in humans. J. Math. Biol. **54**(1), 27–43 (2007)
25. Tyson, J.J.: Modeling the cell division cycle: cdc2 and cyclin interactions. Proc. Nat. Acad. Sci. **88**(16), 7328–7332 (1991)

26. Češka, M., Dannenberg, F., Kwiatkowska, M., Paoletti, N.: Precise parameter synthesis for stochastic biochemical systems. In: Mendes, P., Dada, J.O., Smallbone, K. (eds.) CMSB 2014. LNCS, vol. 8859, pp. 86–98. Springer, Heidelberg (2014)
27. Yordanov, B., Belta, C.: Parameter synthesis for piecewise affine systems from temporal logic specifications. In: Egerstedt, M., Mishra, B. (eds.) HSCC 2008. LNCS, vol. 4981, pp. 542–555. Springer, Heidelberg (2008)
28. Zi, Z., Klipp, E.: SBML-PET: a Systems Biology Markup Language-based parameter estimation tool. Bioinformatics 22(21), 2704–2705 (2006)

Derivation of Qualitative Dynamical Models from Biochemical Networks

Wassim Abou-Jaoudé[2], Jérôme Feret[1][(✉)], and Denis Thieffry[2]

[1] DI-ENS, INRIA/ÉNS/CNRS/PSL, Paris, France
`feret@ens.fr`
[2] IBENS, ÉNS/CNRS/Inserm/PSL, Paris, France
`wassim@biologie.ens.fr`, `thieffry@ens.fr`

Abstract. As technological advances allow a better identification of cellular networks, more and more molecular data are produced allowing the construction of detailed molecular interaction maps. One strategy to get insights into the dynamical properties of such systems is to derive compact dynamical models from these maps, in order to ease the analysis of their dynamics. Starting from a case study, we present a methodology for the derivation of qualitative dynamical models from biochemical networks. Properties are formalised using abstract interpretation. We first abstract states and traces by quotienting the number of instances of chemical species by intervals. Since this abstraction is too coarse to reproduce the properties of interest, we refine it by introducing additional constraints. The resulting abstraction is able to identify the dynamical properties of interest in our case study.

1 Introduction

As technological advances allow a better identification of cellular networks, more and more molecular data are produced allowing the construction of detailed molecular interaction maps. These maps form large and complex intertwined biochemical networks, which dynamical functioning is very hard to decipher. One approach to unravel the dynamical properties of such systems relies on the derivation of qualitative dynamical models from these maps, in order to ease the analysis of their dynamics [9,12].

Automatic methods for such derivations still lack of convenient trade-off between efficiency and accuracy. Some abstractions consist only in partitioning the state space (as in the Boolean semantics of BIOCHAM [6]). These abstractions are usually too conservative and fail in detecting properties of interest.

This material is based upon works sponsored by the "École normale supérieure" (ÉNS) under an incitative action and by the Defense Advanced Research Projects Agency (DARPA) and the U.S. Army Research Office under grant number W911NF-14-1-0367. The views, opinions, and/or findings contained in this article are those of the authors and should not be interpreted as representing the official views or policies, either expressed or implied, of the "École normale supérieure", the Defense Advanced Research Projects Agency, or the U.S. Department of Defense.

O. Roux and J. Bourdon (Eds.): CMSB 2015, LNBI 9308, pp. 195–207, 2015.
DOI: 10.1007/978-3-319-23401-4_17

They have to be refined by integrating an approximated quantitative description of the dynamics of the model in each partition class, as done in tropical approximations [13] and piecewise affine systems [10]. Yet, the latter methods provide no explicit bounds for numerical errors (at best an asymptotic estimation of them).

Our motivation is twofold: not only we want to design an automatic tool to derive accurate logical models from reaction networks, but also we want to better understand the process of logical modelling and its underlying implicit assumptions. To achieve these goals, we use abstract interpretation [5]. Abstract interpretation is a mathematical framework to formally relate the behaviour of programs or models, seen at different levels of abstraction. It can be used not only to establish formal comparisons between abstraction techniques, but also to derive new abstractions of the behaviour of programs or models.

Our approach is the following. In Sect. 3, we formalise the behaviour of reaction networks by keeping the exact number of instances of chemical species. In Sect. 4, we propose an abstraction in which the number of instances is sampled within a finite set of intervals. In Sect. 5, we refine this abstraction by taking into account three kinds of properties: we deal with mass preservation invariants in Sect. 5.1; we detect when the number of instances of a given chemical species cannot cross its sampling intervals in Sect. 5.2; we enrich the description of the behaviour of the models with information about the reaction rates and take into account the separation between time-scales in Sect. 5.3. More details are provided in an extended version of this article [1].

2   Case Study

Let us start with a case study.

We consider a model with three kinds of proteins A, B, C. We assume that the protein B is a scaffold between the proteins A and C, that is to say that each instance of B can bind to an instance of A and/or to an instance of C. We wonder what is the influence of the initial concentration of the protein B on the concentration of the trimer ABC. Intuitively, the more Bs we put in the model, the more ABCs will be formed. Yet this is not the case, since at high concentration, the protein B prevents the proteins A and C to meet since almost each instance of A (resp. C) belongs to a dimer AB (resp. BC), and thus there is no available As (resp. Cs) to form the trimers ABC. Thus, at high concentration, by sequestration effect, the scaffold prevents the formation of trimers ABC.

Figure 1 lists the reactions of the model (Fig. 1(a)), the system of equations (under the assumptions of the law of mass action) (Fig. 1(b)), and the concentration of the trimer ABC at steady state with respect to the initial concentration of the protein B (Fig. 1(c)). We notice that at low concentration of the protein B, the concentration of the trimer ABC at steady state grows linearly, whereas it drops following an homographic function at high concentration of the protein B. Interestingly, this sequestration effect has also been observed *in vivo* [4].

$$A + B \xrightarrow{k_1} AB$$

$$B + C \xrightarrow{k_2} BC$$

$$AB + C \xrightarrow{k_3} ABC$$

$$A + BC \xrightarrow{k_4} ABC$$

(a) Reactions.

$$\begin{cases} \dfrac{d[A]}{dt} = -[A](k_1[B] + k_4[BC]) \\[1mm] \dfrac{d[B]}{dt} = -[B](k_1[A] + k_2[C]) \\[1mm] \dfrac{d[C]}{dt} = -[C](k_2[B] + k_3[AB]) \\[1mm] \dfrac{d[AB]}{dt} = k_1[A][B] - k_3[AB][C] \\[1mm] \dfrac{d[BC]}{dt} = k_2[B][C] - k_4[A][BC] \\[1mm] \dfrac{d[ABC]}{dt} = k_4[A][BC] + k_3[AB][C] \end{cases}$$

(b) Equations.

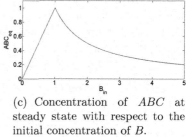

(c) Concentration of ABC at steady state with respect to the initial concentration of B.

Fig. 1. A model with a sequestration effect. We plot the concentration of the trimer ABC at steady state with respect to the initial concentration of the protein B, with all reaction rates equal to 1 and with an initial concentration of 1 for the proteins A and C and of 0 for the complexes AB, BC, and ABC.

This example is well suited for testing the accuracy of our approach, since two different dynamical behaviours may emerge according to the relative position of the quotient between the initial concentration of the protein B and the initial concentrations of the proteins A and C with respect to a semi-quantitative threshold. There is no need to know precisely the rates of the reactions. These quantitative details shift the threshold but have no impact on its existence (unless one of the reaction rate is set to 0). Although we have shown this phenomenon on the deterministic (differential) semantics, considering only forward reactions, it also occurs with a stochastic semantics and/or reversible reactions. A fine description of this model should account for complex properties such as concurrency and sequestration phenomena (when an instance of the protein A is bound to an instance of the protein B, it is no longer available to bind with an instance of the dimer BC), as well as for the race between competing reactions (if there is many instances of B and few instances of BC, an instance of A is more likely to bind to a protein B, than to a complex BC).

3   Trace Semantics

We want to design a framework to automatically abstract logical models from reaction networks. Following a formal approach, we will relate the behaviour of the abstract model with the behaviour of the reaction network. Thus, the first task is to provide a formal definition for the behaviour of reaction networks. In this section, we describe this behaviour qualitatively in terms of a set of traces. Partial information about reaction kinetics will be taken into account in Sect. 5.3.

Firstly, we give the definition of a reaction network.

Definition 1 (Reaction Network). *A network R of n reactions is a pair $(\nu, (M_r, V_r)_{1 \leq r \leq n})$, where: (i) ν is a set of chemical species; (ii) for each integer r between 1 and n, (a) $M_r : \nu \longrightarrow \mathbb{N}$ is a multi-set of chemical species, and (b) $V_r : \nu \longrightarrow \mathbb{Z}$ is a reaction vector, such that $M_r(x) + V_r(x) \geq 0$ for any $x \in \nu$.*

In Definition 1, a pair (M_r, V_r) is called a reaction. In a reaction (M_r, V_r), the multi-set M_r encodes the set of the reactants (with their multiplicities) whereas the vector V_r denotes how many chemical species of each kind is produced and consumed at each application of the reaction.

We can now formally define the set of transitions of a reaction network.

Definition 2 (Transition System). *A reaction network $R \triangleq (\nu, (M_r, V_r)_{1 \leq r \leq n})$ induces a transition system $(\mathcal{Q}_R, T_R)$ where: (i) $\mathcal{Q}_R$ is the set $\mathbb{N}^\nu$ of the functions between ν and $\mathbb{N}$; (ii) T_R is the subset of $\mathbb{N}^\nu \times [\![1,n]\!] \times \mathbb{N}^\nu$ that contains all the triple (q, r, q') such that, for all chemical species $x \in \nu$, (a) $M_r(x) \leq q(x)$ and (b) $q'(x) = q(x) + V_r(x)$.*

In Definition 2, the notation $[\![1, n]\!]$ denotes the set of the integers between 1 and n. The set $\mathcal{Q}_R$ denotes all the potential states of the system. At this level of abstraction, the state of the system describes the number of instances of each kind of chemical species. The elements of T_R are called the transitions of the system. Transitions define the result of the applications of reactions. More precisely, a triple $(q, r, q') \in T_R$ denotes the fact that the system can jump from the state q to the state q' by applying the rule indexed by the integer r. Condition (2a) ensures that enough reactants are available, whereas condition (2b) encodes the consumption/production of the chemical species. We notice that the resulting transition system is equivalent to a Petri net [3], in which each kind of chemical species is denoted by a placeholder and each instance by a token.

Before defining the traces of a reaction network, we introduce some notations. For any two sets A and Σ, and any subset T of the set $A \times \Sigma \times A$, we call a pretrace of elements of A and transitions in T, any element of the set $A \times T^\star$. In a pretrace $\tau \triangleq (a_0', (a_i, \lambda_i, a_i')_{1 \leq i \leq k})$, the element a_0' (resp. a_k') is called the initial (resp. final) state of the pretrace τ and is denoted as $first(\tau)$ (resp. $final(\tau)$). The second element of a pretrace is a (potentially empty) sequence of triples in T. We call a trace any pretrace $(a_0', (a_i, \lambda_i, a_i')_{1 \leq i \leq k})$ such that $a_i = a_{i-1}'$ for any integer i between 1 and k. Lastly, given a triple $(a_{k+1}, \lambda_{k+1}, a_{k+1}')$ in T, we define by $\tau \frown (a_{k+1}, \lambda_{k+1}, a_{k+1}')$ the pretrace $(a_0', (a_i, \lambda_i, a_i')_{1 \leq i \leq k+1})$.

We can now properly define the trace semantics of a reaction network.

Definition 3 (Trace Semantics). *The set of traces that is induced by a reaction network R and a set of initial states $\mathcal{Q}_{R,0} \subseteq \mathcal{Q}_R$ is defined as the set of the traces τ of elements of $\mathcal{Q}_R$ and transitions in T_R such that $first(\tau) \in \mathcal{Q}_{R,0}$.*

We denote by $\mathcal{T}_{R, \mathcal{Q}_{R,0}}$ the set of traces that is induced by the reaction network R and the set of the initial states $\mathcal{Q}_{R,0}$.

Following the abstract interpretation framework [5], we can also express the trace semantics as the least fixpoint of a monotonic function over the powerset $\wp(\mathcal{Q}_R \times T_R^\star)$. Let $\mathbb{F}_{\mathcal{Q}_{R,0}}$ be the function that maps each set X of pretraces into the set of pretraces $\mathcal{Q}_{R,0} \cup \{\tau \frown (q, r, q') \mid \tau \in X \wedge (q, r, q') \in T_R \wedge q = final(\tau)\}$. $\mathbb{F}_{\mathcal{Q}_{R,0}}$ is a monotonic function (i.e. $X \subseteq Y \Rightarrow \mathbb{F}_{\mathcal{Q}_{R,0}}(X) \subseteq \mathbb{F}_{\mathcal{Q}_{R,0}}(Y)$) over a powerset, thus it has a least fixpoint [16]. This least fixpoint, $lfp\ \mathbb{F}_{\mathcal{Q}_{R,0}}$, is indeed the set of all the traces of the reaction network R (i.e. $lfp\ \mathbb{F}_{\mathcal{Q}_{R,0}} = \mathcal{T}_{R, \mathcal{Q}_{R,0}}$). Moreover, the function $\mathbb{F}_{\mathcal{Q}_{R,0}}$ is $\cup$-continuous (i.e. $\mathbb{F}_{\mathcal{Q}_{R,0}}(\cup\{X_j \mid j \in J\}) = \cup\{\mathbb{F}_{\mathcal{Q}_{R,0}}(X_j) \mid j \in J\}$ for any family $(X_j)_{j\in J}$ of sets of pretraces). It follows from [11] that the least fixpoint of $\mathbb{F}_{\mathcal{Q}_{R,0}}$ can also be expressed as the limit of the finite iterates of the function $\mathbb{F}_{\mathcal{Q}_{R,0}}$ (i.e. $\mathcal{T}_{R,\mathcal{Q}_{R,0}} = \cup\{\mathbb{F}^i_{\mathcal{Q}_{R,0}}(\varnothing) \mid i \in \mathbb{N}\}$), which provides an iterative algorithm to enumerate the traces of the network R.

4   Derivation of a Coarse-Grained Qualitative Semantics

The semantics described in Sect. 3 is too fine grained. In particular, each instance of a protein is taken into account. Usually, in a qualitative model, the number of instances of proteins is sampled within a finite number of intervals. In this section, we will use the abstract interpretation framework to derive such an abstraction. Abstract interpretation [5] is a unifying framework for the approximation of mathematical structures. It offers formal tools to relate the observations of the behaviour of a system at different levels of details. It can also be used to systematically derive static analysers (that provide effective definitions of semantics at coarser levels of abstraction).

We use a simple version of the abstract interpretation framework that consists in removing some information from values, states and traces. Our abstraction is twofold. Firstly, we sample the number of instances of chemical species within a finite number of intervals. Secondly, we remove in traces the transitions for which the number of instances of each chemical species remain in the same interval. To sample the number of instances and later the rate of reactions (see Sect. 5.3), we partition the set $\mathbb{R}^+$ over the $p+1$ intervals $[0, \delta[$, $[\delta^i, \delta^{i+1}[$ for each integer i between 1 and $p-1$, and $[\delta^p, \infty[$, where p and δ are integer parameters such that $\delta \geq 2$. We introduce a function β^v to sample positive real numbers over this partition as follows:

Definition 4 (Abstract Values). *We define the function β^v between the set $\mathbb{R}^+$ and the set $[0, p]$ that maps each positive real number $v \in \mathbb{R}^+$ into the least integer in the set $\{p\} \cup \{k \in [0, p] \mid v < \delta^{k+1}\}$.*

Then we lift the function β^v over transition systems.

Definition 5 (Abstract Transition System). *A reaction network R induces an abstract transition system $(\mathcal{Q}^\sharp_R, T^\sharp_R)$ where: (i) $\mathcal{Q}^\sharp_R$ is the set $[0, p]^\nu$ of the functions between the set of the chemical species ν and the integer interval $[0, p]$; (ii) $T^\sharp_R$ is the subset of $[0, p]^\nu \times [1, n] \times [0, p]^\nu$ that is defined by $(q^\sharp, r, q^{\sharp\prime}) \in T^\sharp_R$ if and only if there exist $(q, r, q') \in T_R$ such that $q^\sharp = \beta^v \circ q$ and $q^{\sharp\prime} = \beta^v \circ q'$.*

Thus, the abstract transition system is obtained by applying component-wise the function β^v in the states of the transition system and in the states that occur in transitions. We denote by β^s the function mapping each state $q \in \mathcal{Q}_R$ into the abstract state $\beta^v \circ q \in \mathcal{Q}^\sharp_R$. Then we lift the abstraction β^s to pretraces and traces. We call an abstract pretrace (resp. trace) any pretrace (resp. trace) of elements of $\mathcal{Q}^\sharp_R$ and transitions in $T^\sharp_R$. We denote by β^t_1 the function between the set $\mathcal{Q}_R \times T^\star_R$ and the set $\mathcal{Q}^\sharp_R \times T^{\sharp\star}_R$ that maps each (concrete) pretrace $(q'_0, (q_i, r_i, q'_i)_{1\leq i\leq k})$ to the (abstract) pretrace $(\beta^s(q_0'), (\beta^s(q_i), r_i, \beta^s(q_i'))_{1 \le i \le k})$. We notice that there exists some abstract transitions $(q^\sharp, r, q^{\sharp\prime}) \in T_R^\sharp$ such that $q^\sharp = q^{\sharp\prime}$. Indeed, even if a concrete transition changes the number of instances of a chemical species, this does not always make it exit its sampling interval. We call such transitions silent and we denote by $T_{R/\varepsilon}^\sharp$ the set of the non silent abstract transitions. In order to remove silent transitions, we define the function β_2^t between the set $\mathcal{Q}_R^\sharp \times T_R^{\sharp\star}$ and the set $\mathcal{Q}_R^\sharp \times T_{R/\varepsilon}^{\sharp\star}$, which maps each abstract pretrace $(q_0^\sharp, (q_i^\sharp, r_i, q_i^{\sharp\prime}))$ to the abstract pretrace $(q_0^{\sharp\prime}, (q_{\sigma(i)}^\sharp, r_{\sigma(i)}, q_{\sigma(i)}^{\sharp\prime}))$, where $\sigma(i)$ ranges over the set $\{i \in [\![1, k]\!] \mid q_i^\sharp \ne q_i^{\sharp\prime}\}$ in increasing order.

We denote by β^t the composition of the function β_2^t and β_1^t and use this function to abstract the computation of the trace semantics. Given a set of initial states $\mathcal{Q}_{R,0} \subseteq \mathcal{Q}_R$, we introduce the function $\mathbb{F}_{\mathcal{Q}_{R,0}}^\sharp$ over the set $\wp(\mathcal{Q}_R^\sharp \times T_{R/\varepsilon}^{\sharp\star})$ that is defined as $\alpha^t \circ \mathbb{F}_{\mathcal{Q}_{R,0}} \circ \gamma^t$, where: (i) the function α^t maps each subset X of $\mathcal{Q}_R \times T_R^\star$ into the subset $\{\beta^t(x) \in \mathcal{Q}_R^\sharp \times T_{R/\varepsilon}^{\sharp\star} \mid x \in X\}$ of $\mathcal{Q}_R^\sharp \times T_{R/\varepsilon}^{\sharp\star}$ (ii) and conversely, the function γ^t maps each subset Y of $\mathcal{Q}_R^\sharp \times T_{R/\varepsilon}^{\sharp\star}$ into the subset $\{x \in \mathcal{Q}_R \times T_R^\star \mid \beta^t(x) \in Y\}$ of $\mathcal{Q}_R \times T_R^\star$. Given two subsets $X \subseteq \mathcal{Q}_R \times T_R^\star$ and $Y \subseteq \mathcal{Q}_R^\sharp \times T_{R/\varepsilon}^{\sharp\star}$, the property $\alpha^t(X) \subseteq Y$ is equivalent to the property $X \subseteq \gamma^t(Y)$. Such a pair of functions is called a Galois connection [5]. Intuitively, the parts of the set $\mathcal{Q}_R^\sharp \times T_{R/\varepsilon}^{\sharp\star}$ denote properties about the elements in $\mathcal{Q}_R \times T_R^\star$. The function α^t abstracts each set of elements in $\mathcal{Q}_R \times T_R^\star$ into the most precise property they satisfy (the fact that (α^t, γ^t) is a Galois connection entails that the most precise property always exists). Conversely, the function γ^t concretizes a property in $\wp(\mathcal{Q}_R^\sharp \times T_{R/\varepsilon}^{\sharp\star})$ into the set of the elements which satisfy this property. We define the Galois connections (α^v, γ^v) (resp. (α^s, γ^s)) between sets of concrete values (resp. states) and sets of abstract values (resp. states) the same way.

The function $\mathbb{F}_{\mathcal{Q}_{R,0}}^\sharp$ is monotonic. Thus, by [16], it has a least fixpoint.

Definition 6 (Abstract Trace Semantics). *The set of abstract traces $\mathcal{T}_{\mathcal{Q}_{R,0}}^\sharp$ that is induced by a reaction network R and a set of initial states $\mathcal{Q}_{R,0} \subseteq \mathcal{Q}_R$ is defined as the least fixpoint lfp $\mathbb{F}_{\mathcal{Q}_{R,0}}^\sharp$ of the function $\mathbb{F}_{\mathcal{Q}_{R,0}}^\sharp$.*

The Galois connection (α^t, β^t) can be used to transfer the computation of the concrete fixpoint $\mathcal{T}_{R,\mathcal{Q}_{R,0}} = lfp\, \mathbb{F}_{\mathcal{Q}_{R,0}}$ in the abstract.

Theorem 1 (Fixpoint Transfer). *For any reaction network R and any set of initial states $\mathcal{Q}_{R,0} \subseteq \mathcal{Q}_R$, the set lfp $\mathbb{F}_{\mathcal{Q}_{R,0}}$ is a subset of the set $\gamma^t(lfp\, \mathbb{F}_{\mathcal{Q}_{R,0}}^\sharp)$.*

We have used the Galois connection (α^t, γ^t) so as to abstract the trace semantics. Theorem 1 ensures that our abstraction is conservative, i.e. all the traces of the concrete semantics are taken into account. Moreover, the set of abstract traces can be computed by iterating the function $\alpha^t \circ \mathbb{F}_{\mathcal{Q}_{R,0}} \circ \gamma^t$. This consists in, at each step, (a) computing the concretization of the set of traces, (b) making the computation in the concrete, and (c) abstract the result.

The following property provides a direct way to make this computation without going back and forth in the concrete and provides more intuition about what information is lost with our abstraction. We introduce few notations: we denote by V_∞ (resp. M_∞) the greatest element of the set $\{|V_i(x)| \mid i \in [\![1,n]\!],\ x \in \nu\}$ (resp. of the set $\{M_i(x) \mid i \in [\![1,n]\!],\ x \in \nu\}$); for any integer $z \in \mathbb{Z}$, we define the sign $sign(z)$ of z as: (a) $sign(0) \triangleq 0$, and (b) $sign(z) \triangleq z/|z|$ if $z \neq 0$; and for any function f between two sets A and B and any elements $y \in A$ and $v \in B$, we define $f[y \mapsto v]$ as the function between A and B mapping the element y to the element v, and any element $x \in A \smallsetminus \{y\}$ to the element $f(x)$.

Property 1. The following assertions hold:

1. For any part $Y \subseteq \mathcal{Q}_R^\sharp \times T_{R/\varepsilon}^{\sharp *}$, the set $\mathbb{F}_{\mathcal{Q}_{R,0}}^\sharp(Y)$ is equal to the set $\alpha^s(\mathcal{Q}_{R,0}) \cup \{\tau^\sharp \frown (q^\sharp, r, q^{\sharp\prime}) \mid \tau^\sharp \in Y \ \wedge \ (q^\sharp, r, q^{\sharp\prime}) \in T_{R/\varepsilon}^\sharp \ \wedge \ final(\tau) = q^\sharp\}$.
2. For any abstract transition $(q^\sharp, r, q^{\sharp\prime}) \in T_R^\sharp$, if $\delta > V_\infty$, then the value $q^{\sharp\prime}(x)$ is either equal to $q^\sharp(x)$ or to $q^\sharp(x) + sign(V_r(x))$.
3. For any rule r and any abstract state $q^\sharp \in \mathcal{Q}_R^\sharp$, if $\delta > M_\infty$, then, for any chemical species $y \in \nu$ such that $V_r(y) \neq 0$ and $0 \leq q^\sharp(y) + sign(V_r(y)) \leq p$, we have $(q^\sharp, r, q^\sharp[y \mapsto q^\sharp(y) + sign(V_r(y))]) \in T_R^\sharp$.

Property 1.(1) provides an inductive definition to compute the set of the abstract traces directly, without having to concretize the states. Property 1.(2) establishes the fact that it is not possible to cross a whole interval in a single transition. As formalised in Theorem 1, the abstract trace semantics is a safe over-approximation of the concrete trace semantics. Yet, this semantics introduces spurious behaviours. In particular, Property 1.(3) establishes that it is always possible to change the interval of a chemical species $x \in \nu$ in the direction given by the sign of $V_r(x)$, when applying the rule that is indexed with the integer r, unless the chemical species $x \in \nu$ is already in the first or in the last interval of the partition. This is a very coarse abstraction, which will be refined in the next section.

5   Refinements

As we have noticed in Sect. 4, the abstraction $\mathcal{T}_{\mathcal{Q}_{R,0}}^\sharp$ is very coarse. In particular, it does not exploit the following three kinds of situations. Firstly, the number of instances of chemical species may be entangled by some mass preservation invariants. Secondly, when the number of instances of a chemical species enters a new interval, it is sometimes possible to prove that there are not enough resources in the system to make this number reach the next interval. Thirdly, our concrete semantics is purely qualitative. We propose to add kinetic rates and abstract them accurately in order to account for the potential races between reactions.

In this section, we propose three refinements of the abstract semantics to formalise three corresponding classes of reasoning. These refinements are orthogonal: they can be combined by the means of a reduced product [5].

5.1   Mass Invariants

In the concrete semantics, the number of instances of the chemical species may be related by some mass conservation equations. For instance, in our case study, the overall numbers of As, of Bs, and of Cs are constant. Thus the expressions $q(A) + q(AB) + q(ABC)$, $q(B) + q(AB) + q(BC) + q(ABC)$, and $q(C) + q(BC) + q(ABC)$ keep the same values along a trace.

We are interested in constraints of the form $\sum \alpha_x q(x) = b$ for $(\alpha_x)_{x \in \nu} \in \mathbb{N}^\nu$ and $b \in \mathbb{N}$ (i.e. semi-positive constraints). There are several ways to obtain the semi-positive constraints that are satisfied in a network. An algorithm that computes a basis of the set of the semi-positive relationships in a reaction network is proposed in [15]. Less costly but incomplete approaches can also be used: if we know the protein composition of chemical species, we can detect by scanning the set of the proteins and the set of the reactions which proteins are preserved by each reaction, and infer the corresponding mass preservation invariants.

Mass preservation invariants are particular cases of trace invariants and can thus be used to refine our abstraction. Let $inv \subseteq \mathcal{Q}_R \times T_R^\star$ be a trace invariant (formally, this means that $\mathbb{F}_{\mathcal{Q}_{R,0}}(inv) \subseteq inv$). By [16], the concrete semantics is the most precise of the trace invariants (i.e. $\mathcal{T}_{R, \mathcal{Q}_{R,0}} = \bigcap \{X \mid \mathbb{F}_{\mathcal{Q}_{R,0}}(X) \subseteq X\}$). In particular, $\mathcal{T}_{R, \mathcal{Q}_{R,0}} \subseteq inv$. It follows that $lfp\, \mathbb{F}_{\mathcal{Q}_{R,0}} = lfp\, \mathbb{F}^{\text{INV}}_{\mathcal{Q}_{R,0}, inv}$, where $\mathbb{F}^{\text{INV}}_{\mathcal{Q}_{R,0}, inv}$ is defined as the function over the powerset $\wp(\mathcal{Q}_R \times T_R^\star)$ mapping each subset $X \subseteq \mathcal{Q}_R \times T_R^\star$ to the set $\mathbb{F}_{\mathcal{Q}_{R,0}}(X) \cap inv$. The least fixpoints of both functions $\mathbb{F}_{\mathcal{Q}_{R,0}}$ and $\mathbb{F}^{\text{INV}}_{\mathcal{Q}_{R,0}, inv}$ are equal, but the abstraction of the iterates of the latter may be more precise. Let $\mathbb{F}^{\text{INV}\,\sharp}_{\mathcal{Q}_{R,0}, inv}$ be the function $\alpha^t \circ \mathbb{F}^{\text{INV}}_{\mathcal{Q}_{R,0}, inv} \circ \gamma^t$. The function α^t is $\cap$-complete, so the function $\mathbb{F}^{\text{INV}\,\sharp}_{\mathcal{Q}_{R,0}, inv}$ is equal to $[Y \mapsto \mathbb{F}^\sharp_{\mathcal{Q}_{R,0}}(Y) \cap \alpha^t(inv)]$. The iterates of the function $\mathbb{F}^{\text{INV}\,\sharp}_{\mathcal{Q}_{R,0}, inv}$ provide another effective way, more precise but still sound, to abstract the trace semantics:

Theorem 2 (Abstract Trace Semantics with Invariants). *Let $\mathcal{Q}'_{R,0}$ be a subset of $\mathcal{Q}_{R,0}$ and inv be a part of $\mathcal{T}_{R, \mathcal{Q}_{R,0}}$ such that $\mathbb{F}_{\mathcal{Q}'_{R,0}}(inv) \subseteq inv$.*

Then, we have: $\mathcal{T}_{R, \mathcal{Q}'_{R,0}} \subseteq \gamma^t(lfp\,[Y \mapsto \mathbb{F}^\sharp_{\mathcal{Q}'_{R,0}}(Y) \cap \alpha^t(inv)])$.

In Theorem 2, we have partitioned the traces [2,14] to separate the computation of their abstraction according to their initial states. This leads to a more accurate abstraction whenever some pairs of initial states do not share the same invariants.

When the trace invariant is a set of semi-positive constraints, the following property gives an explicit definition for the term $\alpha^t(inv)$.

Property 2 (mass invariant separation). Let $(a_x)_{x \in \nu} \in \mathbb{N}^\nu \setminus \{0\}^\nu$ be a family of positive integer coefficients (with at least one not equal to 0), $b \in \mathbb{N}$ be a positive integer coefficient, and $q^\sharp$ be an abstract state in $\mathcal{Q}_R^\sharp$. We denote by S the sum of the coefficients a_x for any chemical species $x \in \nu$ and we introduce, for any abstract state $q^\sharp$, $q^\sharp_{max}$ as the maximum element of the set $\{k \in [\![0, p]\!] \mid \exists x \in \nu,\, a_x > 0 \ \wedge\ k = q^\sharp(x)\}$. We further assume that $S < \delta$. Then, if either $S\delta^{\beta^v(b)} \leq b$ or $\beta^v(b) = 0$, the set $\alpha^s(\{q \in \mathcal{Q}_R \mid b = \sum_{x \in \nu} \alpha_x q(x)\})$ is equal to the set $\{q^\sharp \in \mathcal{Q}_R^\sharp \mid q^\sharp_{max} = \beta^v(b)\}$. Otherwise, it is equal to the set $\{q^\sharp \in \mathcal{Q}_R^\sharp \mid q^\sharp_{max} \in \{\beta^v(b) - 1, \beta^v(b)\}\}$.

Property 2 has a flavour of tropical algebræ [13]. In particular, whenever the affine constants of mass preservation invariants are far enough from the lower bound of their sampling interval, the abstraction of the number of instances of a protein is equal to the abstraction of the number of instances of the most abundant chemical species containing this protein.

5.2   Watching Interval Boundaries

So far, we have approximated the number of instances of each chemical species by means of intervals. This is a quite coarse abstraction. Indeed, when the number of instances of a chemical species enters a new interval, there is no way to predict whether or not there may be enough resources in the system so that it may reach and enter the next interval. For instance, in our case study, when the system is in a state $q \in \mathcal{Q}_R$ such that $\beta^s(q)(A)$, $\beta^s(q)(C)$ and $\beta^s(q)(ABC)$ are all equal to 0, it may be possible to reach a state q' such that $\beta^s(q')(ABC) = 1$, because (1) the number of instances of ABC may be close to δ, and (2) there may be enough instances of A to cross this threshold. But, after this, there will be not enough instances of A to reach a state q'' such that $\beta^s(q'')(ABC) > 1$. We formalise this kind of reasoning and refine our abstraction accordingly.

We focus on proving that the number of instances of some chemical species cannot cross their current interval upwards (the dual case can be dealt with the same way). We assume that $\delta > 2V_\infty$. Given a state $q \in \mathcal{Q}_R$ and a chemical species $x \in \nu$, we write $q \models x_\dagger$ if either the value $q(x)$ is in the interval $[\delta^{\beta^v(q(x))}, \delta^{\beta^v(q(x))} + V_\infty]$ or if there is no concrete trace τ in $\mathcal{T}_{R,\{q\}}$ such that $\beta^v(final(\tau)(x)) > \beta^v(q(x))$. We denote by $\mathcal{C}_\dagger$ the set $\{x_\dagger \mid x \in \nu\}$.

We update the definitions of abstract states and abstract traces to take into account the constraints in $\mathcal{C}_\dagger$. Formally, an abstract state is now an element of $\mathcal{Q}_R^\natural \times \wp(\mathcal{C}_\dagger)$. The first component encodes the intervals for the number of instances of chemical species, whereas the second component is a set of constraints that specifies which chemical species may eventually cross their current intervals upwards. We also define a refined abstraction function $\beta_\dagger^s$ by $\beta_\dagger^s(q) \triangleq (\beta^s(q), \{c \in \mathcal{C}_\dagger \mid q \models c\})$. We denote by $\beta_\dagger^t$ the function mapping each concrete trace $\tau \in \mathcal{T}_{R,\mathcal{Q}_{R,0}}$ to the trace obtained by firstly replacing in the trace τ every state q with its abstraction $\beta_\dagger^s(q)$ and by secondly removing silent moves. The Galois connection that is induced by $\beta_\dagger^s$ (resp. $\beta_\dagger^t$) is denoted as $(\alpha_\dagger^s, \gamma_\dagger^s)$ (resp. $(\alpha_\dagger^t, \gamma_\dagger^t)$).

Iterating the most precise counterpart $\alpha_\dagger^t \circ \mathbb{F}_{\mathcal{Q}_{R,0}} \circ \gamma_\dagger^t$ to the function $\mathbb{F}_{\mathcal{Q}_{R,0}}$ would be very costly. Thus we iterate an over-approximation of it instead. We define esc as the set of the triples $(q^\natural, x_\dagger, r) \in \mathcal{Q}_R^\natural \times \mathcal{C}_\dagger \times [1, n]$ such that there is a concrete trace $\tau \in \mathcal{T}_{R,\mathcal{Q}_R}$ which satisfies: (i) $\beta^s(first(\tau)) = q^\natural$, (ii) $first(\tau) \models x_\dagger$, (iii) $\beta^s(first(\tau))(x) < \beta^s(final(\tau))(x)$, (iv) $V_r(x) > 0$, (v) there exists a transition in τ with the label r. Intuitively, the set esc contains all the triples $(q^\natural, x_\dagger, r)$ such that, whenever the system is in a state $q \in \gamma^s(\{q^\natural\})$ satisfying $q \models x_\dagger$, the number of instances of the chemical species x may eventually cross the upper bound of its current interval, in a trace that contains at least one application of the rule indexed by the integer r. So as to offer a choice of trade-off between accuracy and efficiency, we introduce a superset $esc^\natural$ of esc, considered as a parameter of our abstraction. Intuitively, whenever a triple $(q^\sharp, x_\ddagger, r) \in esc^\sharp$, it means that our approximation has failed in proving that the number of instances of the species x will never cross its current interval upwards.

We can now refine the set of the transitions of the abstract semantics.

Definition 7 (Abstract Transitions). *We denote by* $T^{\text{CROSS}\,\sharp}_{esc\,\sharp}$ *the set of the triples* $((q^\sharp, C), r, (q^{\sharp\prime}, C'))$ *in* $(\mathcal{Q}^\sharp_R \times \wp(\mathcal{C}_\ddagger)) \times [\![1, n]\!] \times (\mathcal{Q}^\sharp_R \times \wp(\mathcal{C}_\ddagger))$ *such that:*

1. *either (a)* $(q^\sharp, r, q^{\sharp\prime}) \in T^\sharp_R$, *(b)* $\forall x_\ddagger \in C, q^{\sharp\prime}(x) \leq q^\sharp(x)$, *and (c)* $C' = (C \cup \{x_\ddagger \mid x \in \nu \wedge q^{\sharp\prime}(x) > q^\sharp(x)\}) \setminus \{x_\ddagger \mid x \in \nu \wedge q^{\sharp\prime}(x) < q^\sharp(x)\}$,
2. *or (a)* $q^{\sharp\prime} = q^\sharp$ *and there exists a constraint* $c \in \mathcal{C}_\ddagger$ *such that: (b)* $C' = C \setminus \{c\}$, *and (c)* $(q^\sharp, c, r) \in esc^\sharp$.

We distinguish between two kinds of transitions in Definition 7. The first ones consist in regular computation steps: they apply reactions that are allowed and do not violate the constraints about the capability of the chemical species to cross their intervals. After such reactions, the set of the chemical species that have just entered a new interval from below (resp. above) is recorded in (resp. removed from) the set of the constraints. The second kind of transitions consists in removing a constraint where we are unable to prove that the corresponding chemical species will never cross its current interval upwards.

Let $T^{\text{CROSS}\,\sharp}_{esc\,\sharp}$ be the set of pretraces of elements of $\mathcal{Q}^\sharp_R \times \wp(\mathcal{C}_\ddagger)$ and transitions in $T^{\text{CROSS}\,\sharp}_{esc\,\sharp}$. Given a set of initial states $\mathcal{Q}_{R,0} \subseteq \mathcal{Q}_R$, we consider the function $\mathbb{F}^{\text{CROSS}\,\sharp}_{\mathcal{Q}_{R,0},esc\,\sharp}$ over the set $\wp(T^{\text{CROSS}\,\sharp}_{esc\,\sharp})$ mapping each subset Y of $T^{\text{CROSS}\,\sharp}_{esc\,\sharp}$ to the subset $\alpha^s_\ddagger(\mathcal{Q}_{R,0}) \cup \{\tau^\sharp \frown (q^\sharp, r, q^{\sharp\prime}) \mid \tau^\sharp \in Y \wedge (q^\sharp, r, q^{\sharp\prime}) \in T^{\text{CROSS}\,\sharp}_{esc\,\sharp} \wedge final(\tau^\sharp) = q^\sharp\}$. The function $\mathbb{F}^{\text{CROSS}\,\sharp}_{\mathcal{Q}_{R,0},esc\,\sharp}$ is monotonic and satisfies $[\alpha^t_\ddagger \circ \mathbb{F}_{\mathcal{Q}_{R,0}} \circ \gamma^t_\ddagger](Y) \subseteq \mathbb{F}^{\text{CROSS}\,\sharp}_{\mathcal{Q}_{R,0},esc\,\sharp}(Y)$ for any subset Y of $T^{\text{CROSS}\,\sharp}_{esc\,\sharp}$. By [5], it follows that our approach is sound:

Theorem 3 (Soundness). *The function* $\mathbb{F}^{\text{CROSS}\,\sharp}_{\mathcal{Q}_{R,0},esc\,\sharp}$ *has a least fixpoint. Moreover, we have:* $lfp\,\mathbb{F}_{\mathcal{Q}_{R,0}} \subseteq \gamma^t_\ddagger(lfp\,\mathbb{F}^{\text{CROSS}\,\sharp}_{\mathcal{Q}_{R,0},esc\,\sharp})$.

The following property proposes a trade-off for the definition of the primitive $esc^\sharp$, based on a linear integer decision procedure.

Property 3. Let $(q^\sharp, x_\ddagger, r) \in esc^\sharp$. We have $q^\sharp(x) \neq p$ and there exists a function $w \in \mathbb{N}^{[\![1,n]\!]}$ such that: (i) $w(r) > 0$, (ii) $\delta^{q^\sharp(x)} + V_\infty + V_w(x) \geq \delta^{q^\sharp(x)+1}$, and (iii) $\forall x' \in \nu, q^\sharp(x') \neq p \Rightarrow \delta^{q^\sharp(x')+1} + V_w(x) > 0$, where for any chemical species $x' \in \nu$, $V_w(x)$ denotes the value of the expression $\sum_{1 \leq r' \leq n} w(r')V_{r'}(x')$.

5.3 Scales Separation

In our case study, when there are a lot of Bs and only a few BCs in the system, so as to capture the sequestration effect properly, we have to neglect the binding reaction between the chemical species A and BC. Thus we have to exploit the separation between different time scales. According to the modelling paradigm, several methods are used for the formalisation of the separation between time scales. In the logical approach, we usually assume that a reaction preempts any other much slower reactions. In the tropical approach, special care is taken not to neglect the reactions which are involved in large time relaxations of fast cycles [13]. Another approach consists in encoding scales separation by the means of fairness hypotheses that bound the frequencies of slow reaction steps.

In this section, we propose (1) a generic method to formalise assumptions about time scale separation and (2) a systematic way to lift these assumptions to the abstract semantics. We start from a given scheduler $\mathcal{S}$. Formally, $\mathcal{S}$ is a function between the set $\mathcal{Q}_R \times T_R^\star$ and the set $\wp(\llbracket 1,n \rrbracket)$. Intuitively, the scheduler restricts the set of the reactions which can be computed immediately after a (pre)trace. We refine the concrete semantics accordingly: we define $T_{\mathcal{R},\mathcal{Q}_{R,0},\mathcal{S}}^{\text{TIME}}$ as the least fixpoint of the monotonic function $\mathbb{F}_{\mathcal{Q}_{R,0},\mathcal{S}}^{\text{TIME}}$, which maps any set $X \subseteq \mathcal{Q}_R \times T_R^\star$ of pretraces to the set $\mathcal{Q}_{R,0} \cup \{\tau \frown (q,r,q') \mid \tau \frown (q,r,q') \in \mathbb{F}_{\mathcal{Q}_{R,0}}(X) \ \wedge \ r \in \mathcal{S}(\tau)\}$.

Now we lift the action of the scheduler $\mathcal{S}$ to the abstract semantics. For this end, we introduce, as a parameter of our analysis, a function $\mathcal{S}^\sharp$ between the set $\mathcal{Q}_R^\sharp \times T_{R/\varepsilon}^{\sharp\star}$ and $\wp(\llbracket 1,n \rrbracket)$ such that for any concrete trace $\tau \in T_{\mathcal{R},\mathcal{Q}_{R,0},\mathcal{S}}^{\text{TIME}}$ and any transition $(q,r,q') \in T_R$ that satisfy (i) $\text{final}(\tau) = q$, (ii) $\beta^s(q) \neq \beta^s(q')$, and (iii) $r \in \mathcal{S}(\tau)$, we have $r \in \mathcal{S}^\sharp(\beta^t(\tau))$. Intuitively, a reaction r is in the set $\mathcal{S}^\sharp(\tau^\sharp)$ whenever our approximation fails in proving that no trace $\tau \in \gamma^t(\tau^\sharp)$ can be continued by applying the reaction r while changing the sampling interval of at least one chemical species. We introduce the function $\mathbb{F}_{\mathcal{Q}_{R,0},\mathcal{S}^\sharp}^{\text{TIME}\,\sharp}$ over $\wp(\mathcal{Q}_R^\sharp \times T_{R/\varepsilon}^{\sharp\star})$ that maps each subset Y of $\mathcal{Q}_R^\sharp \times T_{R/\varepsilon}^{\sharp\star}$ to the subset $\alpha^s(\mathcal{Q}_{R,0}) \cup \{\tau^\sharp \frown (q^\sharp,r,q^{\sharp\prime}) \mid \tau^\sharp \frown (q^\sharp,r,q^{\sharp\prime}) \in \mathbb{F}_{\mathcal{Q}_{R,0}}^\sharp(Y) \ \wedge \ r \in \mathcal{S}^\sharp(\tau^\sharp)\}$. The function $\mathbb{F}_{\mathcal{Q}_{R,0},\mathcal{S}^\sharp}^{\text{TIME}\,\sharp}$ is monotonic and satisfies $[\alpha^t \circ \mathbb{F}_{\mathcal{Q}_{R,0},\mathcal{S}}^{\text{TIME}} \circ \gamma^t](Y) \subseteq \mathbb{F}_{\mathcal{Q}_{R,0},\mathcal{S}^\sharp}^{\text{TIME}\,\sharp}(Y)$, for any subset $Y \subseteq \mathcal{Q}_R^\sharp \times T_{R/\varepsilon}^{\sharp\star}$. By [5], it follows that our approach is sound:

Theorem 4. (Soundness). *The function* $\mathbb{F}_{\mathcal{Q}_{R,0},\mathcal{S}^\sharp}^{\text{TIME}\,\sharp}$ *has a least fixpoint. Moreover, we have:* $\text{lfp}\,\mathbb{F}_{\mathcal{Q}_{R,0},\mathcal{S}}^{\text{TIME}} \subseteq \gamma^t(\text{lfp}\,\mathbb{F}_{\mathcal{Q}_{R,0},\mathcal{S}^\sharp}^{\text{TIME}\,\sharp})$.

Let us instantiate our framework. For the sake of simplicity, we opt for the assumptions used in logical modelling, all the more so since there are no large time relaxation of fast cycles in our example. To each integer $r \in \llbracket 1,n \rrbracket$, we associate a kinetic function k_r between the set $\mathcal{Q}_R$ and the set $\wp(\mathbb{R}^+) \setminus \{\varnothing\}$. The set $k_r(q)$ denotes the potential propensity of the reaction indexed by r in the state q according to the (maybe partial) information that we may have about the rate of this reaction. The separation between time scales is encoded by a subset Sep of $(\mathbb{R}^+)^2$ satisfying: (i) for any $(x,y) \in Sep$, $x < y$; (ii) for any $x,y,x',y' \in \mathbb{R}^+$, if $(x,y) \in Sep$, $x' \leq x$, and $y \leq y'$, then $(x',y') \in Sep$. Intuitively, a pair (x,y) belongs to the set Sep when the value y is much greater than the value x. We define the concrete scheduler $\mathcal{S}$ as the function mapping each pretrace $\tau \in \mathcal{Q}_R \times T_R^\star$ to the set of the reactions r such that for all reactions r', we have $k_r(\text{final}(\tau)) \times k_{r'}(\text{final}(\tau)) \notin Sep$, meaning that the reaction r may be fast enough to exclude preemption by any other reaction.

In Property 4, we abstract away the dependency with respect to the concrete state $\text{final}(\tau)$ so as to get an effective instantiation for the parameter $\mathcal{S}^\sharp$.

Property 4. Let $(q^\sharp, r)$ be a pair in $\mathcal{Q}_R^\sharp \times [\![1,n]\!]$. For any integer $r' \in [\![1,n]\!]$, we denote by $k_{r'}^\sharp(q^\sharp)$ the set of real values $\alpha^v(\bigcup\{k_{r'}(q) \mid q \in \gamma^s(\{q^\sharp\})\})$. If $(max(k_r^\sharp(q^\sharp)), min(\bigcup\{k_{r'}^\sharp(q^\sharp) \mid r' \in [\![1,n]\!]\})) \in Sep$, then for any integer $r' \in [\![1,n]\!]$ and any state $q \in \gamma^s(\{q^\sharp\})$, we have $(max(k_r(q)), min(k_{r'}(q))) \in Sep$.

6   Conclusion

We have designed a formal framework to derive qualitative dynamical models from reaction networks. The results of the analysis of our case study is detailed in an extended version [1]. Interestingly, we can capture the sequestration effect: we can prove that when the number of instances of the protein B is very high (level 4) and those of the proteins A and C are low (level 2) in the initial state, then the number of instances of the complex ABC remains very low (levels 0, 1).

Our methodology offers a new trade-off between complexity and accuracy. It captures interesting properties that are beyond the scope of purely qualitative abstractions [6] and avoids the integration of numerical equations [10,13]. Our framework is purely formal and provides a better understanding of the qualitative modelling process, by clarifying the underlying assumptions. Interestingly, we notice that our approach often requires more intervals than in tropical approaches [13]. This is not so surprising, since in tropical approaches two consecutive intervals are assumed to be infinitely far from each other, whereas in our approach they contain arbitrarily close elements. One current limitation of our method is that we use one variable per chemical species, leading to a combinatorial explosion of the dynamics as the model size increases. To cope with this limitation, we plan to extend our framework to the reduced reaction networks obtained by the fragmentation [7,8] of the models written in the kappa language.

References

1. http://www.di.ens.fr/~feret/publications/CMSB2015/
2. Bourdoncle, F.: Abstract interpretation by dynamic partitioning. J. Funct. Prog. **2**(4), 407–435 (1992)
3. Chaouiya, C.: Petri net modelling of biological networks. Brief. Bioinfo. **8**(4), 210–219 (2007)
4. Chapman, S.A., Asthagiri, A.R.: Quantitative effect of scaffold abundance on signal propagation. Mol. Syst. Biol. **5**(1), 313 (2009)
5. Cousot, P., Cousot, R.: Abstract interpretation: a unified lattice model for static analysis of programs by construction or approximation of fixpoints. In: POPL (1977)
6. Fages, F., Soliman, S.: Formal cell biology in biocham. In: Bernardo, M., Degano, P., Zavattaro, G. (eds.) SFM 2008. LNCS, vol. 5016, pp. 54–80. Springer, Heidelberg (2008)
7. Feret, J., et al.: Internal coarse-graining of molecular systems. PNAS **106**, 6453–6458 (2009)

8. Feret, J., et al.: Stochastic fragments: a framework for the exact reduction of the stochastic semantics of rule-based models. IJSI **7**(4), 527–604 (2013)
9. Grieco, L., et al.: Integrative modelling of the influence of MAPK network on cancer cell fate decision. PLoS Comput. Biol. **9**(10), e1003286 (2013)
10. de Jong, H., et al.: Qualitative simulation of genetic regulatory networks using piecewise-linear models. Bull. Math. Biol. **66**(2), 301–340 (2004)
11. Kleene, S.C.: Introduction to Mathematics. ISHI Press International, New York (1952)
12. Niarakis, A., et al.: Computational modeling of the main signaling pathways involved in mast cell activation. Curr. Top. Microbio. Immun. **382**, 69–93 (2014)
13. Radulescu, O., et al.: Reduction of dynamical biochemical reactions networks in computational biology. Front. Genet. **3**, Article 131, 17 p. (2012)
14. Rival, X., Mauborgne, L.: The trace partitioning abstract domain. TOPLAS **29**(5), Article 26, 51 p. (2007)
15. Schuster, S., Höfer, T.: Determining all extreme semi-positive conservation relations in chemical reaction systems: a test criterion for conservativity. J. Chem. Soc. **87**, 2561–2566 (1991)
16. Tarski, A.: A lattice-theoretical fixpoint theorem and its applications. Pacific J. Math. **5**(2), 285–309 (1955)

Model-Based Investigation of the Effect of the Cell Cycle on the Circadian Clock Through Transcription Inhibition During Mitosis

Pauline Traynard[✉], François Fages, and Sylvain Soliman

Inria Paris-Rocquencourt, Team Lifeware, Rocquencourt, France
pauline.traynard@inria.fr

Abstract. Experimental observations have put in evidence autonomous self-sustained circadian oscillators in most mammalian cells, and proved the existence of molecular links between the circadian clock and the cell cycle. Several models have been elaborated to assess conditions of control of the cell cycle by the circadian clock, in particular through the regulation by clock genes of Wee1, an inhibitor of the mitosis promoting factor, responsible for a circadian gating of mitosis and cell division period doubling phenomena. However, recent studies in individual NIH3T3 fibroblasts have shown an unexpected acceleration of the circadian clock together with the cell cycle when the milieu is enriched in FBS, the absence of such acceleration in confluent cells, and the absence of any period doubling phenomena. In this paper, we try to explain these observations by a possible entrainment of the circadian clock by the cell cycle through the inhibition of transcription during mitosis. We develop a differential model of that reverse coupling of the cell cycle and the circadian clock and investigate the conditions in which both cycles are mutually entrained. We use the mammalian circadian clock model of Relogio et al. and a simple model of the cell cycle by Qu et al. which focuses on the mitosis phase. We show that our coupled model is able to reproduce the main observations reported by Feillet et al. in individual fibroblast experiments and use it for making some predictions.

1 Introduction

Most organisms, from bacteria to plants and animals, have a circadian clock present in each cell, generally in the form of a self-sustained genetic oscillator entrained by the day/night cycle through various mechanisms. This circadian clock has many effects on the cell including its metabolism [13]. Experimental results have also shown a regulation of the cell division cycle by the circadian clock [2,16,23], with possible applications to cancer chronotherapies [1,7]. Molecular links between these two cycles have been exhibited to explain this regulation. In particular the regulation of Wee1, an inhibitor of the mitosis promoting factor, by the clock genes, induces a circadian gating of mitosis to particular clock

© Springer International Publishing Switzerland 2015
O. Roux and J. Bourdon (Eds.): CMSB 2015, LNBI 9308, pp. 208–221, 2015.
DOI: 10.1007/978-3-319-23401-4_18

phases and can result in a synchronization of cell division with a 24 h period or 48 h period with period doubling phenomena [8]. Other similar molecular links going in the same direction, through p21 [14] and cMyc [17], have been shown in the literature. A few models have also been developed to further investigate those hypotheses, by coupling a model of the cell cycle with a model of the circadian clock through those direct molecular links, and analyze the conditions of entrainment in period [6,12].

Several studies using large-scale time-lapse microscopy to monitor circadian gene expression and cell division events in real time and in individual cells during several days have unveiled unexpected behaviours, hinting that the relationship might be more complex. Nagoshi et al. [8] have first shown that circadian gene expression in fibroblasts continues during mitosis, but with a consistent pattern in circadian period variation relatively to the circadian phase at division, leading them to hypothesize that mitosis elicits phase shifts in circadian cycles. However, a more recent study of Bieler et al. [3] relating the same experiments on dividing fibroblasts found the two oscillators synchronized in 1:1 mode-locking leading the authors to hypothesize a predominant reverse coupling in NIH3T3 cells. This is in agreement with another study of Feillet et al. [11] which found several different synchronization states in NIH3T3 fibroblasts in different conditions of culture. It was observed there that enriching the milieu with Foetal Bovine Serum (FBS) not only accelerates the cell division cycle but also the circadian clock. For cells cultured in 10 % FBS, both distributions of the cell cycle length and the circadian clock period are centered around 22 h. For cells cultured in 15 % FBS, both the cell cycle and the circadian clock accelerate, with period distributions centered around 19 h. However, when cells reach confluence and stop dividing, the circadian clock slows down and the period distribution is then centered around 24 h. None of the currently available models coupling the cell cycle and the circadian clock can explain these observations since they are based on an unidirectional influence of the circadian clock on the cell cycle [6,12].

In this paper, we hypothesize that the inhibition of transcription during mitosis in eukaryotes [24] constitutes a reverse interaction from the cell cyle to the circadian clock, which can enable an entrainment of the circadian clock by the cell cycle and can explain the acceleration of the circadian clock in non-confluent cells when the concentration of FBS increases. We develop a differential model of this reverse coupling from the cell cycle to the circadian clock and investigate the conditions in which both cycles are mutually entrained. We use the mammalian circadian clock model of Relogio et al. [19] and a simple model of the cell cycle by Qu et al. [18] which focuses on the mitosis phase. We show that our coupled model is able to reproduce the observations on periods reported by Feillet et al. [11] in individual fibroblast without treatment by Dexamethasone. Furthermore we argue that the complex behaviors observed with high variability after treatment by Dexamethasone, modeled by the induction of a high level of Per and the inhibition of the other clock core genes, can be explained by the perturbation of the clock after this treatment. In our model, the stabilization time after that pulse appears to be greater than the time horizon used in those experiments. Our results are thus compatible with the observations on the periods and phase locking modes of [11], however, the observations on the precise phase shift between the mitosis time and the circadian clock REV-ERB-α protein peaks reported in [11] are not reproduced by our model, nor are they by the other coupled models of [6,12]. This intriguing remaining difficulty is discussed at the end of the paper.

The methodology used to perform these investigations is based on a formal specification of the observed behavior with temporal logic patterns [10,22] which are used in the BIOCHAM modeling environment [5] for parameter search [21] and robustness and sensitivity analysis [20][1].

2   Experimental Observations and Their Specification in Temporal Logic

2.1   Experimental Data

In this section we explain the experiments and analysis performed in [11] and the conclusions drawn by the authors. The reported experiments have been done using cell tracking and time-lapse image analysis of various fluorescent markers of the cell cycle and the circadian clock observed during 72 h in proliferating NIH3T3 mouse fibroblasts.

These cells were modified to include three fluorescent markers of the circadian clock and the cell cycle: the REV-ERB-α::Venus clock gene reporter for measuring the expression of the circadian protein REV-ERB-α, and the Fluorescence Ubiquitination Cell Cycle Indicators (FUCCI), Cdt1 and Geminin, two cell cycle proteins which accumulate during the G1 and S/G2/M phases respectively, for measuring the cell cycle phases.

The cells were left to proliferate in regular medium supplemented with different concentrations of FBS (10 %, 15 % and 20 %). Long-term recording was performed in constant conditions with one image taken every 15 min during 72 h. Cell division times were also measured during the tracking of cell lineages. Cell cycle length was measured as the time interval between two consecutive cell divisions and a piece-wise linear model fitted to both markers of the cell cycle extracted the time of the G1-S transitions.

The expression traces of REV-ERB-α were detrended and smoothed, and spectrum resampling was used to estimate the clock period. Cells with less than two REV-ERB-α peaks within their lifetime, a period length outside the interval between 5 h and 50 h or a relative absolute error (RAE) bigger than 0.25 (showing a confidence interval wider than twice the estimated period) were classified as non-rhythmic and discarded, assuming that they do not have a functioning clock.

Furthermore, a series of experiments were done with a pulse of Dexamethasone (Dex) before recording. This glucocorticoid agonist is known to exert a resetting/synchronizing effect on the circadian molecular clocks in cultured cells

---

[1] The models and the specification used in this paper are available on http://lifeware. inria.fr/wiki/software/cmsb15.

through the induction of *Per1*. In that case the cells were incubated for 2 h in the same medium supplemented with Dex, just before returning to a Dex-free medium for the recording.

The quantitative data on the periods of the cell cycle and the circadian clock are summarised in the Table 1 [11]. Cells non-treated with dexamethasone show a similar period for the cell cycle and the circadian clock both in 10 % and 15 % FBS conditions. Interestingly, increasing FBS significantly decreases both mean periods of the clock and the cell cycle, from 21.9 h to 19.4 h and from 21.3 h to 18.6 h respectively, showing that both oscillators remain unexpectedly in 1:1 mode locking. While the speedup of the cell cycle can be directly attributed to the growth factors in increasing concentration of FBS, it can not account for the speedup of the clock the same way, since confluent cells keep a 24-h period for the circadian clock independently of the FBS concentration.

Table 1. Estimated periods of the circadian molecular clock and the cell division cycle measured in [11] in fibroblast cells for various concentrations of FBS, with and without dexamethasone. The experiment done with 20 % FBS have been clustered by the authors of [11] in two groups with different periods.

Medium	No dexamethasone		Dexamethasone	
	Clock period	Division period	Clock period	Division period
FBS 10	21.9 h ± 1.1 h	21.3 h ± 1.3 h	24.2 h ± 0.5 h	20.1 h ± 0.94 h
FBS 15	19.4 h ± 0.5 h	18.6 h ± 0.6 h	NA	NA
FBS 20	NA	NA	21.25 h	19.5 h
FBS 20	NA	NA	29 h	16 h

The results are more complex in the case of the cells treated with dexamethasone. Cells in 10 % FBS show an increased clock period and a low cell cycle period, with an overall ratio of 5:4. In 20 % FBS the cell lineages are dominated by two groups. The first group shows close periods, i.e. a 1:1 mode-locking similarly to the experiments without dexamethasone. The second group shows a high clock period and a fast cell cycle, with an overall ratio close to 3:2 between the clock and cell cycle, explaining the three-peaks distribution of the circadian phase, as already observed by Nagoshi et al. [8] ten years before. It has to be noted that the 20 % FBS dexamethasone-synchronized experiment was repeated with similar results available in the Supplementary Information of [11], although the distribution of the period ratios for the second group is wider in the interval [1.2 − 2].

In [11], the authors suggest that these observations might be interpreted by the existence of distinct oscillatory stable states coexisting in the cell populations, in particular with 5:4 and 1:1 phase-locking modes for the condition 10 % FBS, and 3:2 and 1:1 phase-locking modes for the condition 20 % FBS, and that the dexamethasone could knock the state out of the 1:1 mode toward other attractors. A mechanistic explanation remains to be found to support this interpretation. In this paper, we investigate a simpler hypothesis of entrainment of the circadian clock by cell divisions through the inhibition of transcription during mitosis and show with a model that this hypothesis can explain the observations on the periods.

2.2   Temporal Logic Specification

For the analysis of the dynamical behavior of the system, we rely on the formalisation of the oscillatory properties in quantitative temporal logic with simple formula patterns [10,22], which allow us to combine qualitative properties of oscillations and quantitative properties on the shapes of the traces such as distances between peaks or peak amplitudes. This is useful to capture the periods on both experimental and simulated traces, even when the traces are noisy. We use flexible constraints on the amplitudes and regularity of the peaks to filter out traces, keeping only sustained oscillations even with small irregularities, as it is the case for example on the Fig. 7.

For instance, the following formula is used to compute the period of REV-ERB-α in a trace:

```
distanceSuccPeaks([RevErb::nucl],[period],[80]) &
Exists([maxdiff1,maxdiff2,maxpeak],
maxDiffDistancePeaks([RevErb::nucl],[maxdiff1],[80])
& maxDiffAmplPeaks([RevErb::nucl],[maxdiff2],[80])
& maxAmplPeaks([RevErb::nucl],[maxpeak],[80])
& 4*maxdiff1<period+errordiff1
& 10*maxdiff2<maxpeak+errordiff2
& maxpeak>0.1+errorampl)
```

The period constraint on the oscillations of REV-ERB-α is expressed by the formula pattern *distanceSuccPeaks*, whose validity domain provides all the values of the distances between peaks of concentrations of REV-ERB-α [10], after a transient time of 80 h to avoid irregularities caused by the initial state.

Moreover, the formula patterns *maxDiffDistancePeaks*, *maxDiffAmplPeaks*, and *maxAmplPeaks* capture several variables characterizing irregularity features of the trace: *errordiff1* for the irregularities in distances between peaks, *errordiff2* for the irregularities in the amplitudes of the peaks, and *errorampl* for a small concentration amplitude. Setting then thresholds on these variables ensures that unwanted traces are filtered out.

These logical formulae can then be used in a modeling environment such as BIOCHAM [5] in a variety of ways for data analysis [9], model parameter search in high dimension and robustness and sensitivity measures [4,20,21].

3   Mathematical Models and Their Coupling

3.1   Model of the Cell Cycle

The cell cycle of somatic cells is composed of four phases: DNA replication (S phase) and chromosome segregation or mitosis (M phase), separated by two gap phases (G1 and G2). At the center of the cell cycle regulation, there is a group of proteins, the cyclin-dependent kinases, which are complexes composed of a kinase and a cyclin partner determining the specificity of the complex. Each phase of the cell cycle is controled by a specific cylin-dependent kinase.

For our purpose, it is sufficient to use a model focusing on the G2-M transition which leads to mitosis. We use a model proposed by Qu et al. [18] in which the cell cycle is divided in two different phases, the G1-S-G2 and M phases. The M phase is triggered by the complex CDC2/CYCLIN B. This complex appears in two forms, an active form called MPF (M-phase Promoting Factor) and a phosphorylated, inactive form called preMPF. MPF is phosphorylated and inactivated by the kinase WEE1, and dephosphorylated and activated by the phosphatase CDC25. Both the kinase and phosphatase activities are themselves regulated by MPF, respectively inactivated and activated by the complex (Fig. 1).

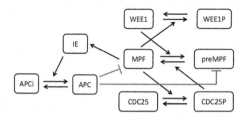

Fig. 1. Schema of the cell cycle model of Qu et al. [18].

The mechanism by which changing the concentration of FBS modulates the cell cycle length is unclear, and probably involves an increase in growth factors. In this model, we assessed the effect of each reaction rate constant on the period of the concentrations and found that two parameters were particularly significant to change the period: *kdie*, the degradation rate of the intermediary enzyme involved in the negative feedback loop between MPF and the proteasome APC, and *kampf*, the activation of MPF by CDC25P. Both are able to change widely the range of the cell cycle period without changing significantly the strength of the coupling, and should thus provide the same effect, so we choose one of them, *kampf*, to modulate the cell cycle period. We shall use the following values for *kampf*: 3.75 for a cell division period of 21.3 h (corresponding to 10 % FBS), 12.1 for a period of 18.6 h (15 % FBS).

More detailed models distinguishing the four phases of the cell cycle of course exist, such as [12] for instance, making possible to represent various regulations from the circadian clock genes, for instance through WEE1 during M-phase and through p21 and Cmyc during the S-phase. However, since the consequences of those regulations have not been observed in the experimental data described in the previous section, we concentrate here on the reverse effect of the cell cycle on the circadian clock by transcription inhibition during mitosis, for which the simpler two phase model of Qu et al. [18] is sufficient.

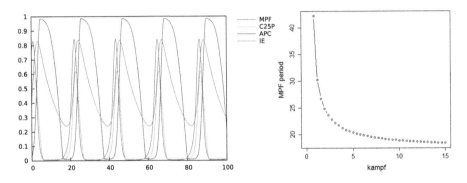

Fig. 2. Left: Simulation of the cell division cycle model of Qu et al. **Right**: Period of the cell division cycle (measured as the distance between successive peaks of MPF) as a function of the parameter *kampf* for MPF activation by CDC25P in the model of Qu et al.

3.2   Models of the Circadian Clock

In many organisms, spontaneous gene expression oscillations with a period close to 24 h have been observed. A biochemical clock present in each cell is responsible for maintaining these oscillations at this period. The central circadian clock in the suprachiasmatic nucleus (SCN) is sensitive to light and entrained by the day-night alternation, allowing molecular clocks in peripheral tissues to be synchronised by central signals. Indeed, Schibler and Nagoshi [8] have shown that in absence of synchronisation by the central clock, autonomous circadian oscillators are maintained in peripheral tissues with the same period, although they are progressively desynchronized.

In mammalian cells, two major proteins are transcribed in a circadian manner, CLOCK and BMAL1 which bind to form a heterodimer responsible for the transcription of several genes involved in intertwined feedback loops such as *Per* (Period), *Cry* (Cryptochrome), *Rev-Erb-α* or *Ror*. The newly-formed proteins then affect their own synthesis as PER and CRY associates to inhibit the activity of the complex CLOCK/BMAL1. REV-ERB-α has a similar effect and these two negative feedback loops give rise to sustained oscillations. Moreover, two positive feedback loops provided by the activation of *Bmal1* by ROR and the activation of *Cry* by REV-ERB-α are believed to bring more robustness to the oscillator.

In this paper we use the circadian clock model of Relogio et al. [19] which has been fitted on suprachiasmatic cells with precise data on the amplitude and phases of the different components. This model is composed of 20 species, 71 parameters, and all the feedback loops described above.

3.3   Coupling from the Cell Cycle to the Circadian Clock by the Inhibition of Transcription During Mitosis

It is known that in eukaryotes, gene transcription is significantly inhibited during mitosis [24]. In particular, the transcription inhibition of clock genes during

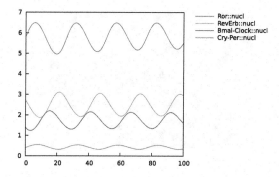

Fig. 3. Simulation of the Circadian Clock model of Relogio et al.

mitosis and its impact on the circadian oscillator by shifting the phase of the circadian cycle has been shown in [15].

In this paper, we model the inhibition of clock genes transcription during mitosis with a negative Hill kinetics for mRNA synthesis taking the ratio between the concentrations of MPF and preMPF as inhibiting factor. The kinetics of mRNA synthesis reactions are thus modified as follows

$$S * \frac{J^n}{J^n + ([MPF]/[preMPF])^n}$$

where S is the original synthesis rate parameter in the model of Relogio et al. [19] and n is taken equal to 4 to mimic the abrupt inhibition of transcription when mitosis occurs. Transcription is thus inhibited when the ratio $[MPF]/[preMPF]$ is high.

This modelling enforces the fact that for quiescent cells, whatever the FBS concentration, the transcription rate will be close to S and therefore the clock close to a period of 24 h.

4   Computational Results

As shown in the right panel of Fig. 2, it is possible to simulate the experimental milieu enrichment with 10 or 15 % FBS by varying the parameter *kampf* of the cell cycle model to obtain the same values for the period of the cell division cycle. The coupling of this model to the circadian clock uses two parameters: the coupling strength J, and the order n of the Hill function. In the results reported in this section, we chose $J = 2$ and $n = 4$, two of the smallest values found through our parameter search procedure.

Figure 4 shows the variation of the period of *REV-ERB-α* when the two parameters *kampf* and J vary. The value of the period is captured with a temporal logic specification as seen in the Subsect. 2.2. Two domains can be distinguished in this parameter space: in the domain on the top left (above the black line) the clock keeps its period constant and close to 24 h, thus it is not entrained. On the

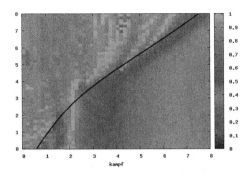

Fig. 4. Periods of REV-ERB-α as a function of *kampf* for varying the cell cycle period, *J*, the strength of transcription inhibition during mitosis. Landscape computed as the satisfaction degree of the formula `distanceSuccPeaks` (`[RevErb::nucl]`,`[period]`,`[transT]`) which defines the period of REV-ERB-α after a transient time transT = 80, and with an objective of 24 h for the *period*. Full satisfaction in yellow indicates a period of 24 h for REV-ERB-α, while the other colours indicate the absolute difference to 24 h (Colour figure online).

Table 2. Measured in the coupled model with different values of *kampf* for modeling the different culture conditions.

kampf	FBS %	Circadian clock period (h)	Cell division period (h)
3.75	10	21.43	21.30
12.1	15	18.60	18.60
1.6	5?	26.16	26.32

contrary, in the domain on the bottom right (below the black line) the clock is entrained to the same period as the cell cycle. One can see that using a different value for *J* would have led to different values for *kampf* in Table 2.

4.1   Comparison to Experimental Data Without Dexamethasone

Table 2 shows the periods of the circadian clock and the cell division cycle in our model with different values of *kampf* corresponding to the different culture conditions. In all cases, the cell division manages to entrain the circadian clock (that has a free period around 24 h) to its period, simply through this mechanism of transcription inhibition, as depicted in Figs. 5 and 6 left panel. These simulation results reproduce quite well the data of Table 1 when there is no treatment by Dex. Note that our model can also have a cell division time higher than 24 h, for instance with *kampf* = 1.6 which might correspond to a concentration of FBS around 5 %. In that case we predict that the cell cycle will still entrain the circadian clock, lowering its period, even if our simulations show a longer transitory period, as depicted in Fig. 6 (right panel).

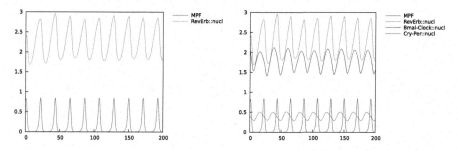

Fig. 5. Entrainment to a period around 21.3 h with $kampf = 3.75$ corresponding to a milieu enriched with 10 % FBS. The same simulation including more clock genes is shown on the right to compare with Fig. 3.

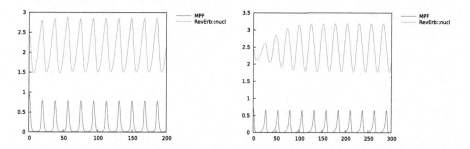

Fig. 6. Left: Entrainment to a period around 18.6 h with $kampf = 12.1$ corresponding to a milieu enriched with 15 % FBS. **Right:** Entrainment to a period around 26.3 h with $kampf = 1.6$ corresponding to a poor milieu (FBS 5 %?), as predicted by our model. Note that the Circadian clock takes more time to adjust to this lengthening of its period.

4.2 Comparison to Data with Dexamethasone

In order to take into account the experiments with Dexamethasone, the model can be extended with an event, lasting for two hours, and inducing *Per* mRNA while inhibiting the other clock genes.

Figure 7 shows that in our models, regardless of the milieu (i.e. of the value of *kampf*), the Dex pulse results in a perturbation of the clock and then returns to the observed entrainment.

These simulations point us to the possibility that the noisy data reported in Table 1 after the Dex pulse might simply be due to the various states in which the pulse happened and to the time necessary for the cells to recover their clock entrainment, rather than to two different oscillatory attractors of the system.

A pulse at time 190 h disrupted only slightly our clock, leading to mostly remaining in mode-locking 1:1, whereas postponing that same pulse by 10 h (corresponding to giving the pulse to a cell in a different state) leads to a bigger disturbance, some peak-to-peak distances close to 24 h, others to 17 h, and even

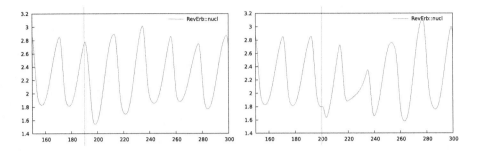

Fig. 7. Effect of a Dexamethasone pulse on the entrainment. The pulse alters the clock before returning to the previously observed entrainment regime. In the left panel the pulse is from time 190 to 192 while on the right it is from 200 to 202. The left panel's peak-to-peak distance remains in the [20.25, 22.3] interval, while the right one is in the [17.9, 24.1] interval. This might correspond to the two groups observed in [11]. The time to recover normal entrainment varies but is often larger than 72 h.

if this is transitory, this might correspond to the type of data observed in the Group 2 of Table 1.

4.3   Remaining Paradox on Phase Data

So far we have considered the periods of the circadian clock and the cell division cycle, but not their phase. The experimental data on the phase between the cell division time and the peak of REV-ERB-α protein in NIH3T3 fibroblast are quite consistent in Bieler et al. [3] and Feillet et al. [11] to indicate that the REV-ERB-α occurs 3–5 h after cell division. However this is not the case in our coupled model where the peak of REV-ERB-α appears 17–20 h after cell division, as shown in Table 3.

Table 3. Phases as time delays, observed experimentally (without Dexamethasone) and by simulation, between the cell division time (peak of MPF in the simulations) and the appearance of a peak of concentration of REV-ERB-α.

Medium	Experimental data	Model simulation
FBS 5	NA	18.6 h
FBS 10	3.82 h	20.7 h
FBS 15	3.98	17.8 h

Interestingly, a similar discrepancy appears in the model of Gerard and Goldbeter [12] which models the reverse effects of the circadian clock genes on the cell cycle, through Wee1, p21 and Cmyc, and shows mitosis gating. In their simulations, the peak of REV-ERB-α appears around 16 h after cell division. We do not have explanations for these discrepancies between the computational models and the recent data which now permit to fit the models in phase in addition to periods.

In the circadian clock model of Relogio et al. [19], the phases of the different markers of the circadian clock have been precisely fitted to observations made in mice suprachiasmatic nucleus cells, however without data about cell divisions. On the other hand, in the data of Feillet et al. [11], REV-ERB-α is the only marker on the circadian clock and no comparison is thus possible with the other data.

5   Conclusion

Through a simple model for the transcription inhibition during mitosis, we have presented in this article the first mechanistic dynamical model demonstrating the entrainment of the circadian clock by the cell division cycle. This model has been built on the ideas of [3] that the primary coupling between those two oscillators is from the cell division cycle to the circadian clock.

We have demonstrated that such a model is enough to reproduce the recently published biological data of [11] with different medium enrichment leading to different periods for mode-locked oscillators in dividing cells, whereas the quiescent cells still have a 24 h clock. Our model also postulates a different interpretation of some of the results of that article when cells are treated by a 2 h pulse of Dexamethasone: instead of different autonomous cycling regimes, the model predicts temporary perturbations leading to shorter or longer peak-to-peak distances, but returning to the previous entrainment regime after some time, longer than the horizon used in the experiments.

It is noteworthy that in our transcription-inhibition coupled models, the oscillations of the clock's core gene products are much sharper and their peaks closer in time (see for instance Figs. 3 and 5 right panel). Indeed, the peaks get "concentrated" outside of the time of transcription inhibition. A prediction of the model is therefore that in quickly dividing cells, the phase shifts between the different components of the clock are shorter than in quiescent cells where such a phenomenon should not occur.

Finally, though our rather simple model properly fits the data about the periods of the different cycles, the time difference observed between the peaks of MPF and of REV-ERBα is quite different in our model and in the experimental data. A similar discrepancy seems to also appear in the coupled model of [12]. More work is needed now to try to fit these models to the available phase data and probably create new data with several markers of the circadian clock in addition to cell division time.

Acknowledgements. This work has been funded by the ERASysBio+ project C5Sys through the ANR grant 2009-SYSB-002-02. We would like to thank all the partners of this project for fruitful discussions. This work was also granted access to the HPC resources of the CINES under the allocation c2015036437 made by GENCI.

References

1. Ballesta, A., Dulong, S., Abbara, C., Cohen, B., Okyar, A., Clairambault, J., Levi, F.: A combined experimental and mathematical approach for molecular-based optimization of irinotecan circadian delivery. PLOS Comput. Biol. **7**(9), e1002143 (2011)
2. Barnes, J.W., Tischkau, S.A., Barnes, J.A., Mitchell, J.W., Burgoon, P.W., Hickok, J.R., Gillette, M.U.: Requirement of mammalian timeless for circadian rhythmicity. Science **302**(5644), 439–442 (2003)
3. Bieler, J., Cannavo, R., Gustafson, K., Gobet, C., Gatfield, D., Naef, F.: Robust synchronization of coupled circadian and cell cycle oscillators in single mammalian cells. Mol. Syst. Biol. **10**(7), 739 (2014)
4. Calzone, L., Chabrier-Rivier, N., Fages, F., Soliman, S.: Machine learning biochemical networks from temporal logic properties. In: Priami, C., Plotkin, G. (eds.) Transactions on Computational Systems Biology VI. LNCS (LNBI), vol. 4220, pp. 68–94. Springer, Heidelberg (2006)
5. Calzone, L., Fages, F., Soliman, S.: BIOCHAM: An environment for modeling biological systems and formalizing experimental knowledge. Bioinformatics **22**(14), 1805–1807 (2006)
6. Calzone, L., Soliman, S.: Coupling the cell cycle and the circadian cycle. Research Report 5835, INRIA, February 2006
7. De Maria, E., Fages, F., Rizk, A., Soliman, S.: Design, optimization, and predictions of a coupled model of the cell cycle, circadian clock, dna repair system, irinotecan metabolism and exposure control under temporal logic constraints. Theor. Comput. Sci. **412**(21), 2108–2127 (2011)
8. Emi, N., Camille, S., Christoph, B., Thierry, L., Felix, N., Schibler, U.: Circadian gene expression in individual fibroblasts: cell-autonomous and self-sustained oscillators pass time to daughter cells. Cell **119**, 693–705 (2004)
9. Fages, F., Rizk, A.: On temporal logic constraint solving for the analysis of numerical data time series. Theor. Comput. Sci. **408**(1), 55–65 (2008)
10. Fages, F., Traynard, P.: Temporal logic modeling of dynamical behaviors: first-order patterns and solvers. In: del Cerro, L.F., Inoue, K. (eds.) Logical Modeling of Biological Systems, Chapter 8, pp. 291–323. Wiley, New York (2014)
11. Feillet, C., Krusche, P., Tamanini, F., Janssens, R.C., Downey, M.J., Martin, P., Teboul, M., Saito, S., Lévi, F.A., Bretschneider, T., van der Horst, G.T.J., Delaunay, F., Rand, D.A.: Phase locking and multiple oscillating attractors for the coupled mammalian clock and cell cycle. Proc. Nat. Acad. Sci. U.S.A **111**(27), 9833–9928 (2014)
12. Gérard, C., Goldbeter, A.: Entrainment of the mammalian cell cycle by the circadian clock: modeling two coupled cellular rhythms. PLoS Comput. Biol. **8**(21), e1002516 (2012)
13. Glass, L.: Synchronization and rhythmic processes in physiology. Nature **410**(6825), 277–284 (2001)
14. Gréchez-Cassiau, A., Rayet, B., Guillaumond, F., Teboul, M., Delaunay, F.: The circadian clock component bmal1 is a critical regulator of p21WAF1/CIP1 expression and hepatocyte proliferation. J. Biol. Chem. **283**, 4535–4542 (2008)
15. Kang, B., Li, Y.-Y., Chang, X., Liu, L., Li, Y.-X.: Modeling the effects of cell cycle m-phase transcriptional inhibition on circadian oscillation. PLoS Comput. Biol. **4**(3), e1000019 (2008)

16. Matsuo, T., Yamaguchi, S., Mitsui, S., Emi, A., Shimoda, F., Okamura, H.: Control mechanism of the circadian clock for timing of cell division in vivo. Science **302**(5643), 255–259 (2003)
17. Perez-Roger, I.: Myc activation of cyclin e/cdk2 kinase involves induction of cyclin e gene transcription and inhibition of p27(kip1) binding to newly formed complexes. Oncogene **14**(20), 2373–81 (1997)
18. Qu, Z., MacLellan, W.R., Weiss, J.N.: Dynamics of the cell cycle: checkpoints, sizers, and timers. Biophys. J. **85**(6), 3600–3611 (2003)
19. Relógio, A., Westermark, P.O., Wallach, T., Schellenberg, K., Kramer, A., Herzel, H.: Tuning the mammalian circadian clock: robust synergy of two loops. PLoS Comput. Biol. **7**(12), e1002309 (2011)
20. Rizk, A., Batt, G., Fages, F., Soliman, S.: A general computational method for robustness analysis with applications to synthetic gene networks. Bioinformatics **12**(25), il69–il78 (2009)
21. Rizk, A., Batt, G., Fages, F., Soliman, S.: Continuous valuations of temporal logic specifications with applications to parameter optimization and robustness measures. Theor. Comput. Sci. **412**(26), 2827–2839 (2011)
22. Traynard, P., Fages, F., Soliman, S.: Trace simplifications preserving temporal logic formulae with case study in a coupled model of the cell cycle and the circadian clock. In: Mendes, P., Dada, J.O., Smallbone, K. (eds.) CMSB 2014. LNCS, vol. 8859, pp. 114–128. Springer, Heidelberg (2014)
23. Ünsal-Kaçmaz, K., Mullen, T.E., Kaufmann, W.K., Sancar, A.: Coupling of human circadian and cell cycles by the timeless protein. Mol. Cell. Biol. **25**(8), 3109–3116 (2005)
24. Weisenberger, D., Scheer, U.: A possible mechanism for the inhibition of ribosomal rna gene transcription during mitosis. J. Cell Biol. **129**, 561–575 (1995)

Analysis of a Post-translational Oscillator Using Process Algebra and Spatio-Temporal Logic

Christopher J. Banks[1]([⊠]), Daniel D. Seaton[2], and Ian Stark[3]

[1] Systems Genomics,
The Roslin Institute, University of Edinburgh, Edinburgh, UK
`c.banks@ed.ac.uk`
[2] Millar Lab, Synthetic and Systems Biology,
University of Edinburgh, Edinburgh, UK
[3] Laboratory for Foundations of Computer Science,
School of Informatics, University of Edinburgh, Edinburgh, UK

Abstract. We describe the modelling of a post-translational oscillator using a process algebra and the specification of complex properties of its dynamics using a spatio-temporal logic. We show that specifications in the Logic of Behaviour in Context can be seen as hypotheses about oscillations and other biochemical behaviours, to be tested automatically by model-checking software. By using these techniques we show that the theoretical model behaves in a manner in keeping with known properties of biological circadian oscillators.

1 Introduction

In this paper we describe the encoding of a post-translational oscillator (PTO) model in the Continuous Pi-calculus process algebra (cπ) [11,12] and the results of computational experiments made on the model including the use of a novel spatio-temporal logic, the Logic of Behaviour in Context ($\mathcal{LBC}$) [4], to specify and check complex properties of the model. The spatio-temporal logic $\mathcal{LBC}$ can be seen as a formal logical language for expressing properties of biochemical systems. There is a long-standing problem of how to express properties of oscillation in temporal logic and one contribution of this paper is to neatly define a temporal logic specification for both general and more specific oscillatory behaviour.

PTOs generate sustained oscillations in the absence of transcription and translation. Such oscillators are of particular interest in the circadian clock field, where PTOs have recently been postulated to generate endogenous 24-hour rhythms in diverse organisms [15]. Here, we investigate a minimal model of a PTO due to Jolley et al. [10]. This model has a simple structure—it consists only of a kinase, a phosphatase, and a substrate—but can exhibit robust oscillatory behaviour similar to that observed in circadian clocks.

The purpose of our study is to further examine the behavioural properties of the PTO when it is coupled with other PTOs, other reaction pathways, and inhibitors. We examine these properties using both simple computational experiments and more complicated, *higher-order* experiments defined by $\mathcal{LBC}$ properties and performed by model checking. The ultimate goal being to evaluate $\mathcal{LBC}$

© Springer International Publishing Switzerland 2015
O. Roux and J. Bourdon (Eds.): CMSB 2015, LNBI 9308, pp. 222–238, 2015.
DOI: 10.1007/978-3-319-23401-4_19

as a useful logical tool to perform these sorts of analyses and also to draw some conclusions about the behaviour of the theoretical PTO in relation to real circadian oscillators. In particular circadian clock mechanisms must interact with other systems in an organism; this includes the control of metabolic processes and coupling with the classical transcription-translation feedback loop (TTFL) circadian clocks [1]. This potential to robustly interact with other systems is, to date, unexplored for this model.

A key benefit of using process algebra here is the ability to readily build complex models of interaction through combining simpler components. This composition can be challenging, particularly where components are shared or linked; and this is important as such sharing can be source of significant new behaviours [16]. In this study we compose oscillators through shared enzymes.

The advantage of using a formal logical language to specify and check properties of the model, and its composition with other models, is that it gives us a concise and precise means of expressing the hypothesis we wish to test. Model-checking software then gives us the means to test this. This is especially true where we have a mixture of temporal and spatial behaviour we wish to test; e.g. if we wish to know if the introduction of an inhibitor (*spatial* change) has a given effect regardless of the time at which it is applied (*temporal* change).

Using these techniques we show that the Jolley PTO model does indeed exhibit some of the properties that would be expected of a biological PTO. The oscillator is robust when coupled with other oscillators, using different coupling mechanisms, and crucially when coupling at any point in the oscillation cycle. We also show that the oscillator is robust when perturbed by other simple mechanisms. Finally we demonstrate that $\mathcal{LBC}$ has the potential to specify, at least the qualitative aspects of, even more complex properties of oscillators, such as phase response—that is, how the oscillation is affected by a small perturbation at different points in its cycle.

2   The Jolley PTO Model

Jolley, Ode, and Ueda present their model as a set of coupled ODEs. In their paper [10], sets of parameters are identified which give distinct patterns of oscillation in the system. The model aims to provide a framework for analysing and synthesising PTOs and they provide evidence that it is a viable candidate for a minimal circadian clock. However, to date, little further analysis of the properties of the complex behaviour of this oscillator has been done.

The model arose from the observation that PTOs and other oscillatory systems which exist in nature are commonly mediated by multi-site phosphorylation, these include evidence from observations and existing models of the KaiC circadian oscillator [9,14,17], the MAP Kinase signalling pathway [6,13], and others [10]. This motivated the search for the simplest possible phosphorylation-mediated oscillator, to serve as a design principle.

The structure of Jolley's PTO (jPTO), described diagrammatically in Fig. 1(a), is one molecule with two phosphorylation sites. Therefore the molecule has four states (S00, S01, S10, S11) depending on which of its sites are phosphorylated. Two opposing enzymes, a kinase (E) and a phosphatase (F), act to phosphorylate or dephosphorylate a site, respectively.

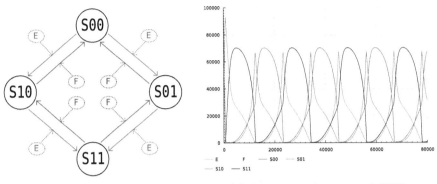

(a) Structure of jPTO, showing the four substrate molecule states, the kinase E, and the phosphatase F.

(b) Oscillatory dynamics of jPTO.

Fig. 1. Structure and dynamics of the Jolley PTO.

The parameters for this model were found by using computational parameter fitting techniques. They then used a clustering algorithm to determine two distinct clusters of parameter sets which produced two different patterns of oscillation. In this study we use one of these; Fig. 1(b) shows the behaviour of the model given these parameters.

3   Process Algebra Model Construction

Model construction in $c\pi$ is species-centric. That is the biochemical species, or reagents, are the focus of the modelling process. We first define each species and its binding sites and actions. We then define how different species can interact with each other. Then we define the initial conditions of our mixture, which species are present and in what concentrations. The model can then be executed to determine the behaviour, using numerical simulation. The remainder of this section gives an overview of the construction and execution of the model in $c\pi$. We omit the finer details of $c\pi$ syntax and semantics as these are described by Kwiatkowski and Stark [12] and in Kwiatkowski's thesis [11].

3.1   Species

The species in our model are the kinase E, the phosphatase F, and the substrate molecule which has four phosphorylation states $S00$, $S01$, $S10$, and $S11$. The simplest of these are the two enzymes; they are defined as follows:

$$E \triangleq e(x).x.E$$
$$F \triangleq f(x).x.F$$

The kinase E has a site e and the phosphatase F has a site f. Each can interact on its site with another molecule, perform some other function which depends on the molecule it is bound to, then return to its original state—from which it can perform the same action again. This directly corresponds to the definition of an enzyme.

In our cπ model we represent each of the four states of the substrate as a distinct species. This is simply to break down the syntactic description into smaller parts. In this model a change of state is essentially a change of species, but to the observer these species can be considered as one. The substrate can be defined as follows:

$$S00 \triangleq (\nu M_{00})\; s00a\langle be\rangle.(u.S00 + ra.S01)$$
$$+ s00b\langle be\rangle.(u.S00 + rb.S10)$$
$$S01 \triangleq (\nu M_{01})\; s01e\langle be\rangle.(u.S01 + r.S11)$$
$$+ s01f\langle bf\rangle.(u.S01 + r.S00)$$
$$S10 \triangleq (\nu M_{10})\; s10e\langle be\rangle.(u.S10 + r.S11)$$
$$+ s10f\langle bf\rangle.(u.S10 + r.S00)$$
$$S11 \triangleq (\nu M_{11})\; s11a\langle bf\rangle.(u.S11 + ra.S01)$$
$$+ s11b\langle bf\rangle.(u.S11 + rb.S10)$$

Here each of the states is defined, each containing a definition of the behaviour at each of the two phosphorylation sites. Each of these definitions is similar in structure, reflecting that they in fact represent distinct states of the same molecule. For example, let us examine the definition of $S01$.

One of the two states where one site is phosphorylated, but not the other, is $S01$. The term begins with a ν-term. The ν-term defines a local affinity network M_{01}; this governs the local interactions of unbinding or reacting in the same way as the global affinity network which will be defined below and defines the internal interaction potential of the complexes formed between substrate and enzyme to unbind (u) or react (r).

The structure of $S01$ is then defined as having two sites $s01e$ and $s01f$, each with some behaviour which follows from another molecule binding on that site. Once we have defined which molecules can interact on which sites (below), $s01e$ will accept the kinase E and $s01f$ will accept the phosphatase F. The behaviour which follows binding is defined by the next part of the term; in this case the bound enzyme can either unbind and the substrate returns to state $S01$ or the reaction can occur, changing the substrate either to state $S00$ or to $S11$, depending on whether F or E is bound.

The definition of each of the other states of the substrate follow the same pattern. Full details of definitions, the affinity networks, and their rates can be found in Appendix A.

3.2 Interactions

Now we have the definitions of the molecules and their interaction sites, we need to define which molecules can bind to which sites and at what rate these reactions occur. This is done by means of an affinity network M:

$$M = \{s00a \leftrightarrow e, \quad s00b \leftrightarrow e, \quad s01e \leftrightarrow e, \quad s10e \leftrightarrow e,$$
$$s01f \leftrightarrow f, \quad s10f \leftrightarrow f, \quad s11a \leftrightarrow f, \quad s11b \leftrightarrow f\}$$

Here we state that each of the substrate sites interacts with either site e of the kinase or site f of the phosphatase. Each of these interactions has a given reaction rate (see Appendix A).

3.3 Mixture

Having now defined the structure and rate parameters of the model, all that remains to be able to execute the model is a definition of the initial conditions we wish to simulate. Here we define a process Π which lists the species present and their initial concentrations.

$$\Pi \triangleq c_S \cdot S00 \parallel c_E \cdot E \parallel c_F \cdot F$$

Here we have some concentration c_S of substrate in its unphosphorylated state $S00$ and likewise some concentrations c_E and c_F of E and F (see Appendix A).

3.4 Validation

From this description the cπ tool generates a vector-space model of species concentrations over time. This is then compiled into a set of model ODEs and an initial value problem, suitable for numerical simulation. In this case the model description generates precisely the set of ODEs which were defined by Jolley et al. and therefore precisely the same behaviour; as shown in Fig. 1(b).

4 Basic Time Series Analysis

In this section we describe a number of computational experiments which were performed, aided by the compositional nature of the cπ description of the model.

4.1 Coupled jPTOs

The first experiment determines the behaviour of two identical jPTOs when coupled. The coupling is achieved by the two jPTOs sharing a pool of enzymes E and F.

We achieve the coupling in our $c\pi$ model in the following way. First we make a copy of the substrate species, call it T as shown in Appendix B. The process term can then be updated to include our new copy:

$$\Pi \triangleq c_S \cdot S00 \parallel c_T \cdot T00 \parallel c_E \cdot E \parallel c_F \cdot F$$

and the global affinity network M can then allow E and F to interact with the sites of T.

The behaviour of the coupled jPTOs can be seen in Fig. 2(a). The result of coupling two identical jPTOs is that the two act in synchrony, but the period is doubled. It is clear that the doubling of the period is due to each jPTO only having half the concentration of enzymes available, the other half of the concentration being sequestered by the other jPTO—each is competing equally for the same pool.

4.2   Weaker Coupling

It is possible to consider other schemes for coupling. For example, if the coupling was made weaker by only sharing one of the enzymes, does synchronisation still occur?

Here we take two jPTOs in a similar manner to above, however we only share the kinase E. This is achieved in the model simply by having a separate phosphatase for each jPTO, F_S and F_T:

$$\Pi \triangleq c_S \cdot S00 \parallel c_T \cdot T00 \parallel c_E \cdot E \parallel c_{F_S} \cdot F_S \parallel c_{F_T} \cdot F_T$$

Here we set $c_{F_S} = c_{F_T} = c_E$ and we then set the global affinity network accordingly. See Appendix C.

We can see, in Fig. 2(b), that indeed the jPTOs still synchronise when coupled less strongly. We can also see that each jPTO, given its own pool of phosphatase, spends more time in the less phosphorylated states as it can dephosphorylate at a greater rate than it can phosphorylate. If the concentration of each enzyme was adjusted accordingly, so $c_{F_S} + c_{F_T} = c_E$, then the system behaves as the coupled jPTOs sharing both kinase and phosphatase (as Fig. 2(a)).

4.3   Coupling Out of Phase

This experiment determines the behaviour of coupling a jPTO with an identical jPTO, but out of phase—that is, when the models are coupled with oscillators beginning at different points in their cycle. To achieve this we take two jPTO models of identical structure, but the second is shifted by a quarter of its cycle, we call this jPTO-90.

When the two models are composed we can see that, after a transient period, the cycles of the two jPTOs begin to synchronise as shown in Fig. 3(a). For comparison we also coupled jPTOs in various phase states. Synchronisation appears to occur in a number of selected phases. This suggests that the synchronisation of two jPTOs is quite robust. Figure 3(b) shows synchronisation when jPTOs are coupled in anti-phase: jPTO-180.

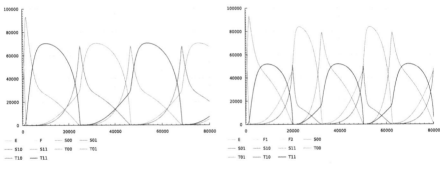

(a) Coupling two identical jPTOs.   (b) Weaker coupling by sharing kinase only.

Fig. 2. jPTO composition dynamics.

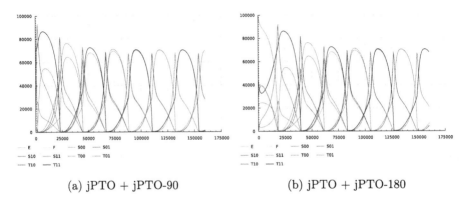

(a) jPTO + jPTO-90   (b) jPTO + jPTO-180

Fig. 3. Coupling of two identical clocks, sharing an enzyme pool, starting out of phase.

4.4   Perturbation

Another useful property of a circadian oscillator is that it is robust to some perturbations—although others may disrupt it. In this experiment we determine the behaviour of the jPTO when perturbed by a pulse of some inhibitor.

To construct a model for this we first construct an inhibitor molecule which rapidly appears in the system and decays rapidly. The mechanism for inhibitor appearing in the system is to have another molecule which is initially present and autonomously becomes the inhibitor. The inhibitor then decays. We will use the inhibitor to bind and sequester components of the jPTO. See Appendix E.

Figure 4(a) shows the effect of inhibiting the kinase; there is a transient period—about as long as the pulse—and then the jPTO settles back into its normal oscillation. This shows that the jPTO is somewhat robust to temporary sequestration of its enzymes.

Figure 4b shows the result of the inhibitor sequestering the doubly phospho-rylated substrate molecule S11. Here the fact that S11 itself is only present in pulses and the fact that the inhibitor does not decay when it is bound to S11 means that the inhibitor remains in the system for longer. We can see echo pulses as the inhibitor binds and unbinds the fluctuating concentration of S11. However, overall, the inhibitor eventually decays and the system stabilises. This shows that the substrate is also robust to this kind of perturbation.

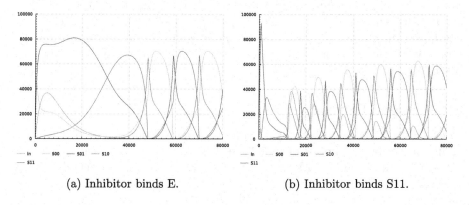

(a) Inhibitor binds E.   (b) Inhibitor binds S11.

Fig. 4. Perturbation of a jPTO with a pulse of inhibitor.

5   Model-Checking Experiments

The experiments in the previous section show a number of properties which are mostly amenable to analysis by conventional techniques. The compositional nature of $c\pi$ models aids greatly in the model construction for models where we are looking at compositions of two models or composition with an inhibitor; something which is much more difficult to do by working directly with ODEs. However the analysis of these models is little more than the inspection of time series for a relatively small set of models and initial conditions.

We will see that we can use the model checking of $\mathcal{LBC}$ specifications to automate the process of inspecting time series for a given behaviour. Moreover, and most importantly, we can define higher-order experiments which require many models and many initial conditions. We gain a means to precisely express a set of computational experiments, which in a conventional setting would require case-specific programming, and to have them automatically checked.

5.1   Behaviour Under Composition

The spatial aspects of $\mathcal{LBC}$ [4] directly take advantage of model compositionality. Specifications about the behaviour of a model when it is composed with another model can be made using the context modality. We can make use of this in analysing the behaviour of coupled oscillators.

A basic assertion in $\mathcal{LBC}$ has the form $P \models \phi$, meaning that system P satisfies property ϕ. A property of the form $Q \rhd \psi$ is a *guarantee*: if Q is introduced to the system, then the resulting mixture satisfies ψ. For example: $PTO1 \models PTO2 \rhd \phi$ states that when we couple $PTO1$ and $PTO2$ we have some behaviour ϕ. Likewise we could state $PTO \models Inhib \rhd \phi$ meaning that our PTO has some behaviour ϕ when we introduce an inhibitor (*Inhib*).

However, the most interesting properties are those which make a statement about introducing something over time. For example: $PTO1 \models \mathbf{G}_t(PTO2 \rhd \phi)$ which uses the temporal operator $\mathbf{G}$ (for *globally* true) to declare $PTO1$ has the property that for any time up to t, if we add $PTO2$ then the system from that point will satisfy ϕ.

5.2 Complex Dynamics

$\mathcal{LBC}$ also has the power to express complex dynamics, such as periodicity and oscillation. Numerous bodies of work have attempted to express oscillation properties in standard temporal logic [2,3,5], but all fall short of a general formula for oscillation. It is possible to express oscillation, however, with some prior knowledge of the type of oscillation. Following the idea in Calzone et al. [5] and extending it to a time-bounded logic, we can express oscillation in the temporal fragment of $\mathcal{LBC}$ as follows:

$$PTO \models \mathbf{G}_{[0,t]}(\mathbf{F}_{[0,p]}(([S]' > 0) \wedge \mathbf{F}_{[0,p]}([S]' < 0)))$$

where $[S]'$ is the first derivative of $[S]$ with respect to time. The formula states that at any time up to t the concentration of S will, within a further time p, be rising and then within another additional time p be falling. This describes a repeated rising and falling with period at most p. Whilst this is not a general formula, it does cover a large class of sustained oscillation. However, its weakness is that it does not distinguish from noise—although noise is not a problem when studying ODE models.

It has been shown that more expressive logics can express more general formulae for oscillation; for example Dluhoš et al. [7] show that one can use a "freeze operator" to do this. In fact, it is possible to give a general formula for sustained—and not necessarily regular—limit cycle periodicity using $\mathcal{LBC}$. The formula:

$$PTO \models \mathbf{F}_{[p_{min},p_{max}]}(\widehat{PTO} \rhd (\mathbf{F}_{[0,s]} \mathbf{G}_{[0,t]}(|[S] - [\widehat{S}]| < \epsilon))) \tag{1}$$

where $\widehat{PTO}$ is a copy of PTO, S is the species being observed, $\widehat{S}$ is the copy of S in $\widehat{PTO}$, and s is a maximum transient period before reaching the limit cycle. The formula states that if we introduce $\widehat{PTO}$ after some period in $[p_{min}, p_{max}]$ then, within s, $[S]$ and $[\widehat{S}]$ will synchronise to within ϵ for at least time t. This essentially takes a copy of the model, shifts it forward in time by $p_{min} \le t \le p_{max}$, and determines if it matches up with the original model. If it does, allowing for some initial transient period, then the model is periodic in species S.

In the context of our case study, we can now check if coupled PTOs still oscillate:

$$PTO1 \models \mathbf{G}_{[0,c]}(PTO2 \triangleright \mathtt{Osc})$$

where c is the end of the first cycle of $PTO1$ and $\mathtt{Osc}$ is one of our oscillation formulae from above. If coupling the PTOs at any time within the first cycle of $PTO1$ gives a system which still oscillates—with some period bounds, as above—then the formula will be true.

5.3 Perturbation Response

$\mathcal{LBC}$ can be used to express properties of a system under perturbation. For example, one might wish to determine if some perturbation causes a greater peak concentration in a species S. The formula:

$$PTO \models \mathbf{F}_{[0,t]}(P \triangleright \mathbf{F}_{[0,r]}([S] > pk))$$

states that some peak value pk is exceeded under some perturbation P, within time t, where r is the maximum expected time of the peak after the perturbation. As the perturbation P could be any model, it could simply be a quantity of some species, a constant amount of inhibitor, a pulse of inhibitor, etc.

Of particular interest in the study of oscillators are the *phase response* [8] characteristics of system. That is, given a short perturbation, at any point in the cycle, what is the effect on the phase of the oscillation? Biologists often plot a *phase response curve*, using a large number of experiments, to visualise the phase response. $\mathcal{LBC}$ cannot give such a precise and quantitative account of phase response as this, however it is certainly possible to formulate some more qualitative—or even semi-quantitative—properties of phase response. For example:

$$PTO \models \widehat{PTO} \triangleright \mathbf{F}_{[c_1,c_2]}(P \triangleright (\mathbf{G}_{[t_1,t_2]}([\widehat{S}]' > 0 \implies \mathbf{F}_{[s_1,s_2]}[S]' > 0)))$$

states that some perturbation P applied within $[c_1, c_2]$ will cause a *forward* phase shift in $[s_1, s_2]$. t_1 is a known max transient period after introducing P, t_2 is a sensible maximum time to simulate for, and the formula assumes that we know the perturbed system still oscillates.

5.4 Results

The following results of verifying the above $\mathcal{LBC}$ properties against the cπ models of Jolley's PTO were obtained by using the reference implementation of the $\mathcal{LBC}$ signal-based model checker[1]. First we show a number of formulae which give the same results as the experiments performed above, albeit without the need to

---

[1] Part of the CPiWorkBench: http://banks.ac/software/.

manually inspect a simulation trace. These results serve to verify the use of the model checker. Finally we show the results of checking formulae which describe higher-order computational experiments, i.e. those which check properties which would require the inspection of a great number of simulation runs.

Oscillation. Our first test was to check for oscillation using Formula 1. We let:

$$\texttt{Osc} = \mathbf{F}_{[p_{min},p_{max}]}(\widehat{jPTO} \triangleright (\mathbf{F}_{[0,s]}\mathbf{G}_{[0,t]}(|[S00] - [\widehat{S00}]| < \epsilon)))$$

where: we know that the period is around 24000 so we set $p_{min} = 23000$ and $p_{min} = 25000$; we know the system will reach limit cycle within $s = 10000$; we must choose an oscillating species and so choose $S = S00$; a reasonable time to simulate for is $t = 80000$—a few cycles; and we choose $\epsilon = 1$ as our concentration accuracy. The copy model $\widehat{jPTO}$ can be constructed in the same manner as the copy model in Sect. 4.1 or by using the appropriate function in the reference implementation.

Upon checking $jPTO1 \models \texttt{Osc}$ we find that it returns *True*. This confirms what we have been able to determine manually from inspecting the simulation traces in Fig. 1(b). Moreover, it shows that the $\mathcal{LBC}$ formula is a succinct and precise means of expressing the oscillation property and the model checker provides an automatic means for testing such a hypothesis.

Coupled Oscillators. The next step is to test coupled oscillators for oscillation, as in Sect. 4.1. First we take identical PTOs: $jPTO1$ and $jPTO2$. Upon checking $jPTO1 \models jPTO2 \triangleright \texttt{Osc}$ using the same formula parameters as above, we find that the result is *False*. This is because, as seen in Fig. 2(a), the period of the coupled oscillators is doubled. Therefore, upon relaxing the desired period range to $p_{min} = 23000$ and $p_{max} = 49000$ we find the formula is satisfied and the checker returns *True*.

Out-of-Phase Coupling. In Sect. 4.3 we showed that for a limited number of out-of-phase couplings of $jPTO1$ and $jPTO2$ the two systems did indeed synchronise and oscillate together after an initial transient period. This however does not confirm whether this is the case for *all* phase shifts.

Using the test $jPTO1 \models \mathbf{G}_{[0,c]}(jPTO2 \triangleright \texttt{Osc})$ we can use the model checker to give a greater guarantee that coupling the oscillators in any phase shift, up to c times the length of one cycle. Here we know that the length of one cycle is no more than, say, $c = 26000$.

Upon checking, again with the above formula parameters and the relaxed period range, we find that the result is *False*. This is because we have not accounted for the lengthened transient period when coupling out of phase. If we increase the parameter s to 120000 we find the formula is now satisfied, the result is *True*. This higher-order property gives a much stronger guarantee that all out-of-phase couplings oscillate than a limited number of manually inspected simulation traces would give.

Phase Response. Another higher-order property is the phase response characteristic. Using the inhibitor pulse model in Sect. 4.4—except using a pulse which lasts for fewer than 2000 time units to match the kind of pulse which would be used to plot a phase response curve—and the formula in Sect. 5.3 we can place some bounds on the phase response characteristics of the model:

$$jPTO \models \widehat{jPTO} \rhd \mathbf{F}_{[c_1,c_2]}(Pulse \rhd (\mathbf{G}_{[t_1,t_2]}([\widehat{S00}]' > 0 \implies \mathbf{F}_{[s_1,s_2]}[S00]' > 0)))$$

where: $[c_1, c_2] = [10000, 34000]$ which is roughly one cycle, this limits the computation; the maximum expected transient period is $t_1 = 10000$; the maximum time to compare oscillations is $t_2 = 80000$; and $[s_1, s_2] = [0, 1000]$ ensures that the whole formula states that: "there is always a forward phase response of no more than 1000 time units".

The model checker confirms that this statement is true for this model. So our small pulse may delay the cycle, but only by a relatively small time; it does not speed up the cycle.

6   Conclusions

We have shown that, using a combination of $c\pi$ and $\mathcal{LBC}$, we can express a variety of complex properties of biochemical models. We have shown that precise and succinct statements of complex properties can be built up in a modular fashion. One can even think of higher-order $\mathcal{LBC}$ properties as precise statements of an experimental hypothesis, to be tested by the model checker.

We have also shown that the Jolley PTO model does indeed interact robustly with other oscillators and inhibitors. This includes showing that the oscillator can be coupled at any point in its cycle and that it shows a robust inhibitor response at any point in its cycle. These latter properties are shown using $\mathcal{LBC}$ statements describing higher-order experiments; these are automatically tested by the model checker without any human intervention nor the necessity to write explicit programs for the necessary inspection of large numbers of simulation runs. Extensions to this work could readily include investigating the results of coupling with oscillators of a different type and of coupling with downstream networks.

A   Basic Jolley Model

The basic Jolley PTO model is constructed in $c\pi$ as follows:

$$E \triangleq e(x).x.E$$
$$F \triangleq f(x).x.F$$

$$
\begin{aligned}
S00 &\triangleq (\nu M_{00})\ s00a\langle be\rangle.(u.S00 + ra.S01) \\
&\quad + s00b\langle be\rangle.(u.S00 + rb.S10) \\
S01 &\triangleq (\nu M_{01})\ s01e\langle be\rangle.(u.S01 + r.S11) \\
&\quad + s01f\langle bf\rangle.(u.S01 + r.S00) \\
S10 &\triangleq (\nu M_{10})\ s10e\langle be\rangle.(u.S10 + r.S11) \\
&\quad + s10f\langle bf\rangle.(u.S10 + r.S00) \\
S11 &\triangleq (\nu M_{11})\ s11a\langle bf\rangle.(u.S11 + ra.S01) \\
&\quad + s11b\langle bf\rangle.(u.S11 + rb.S10)
\end{aligned}
$$

$$
\Pi \triangleq c_S \cdot S00 \parallel c_E \cdot E \parallel c_F \cdot F
$$

where

$$
c_S = 10^5, c_E = 1, c_F = 1.
$$

$$
\begin{aligned}
M_{00} = \{&be \leftrightarrow u : 10.02, \\
&be \leftrightarrow ra : 163.31, \\
&be \leftrightarrow rb : 0\} \\
\\
M_{01} = \{&be \leftrightarrow u : 10.02, \\
&be \leftrightarrow r : 40.83, \\
&bf \leftrightarrow u : 10.02, \\
&bf \leftrightarrow r : 8.17\}
\end{aligned}
$$

$$
\begin{aligned}
M_{10} = \{&be \leftrightarrow u : 10.02, \\
&be \leftrightarrow r : 8.17, \\
&bf \leftrightarrow u : 10.02, \\
&bf \leftrightarrow r : 40.83\} \\
\\
M_{11} = \{&bf \leftrightarrow u : 10.02, \\
&bf \leftrightarrow ra : 0, \\
&bf \leftrightarrow rb : 163.31\}
\end{aligned}
$$

$$
\begin{aligned}
M = \{&s00a \leftrightarrow e : 818.18, \\
&s00b \leftrightarrow e : 0, \\
&s01e \leftrightarrow e : 13.64, \\
&s10e \leftrightarrow e : 4903.17, \\
&s01f \leftrightarrow f : 4903.17, \\
&s10f \leftrightarrow f : 13.64, \\
&s11a \leftrightarrow f : 0, \\
&s11b \leftrightarrow f : 818.18\}
\end{aligned}
$$

B   Coupled jPTOs Model

The coupled model is constructed from the same substrate and enzyme species as the basic model in Appendix A. The second jPTO is a copy of the original substrate, renamed so it forms a distinct species:

$$
\begin{aligned}
T00 &\triangleq (\nu M_{00})\ t00a\langle be\rangle.(u.T00 + ra.T01) \\
&\quad + t00b\langle be\rangle.(u.T00 + rb.T10) \\
T01 &\triangleq (\nu M_{01})\ t01e\langle be\rangle.(u.T01 + r.T11) \\
&\quad + t01f\langle bf\rangle.(u.T01 + r.T00) \\
T10 &\triangleq (\nu M_{10})\ t10e\langle be\rangle.(u.T10 + r.T11) \\
&\quad + t10f\langle bf\rangle.(u.T10 + r.T00) \\
T11 &\triangleq (\nu M_{11})\ t11a\langle bf\rangle.(u.T11 + ra.T01) \\
&\quad + t11b\langle bf\rangle.(u.T11 + rb.T10)
\end{aligned}
$$

The process term is the same as above, but with the addition of the new (copy) substrate:

$$\Pi \triangleq c_S \cdot S00 \parallel c_T \cdot T00 \parallel c_E \cdot E \parallel c_F \cdot F$$

where

$$c_S = 10^5, c_T = 10^5, c_E = 1, c_F = 1,$$

and the global affinity net is then extended to allow the new substrate to interact with the enzymes:

$$M = \{ s00a \leftrightarrow e : 818.18, \qquad\qquad t00a \leftrightarrow e : 181.18,$$
$$s00b \leftrightarrow e : 0, \qquad\qquad t00b \leftrightarrow e : 0,$$
$$s01e \leftrightarrow e : 13.64, \qquad\qquad t01e \leftrightarrow e : 13.64,$$
$$s10e \leftrightarrow e : 4093.17, \qquad\qquad t10e \leftrightarrow e : 4093.17,$$
$$s01f \leftrightarrow f : 4093.17, \qquad\qquad t01f \leftrightarrow f : 4093.17,$$
$$s10f \leftrightarrow f : 13.64, \qquad\qquad t10f \leftrightarrow f : 13.64,$$
$$s11a \leftrightarrow f : 0, \qquad\qquad t11a \leftrightarrow f : 0,$$
$$s11b \leftrightarrow f : 818.18, \qquad\qquad t11b \leftrightarrow f : 818.18\}.$$

C   Weaker Coupled jPTOs

For the weaker coupled model we have a separate phosphatase for each substrate. The model in Appendix B. is extended by replacing species F with the following:

$$F_S \triangleq fs(x).x.F_S$$
$$F_T \triangleq ft(x).x.F_S$$

and the process term is extended:

$$\Pi \triangleq c_S \cdot S00 \parallel c_T \cdot T00 \parallel c_E \cdot E \parallel c_{F_S} \cdot F_S \parallel c_{F_T} \cdot F_T$$

where

$$c_S = 10^5, c_T = 10^5, c_E = 1, c_{F_S} = 1, c_{F_T} = 1,$$

and the affinity net is altered so each substrate only has affinity for one of the phosphatases:

$$M = \{s00a \leftrightarrow e : 818.18, \qquad\qquad t00a \leftrightarrow e : 181.18,$$
$$s00b \leftrightarrow e : 0, \qquad\qquad t00b \leftrightarrow e : 0,$$
$$s01e \leftrightarrow e : 13.64, \qquad\qquad t01e \leftrightarrow e : 13.64,$$
$$s10e \leftrightarrow e : 4093.17, \qquad\qquad t10e \leftrightarrow e : 4093.17,$$
$$s01f \leftrightarrow fs : 4093.17, \qquad\qquad t01f \leftrightarrow ft : 4093.17,$$
$$s10f \leftrightarrow fs : 13.64, \qquad\qquad t10f \leftrightarrow ft : 13.64,$$
$$s11a \leftrightarrow fs : 0, \qquad\qquad t11a \leftrightarrow ft : 0,$$
$$s11b \leftrightarrow fs : 818.18, \qquad\qquad t11b \leftrightarrow ft : 818.18\}.$$

D   Driving Other Reactions

To construct the model which drives another phosphorylation reaction, we first construct P which is the molecule to be phosphorylated:

$$P \triangleq (\nu M_P)\ p\langle x\rangle.(u.P + r.P')$$
$$P' \triangleq \tau_d.P$$

where $d = 10^{-4}$ and $M_P = \{x \leftrightarrow u : 1, x \leftrightarrow r : 1\}$.

The model is then the same as the basic model in Appendix A, but with a new site, which interacts with the P molecule, added to the $S11$ state of the substrate:

$$S11 \triangleq (\nu M_{11})\ s11a\langle bf\rangle.(u.S11 + ra.S01)$$
$$+ s11b\langle bf\rangle.(u.S11 + rb.S10)$$
$$+ s11p(x).x.S11$$

the new molecule added to the process:

$$\Pi \triangleq c_S \cdot S00 \parallel c_E \cdot E \parallel c_F \cdot F \parallel c_P \cdot P$$

where

$$c_S = 10^5, c_E = 1, c_F = 1, c_P = 10^5$$

and the affinity net is extended with

$$M = \{s00a \leftrightarrow e : 818.18,$$
$$s00b \leftrightarrow e : 0,$$
$$s01e \leftrightarrow e : 13.64,$$
$$s10e \leftrightarrow e : 4903.17,$$
$$s01f \leftrightarrow f : 4903.17,$$
$$s10f \leftrightarrow f : 13.64,$$
$$s11a \leftrightarrow f : 0,$$
$$s11b \leftrightarrow f : 818.18,$$
$$s11p \leftrightarrow p : 3 \times 10^{-4}\}.$$

E   Perturbation

To construct the model with a pulse of inhibitor, we take the model in Appendix D and replace the driven species P with an inhibitor In which decays and a species $ProdIn$ which autonomously produces the inhibitor:

$$In \triangleq (\nu M_{In})\ p\langle x \rangle u.In + \tau_d.0$$
$$ProdIn \triangleq \tau_d.P$$

where $M_{In} = \{x \leftrightarrow u : 0.1\}$ and $d = 5 \times 10^{-3}$ and the inhibitor producer added to the process:

$$\Pi \triangleq c_S \cdot S00 \parallel c_E \cdot E \parallel c_F \cdot F \parallel c_P \cdot ProdIn$$

where

$$c_S = 10^5, c_E = 1, c_F = 1, c_P = 10^5$$

In this model the inhibitor binds to the substrate in its $S11$ state. The models where the inhibitor binds to one or the other of the enzymes is constructed in a similar way, with a corresponding new site on the enzyme instead of the substrate. When binding to the enzyme, however the rate should be adjusted from 3×10^{-4} to 5.

References

1. Abraham, U., Granada, A.E., Westermark, P.O., Heine, M., Kramer, A., Herzel, H.: Coupling governs entrainment range of circadian clocks. Mol. Syst. Biol. **6**, 1 (2010)
2. Ballarini, P., Guerriero, M.L.: Query-based verification of qualitative trends and oscillations in biochemical systems. Theor. Comput. Sci. **411**(20), 2019–2036 (2010)

3. Ballarini, P., Mardare, R., Mura, I.: Analysing biochemical oscillation through probabilistic model checking. Electron. Notes Theor. Comput. Sci. **229**(1), 3–19 (2009)
4. Banks, C.J., Stark, I.: A Logic of Behaviour in Context. Inf. Comput. **236**, 3–18 (2014)
5. Calzone, L., Chabrier-Rivier, N., Fages, F., Soliman, S.: Machine learning biochemical networks from temporal logic properties. In: Priami, C., Plotkin, G. (eds.) Transactions on Computational Systems Biology VI. LNCS (LNBI), vol. 4220, pp. 68–94. Springer, Heidelberg (2006)
6. Chickarmane, V., Kholodenko, B.N., Sauro, H.M.: Oscillatory dynamics arising from competitive inhibition and multisite phosphorylation. J. Theor. Biol. **244**(1), 68–76 (2007)
7. Dluhoš, P., Brim, L., Šafránek, D.: On expressing and monitoring oscillatory dynamics. Electron. Proc. Theor. Comput. Sci. **92**, 73–87 (2012)
8. Granada, A., Hennig, R.M., Ronacher, B., Kramer, A., Herzel, H.: Phase response curves elucidating the dynamics of coupled oscillators. Methods Enzymol. **454**, 1–27 (2009)
9. Johnson, C.H., Mori, T., Xu, Y.: A cyanobacterial circadian clockwork. Curr. Biol. **18**(17), R816–R825 (2008)
10. Jolley, C.C., Ode, K.L., Ueda, H.R.: A design principle for a posttranslational biochemical oscillator. Cell Rep. **2**(4), 938–950 (2012)
11. Kwiatkowski, M.: A formal computational framework for the study of molecular evolution. Ph.D. thesis, University of Edinburgh (2010)
12. Kwiatkowski, M., Stark, I.: The continuous π-calculus: a process algebra for biochemical modelling. In: Heiner, M., Uhrmacher, A.M. (eds.) CMSB 2008. LNCS (LNBI), vol. 5307, pp. 103–122. Springer, Heidelberg (2008)
13. Liu, P., Kevrekidis, I.G., Shvartsman, S.Y.: Substrate-dependent control of ERK phosphorylation can lead to oscillations. Biophys. J. **101**(11), 2572–2581 (2011)
14. Nakajima, M., Imai, K., Ito, H., Nishiwaki, T., Murayama, Y., Iwasaki, H., Oyama, T., Kondo, T.: Reconstitution of circadian oscillation of cyanobacterial KaiC phosphorylation in vitro. Science **308**(5720), 414–415 (2005)
15. O'Neill, J.S., van Ooijen, G., Dixon, L.E., Troein, C., Corellou, F., Bouget, F.-Y., Reddy, A.B., Millar, A.J.: Circadian rhythms persist without transcription in a eukaryote. Nature **469**(7331), 554–558 (2011)
16. Seaton, D.D., Krishnan, J.: The coupling of pathways and processes through shared components. BMC Syst. Biol. **5**(1), 103 (2011)
17. van Zon, J.S., Lubensky, D.K., Altena, P.R.H., ten Wolde, P.R.: An allosteric model of circadian KaiC phosphorylation. Proc. Natl. Acad. Sci. U.S.A. **104**(18), 7420–7425 (2007)

Modeling of Resilience Properties in Oscillatory Biological Systems Using Parametric Time Petri Nets

Alexander Andreychenko[1]([✉]), Morgan Magnin[2,3], and Katsumi Inoue[2]

[1] Saarland University, 66123 Saarbrucken, Germany
`alexander.andreychenko@uni-saarland.de`
[2] National Institute of Informatics, 2-1-2, Hitotsubashi,
Chiyoda-ku, Tokyo 101-8430, Japan
[3] IRCCyN UMR CNRS 6597 (Institut de Recherche en Communications et
Cybernétique de Nantes), LUNAM Université, École Centrale de Nantes,
1 Rue de la Noë, B.P. 92101, 44321 Nantes Cedex 3, France

Abstract. Automated verification of living organism models allows us to gain previously unknown knowledge about underlying biological processes. In this paper, we show the benefits to use parametric time Petri nets in order to analyze precisely the dynamic behavior of biological oscillatory systems. In particular, we focus on the resilience properties of such systems. This notion is crucial to understand the behavior of biological systems (e.g. the mammalian circadian rhythm) that are reactive and adaptive enough to endorse major changes in their environment (e.g. jet-lags, day-night alternating work-time). We formalize these properties through parametric TCTL and demonstrate how changes of the environmental conditions can be tackled to guarantee the resilience of living organisms. In particular, we are able to discuss the influence of various perturbations, e.g. artificial jet-lag or components knock-out, with regard to quantitative delays. This analysis is crucial when it comes to model elicitation for dynamic biological systems. We demonstrate the applicability of this technique using a simplified model of circadian clock.

Keywords: Parametric time Petri net · Resilience · Biological oscillators · Model checking

1 Introduction

Understanding the mechanisms involved in oscillatory biological regulation is a fundamental issue to analyze living systems. Time delays play a major role in the sustainability and control of oscillations, as shown for example in phenomena related to the mammalian circadian clock [22]. Taking account of these delays in the modeling process is therefore fundamental to have a precise understanding of the chrono-biological phenomena. A major issue consists in identifying the value of (or the interval associated to) each delay. Some of them cannot be

© Springer International Publishing Switzerland 2015
O. Roux and J. Bourdon (Eds.): CMSB 2015, LNBI 9308, pp. 239–250, 2015.
DOI: 10.1007/978-3-319-23401-4_20

obtained through biological experiments. And most methodologies are not well-suited to capture parametric systems, e.g. simulation is adapted to assess the quality of one (or some) run(s) of the system, but raises hard problems when it comes to an exhaustive analysis. That is why automatic reasoning provided by model-checking techniques is useful to get formal proofs about the evolution of the timed system. The idea to consider time as a discrete variable helped the representation of the sequence of events that punctuate the featuring phases of biological systems. Given that the time delays are generally difficult to determine experimentally, computational approaches to model and infer the precise delays value *in silico* are fundamental.

To refine the quality of the given biological model, the bridge has to be made between the observed system properties and the dynamic behaviors of the model. This means that we need to perform a model elicitation procedure (using for example a model-checking approach) with regard to a relevant class of properties, which are to be expressed through modal logics, especially with regard to the extensions of LTL [29] and CTL [10] logics.

TCTL is one of these logics, aiming at the verification of properties with quantitative timing information [2]. As TCTL model-checking is undecidable for the general classes of timed extensions of Petri nets or automata, the main challenge is to identify the relevant subclasses of models (or properties) where decidability can be settled and the associated complexity can be handled in an efficient way. Recently, the authors of [19] identified a subclass of parametric timed automata that can benefit from efficient analysis of TCTL model-checking.

In our context, we were looking for a model expressive enough to capture the timed behavior of biological systems, easy-to-understand for biological collaborators, and with existing tools to perform parametric model-checking. Extending time Petri nets [27] with strong semantics, the framework of (bounded) parametric time Petri nets [35] with parametric intervals associated to transitions, meets these requirements thus motivating our choice.

1.1 Petri Nets to Model Dynamical Biological Systems

Concurrence between different components, either at a micro or macro scale, is central to biological systems. Petri nets are capable to concisely represent this this concept of concurrence and to simulate the behavior of concurrent systems biology models [9]. This framework is associated with a number of extensions, including *stochastic* Petri nets that allow to represent stochastic behavior or *time* Petri nets to include quantitative timing information.

Stochastic extensions of Petri nets are effective, especially for modeling biochemical systems. The main work on stochastic networks involve Markov models [17] for which the model checking techniques are well-established. But the formalization of oscillatory properties is a challenging task. To address it, the authors of [4,32] add the observer automata to the system such that it allows to precisely describe the noisy oscillatory trends.

During the last decade, some work (especially [20] and [8]) demonstrated how Petri nets could be used for both qualitative and quantitative analysis of biological systems. The unifying framework to conduct model checking tasks using Petri nets is given in [14]. In [8], the authors defined a systematic rewriting of Boolean models of logical regulatory networks into a standard PN formalism. In [9], they extended this previous work to multi-level logical models, but without incorporating delays. On the other hand, hybrid modeling have been studied in the context of the much expressive hybrid automata [1]. Taking inspiration of these works, we propose here to translate the multi-valued logical models into Petri nets, associating time intervals to transitions to capture the quantitative delays between discrete events.

1.2   Resilience Properties

Taking its inspiration from studies in ecological systems [18], *resilience* recently raised a growing interest among the research community [16,25,33]. This notion is critical to design a system reactive enough to face major changes in its environment: at an organizational level, this can be the security logistics in case of an earthquake; in biology, the functionality of circadian rhythm confronted to a wide range of perturbations. Resilience encompasses a family of four core properties, which are *resistance, recoverability, functionality* and *stability* [31]. The main difference to the design of critical systems lies in the fact that resilient systems may experience changes to its very nature, adapt and maintain some properties [23]. In this paper, we are investigating resilience properties in oscillatory models, more specifically in a biological context. While most existing works around resilience are limited to chronological models, we focus on an analysis based on quantitative timing information.

1.3   Modeling of the Mammalian Circadian Clock

Circadian rhythms control numerous biological mechanisms in various species. These endogenous oscillators are entrained [15] by environmental factors (Zeitgeber) such as light and temperature conditions. One of the main oscillatory mechanisms in mammals [13] is associated with so-called suprachiasmatic nucleus (SCN) that serves as the master clock for cellular clocks in peripheral tissues. This effect of signal propagation triggered by the oscillatory trend in the master clock is known as coupling of oscillators.

One of the first models of mammalian circadian clock formulated using differential equations is given in [22]. We consider its simplified version presented in [11], where Comet et al. applied a series of transformations to obtain the minimal discrete-state model with delays which allows to show important behavioral patterns. We converted this model into parametric time Petri net where the state of each gene is encoded by a place, whose (safe) marking corresponds to the Boolean status of the gene.

Previous research in the literature includes the earlier hybrid Petri net representation of circadian clock by [26], which has been analyzed using simulation. In our paper we aim to provide a more systematic method to study the dynamic properties of the gene regulatory network behind circadian rhythm. For the sake of simplicity, we chose to stick to the model from [11], but connections with the model of [26] has to be investigated in future works.

1.4   Our Contribution

In this paper, we propose a methodology based on parametric time Petri nets to assess the resilience of the gene regulatory network controlling the mammalian circadian clock system. Analyzing the literature, we formalize the corresponding properties in the TCTL logic and apply them on a simplified version of the circadian rhythm [11]. In particular, thanks to these resilience properties, we are able to perform model elicitation and gain the information about the delays involved in the regulations of this system. The same kind of approach could be applied to larger models of the circadian clock oscillator, by changing the input model (assuming we get it from further collaboration with biologists) and translating it into the framework of parametric time Petri nets.

1.5   Outline of the Paper

The rest of the paper is organized as follows: in Sect. 2 we introduce the notions of parametric time Petri net and *TPN-TCTL* logic. In Sect. 2.1 we describe the model of mammalian circadian clock and its representation as time Petri net. In Sect. 3 we state the properties that address the basic properties of biological systems. The extended discussion of the properties that use observers together with the resilience of oscillatory systems in given in the Appendix. Our contribution is summarized in the final section of the paper.

2   Logical Characterization of Circadian Clock Model

Timed models are capable of describing the complex behavioral patterns of biological systems. One of the ways to gain new insights about the underlying processes is to analyze the traces of execution of the given model. Here we apply the model-checking approach that verifies the model versus the given logical characterization of certain behavioral pattern. This approach is also capable to provide bounds on the time parameters of the model. It gives the additional information on how the model can be modified in order to satisfy the desired specification. In this paper we describe how oscillatory behavior is formalized in our framework using parametric temporal logics *PTPN-TCTL*.

In addition to the logical framework given in Appendix A, it is possible to define and use observers to model-check additional properties in TPNs and P-TPNs. It consists in adding to the Petri net - in a non-intrusive manner - places and transitions to model the property to check. The property is encoded as a marking on the extended Petri net and we check its reachability [34].

The main drawbacks of observers are two folds: first, there is no automatic procedure to build them; second, the observer can dramatically increase the

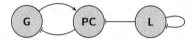

Fig. 1. Gene regulatory network of circadian clock

size of the state space to be explored by the model-checking procedure. To conduct the model checking of P-TPN models we use the tool ROMÉO [24][1] that is capable of compute the state space of the model (using state class graph) and analyze reachability and TCTL properties. The performance of the tool highly depends on the number of simultaneously enabled transitions and on the number of parameters that should remain rather small.

2.1 Circadian Clock Model

We consider the simplified model of the mammalian circadian clock proposed in [11]. This model reveals to be expressive enough to mimic the important behaviors of the circadian clock mechanism having the smallest amount of components, namely the abstract set of controlling genes (G), the protein complex PER–CRY in the nucleus (PC) and the external light condition (L), where each component is Boolean (shown in Fig. 1). It describes the main feedback loop (G $\leftrightarrow$ PC) that generates the oscillations. This non-deterministic model has the asynchronous semantics so that exactly one variable may be changed by any transition.

We consider this model as a multi-valued network [30] $(\mathbf{G}_{\mathcal{N}}, \mathbf{F})$ where the set of nodes $\mathbf{G}_{\mathcal{N}} = \{\text{G}, \text{PC}, \text{L}\}$ and each node has two qualitative states (0 and 1). The state transitions are given by the function $\mathbf{F}$ shown in Table 1. Each row describes the transition(s) $(\text{L}, \text{G}, \text{PC}) \rightarrow (\text{L}', \text{G}', \text{PC}')$, where $(\text{L}, \text{G}, \text{PC})$ is the state of variables before the transition and $(\text{L}', \text{G}', \text{PC}')$ is the state of variables after the transition. The symbol $*$ refers to any value (0 and 1) of the corresponding variable before and after transition. The authors [11] extended the gene regulatory network with delays $\tau_a, \ldots, \tau_{on}$ defined with respect to the knowledge about the temporal behavior of circadian cycle. The values of delays are provided in Table 1 (except for the unknown delay τ_g).

Table 1. Transitions in circadian clock model

t	$(L,G,PC) \rightarrow (L',G',PC')$	τ_t	t	$(L,G,PC) \rightarrow (L',G',PC')$	τ_t
a	$(0,0,1) \rightarrow (0,0,0)$	7	e	$(1,0,1) \rightarrow (1,0,0)$	1
b	$(*,0,0) \rightarrow (0,1,0)$	5	g	$(1,1,1) \rightarrow (1,1,0)$	τ_g
c	$(0,1,0) \rightarrow (0,1,1)$	7	off	$(1,*,*) \rightarrow (0,*,*)$	12
d	$(*,1,1) \rightarrow (0,0,1)$	5	on	$(0,*,*) \rightarrow (1,*,*)$	12

---

[1] http://romeo.rts-software.org/.

2.2 Translation of Gene Regulatory Network to Time Petri Net

In this subsection, we give the principle of our translation of gene regulatory networks into Petri nets, which is similar to [8], where the number of the places corresponds to the qualitative expression levels of a biological component (such transformation results in safe Petri net). We consider gene regulatory network $(\mathbf{G}_{\mathcal{N}}, \mathbf{F})$ as an input. It consists of a set of nodes $\mathbf{G}_{\mathcal{N}} = \{g_1, \ldots, g_{n_0}\}$ representing chemical species of a biological network (regulatory entities), where each entity $g \in \mathbf{G}_{\mathcal{N}}$ has a finite number of qualitative states, i.e. $\hat{g} \in S_g$, $S_g = \{0, 1, \ldots, k_g\}$, and a set of entities $\Omega(g) \subseteq \mathbf{G}_{\mathcal{N}}$ that affect g. The state s of the network is given by the discrete states of each entity in $\mathbf{G}_{\mathcal{N}}$, i.e. $s = (\hat{g}_1, \ldots, \hat{g}_{n_0})$. The next state s' is defined by the function $\mathbf{F} : S_{\mathcal{N}}^{\mathbf{G}} \times H \times (\mathbb{N} \cup \bot) \mapsto S_{\mathcal{N}}^{\mathbf{G}}$, where the state $s \in S_{\mathcal{N}}^{\mathbf{G}}$, $S_{\mathcal{N}}^{\mathbf{G}} = S_{g_1} \times \ldots \times S_{g_{n_0}}$, the set $H \subseteq \mathbf{G}_{\mathcal{N}}$ shows which entries affect the change and the delays of transitions $\tau \in (\mathbb{N} \cup \bot)$, where $\bot$ corresponds to the unknown delay. Each mapping in $\mathbf{F}$ changes one entity of a gene regulatory network such that the new state is given by $\hat{g}' \in \{\hat{g} - 1, \hat{g} + 1\}$ if $\hat{g} \in \{1, \ldots, k_g - 1\}$, and $\hat{g}' = 1$ for $\hat{g} = 0$, $\hat{g}' = k_g - 1$ for $\hat{g} = k$. The change of the state uses the asynchronous update semantics [12].

Given a gene regulatory network $(\mathbf{G}_{\mathcal{N}}, \mathbf{F})$, we construct a time Petri net with read arcs as follows:

- the set of places $\mathbf{P}$ is given by $\mathbf{P} = P_1 \cup \ldots \cup P_{n_0}$, where the set $P_i = \{p_{i,0}, p_{i,1}, \ldots, p_{i,k_{g_i}}\}$ corresponds to the qualitative levels of i-th entity. This correspondence is defined by the mapping $Pl_i : S_{g_i} \mapsto P_i$.
- each mapping in $f \in F$ corresponds to the transition $t_f \in \mathbf{T}$, where f changes the state of entity g_i from k to l. The transition t_f is defined by $^\bullet t_f = p_{i,k}$, $t_f^\bullet = p_{i,l}$ and $^\Box t_f = \{Pl_h(\hat{g}_h), \ldots, Pl_j(\hat{g}_j)\}$, where $\{g_h, \ldots, g_j\} \in H/g_i$.
- each mapping in $f \in F$ is associated with the delay $\delta \in (\mathbb{N} \cup \bot)$. If the delay δ is known $\delta = q$, $q \in \mathbb{N}$ then the firing interval t_f is defined as $J_s(t_f) = [q, q]$. Otherwise the corresponding firing interval is parametric with parameter τ_f, $J_s(t_f) = [\tau_f, \tau_f]$.
- initial state $s_0 = \{\hat{g}_1, \ldots, \hat{g}_{n_0}\}$ of the gene regulatory network defines the initial marking M_0 such that $M_0(Pl_i(\hat{g}_i)) = 1$ for $i \in \{1, \ldots, n_0\}$ and 0 otherwise.

It is important to notice that this transformation produces a safe Petri net.

There is always one token for each group of places P_i, $\sum_{j=0}^{k_{g_i}} M(p_{i,j}) = 1$. The time Petri net model of circadian clock that we use further is shown in Fig. 2, where read arcs are shown with white rectangles and each transition is annotated with the corresponding firing interval. We restrict ourselves to a Boolean representation, as the model in question in intrinsically Boolean. We add an additional restriction $\gamma = \{\tau_g \geq 1\}$ to emphasize that the delay of the corresponding biological process is not instantaneous. Here, places L0 and L1 correspond to the the absence and presence of the light. The state of the set of genes (inactive or active) is encoded by places G0 and G1 and the presence of the protein complex PER–CRY is given by places PC0 and PC1. Starting from the initial state $(\mathtt{L}, \mathtt{G}, \mathtt{PC}) = (1, 0, 1)$, the regular oscillation behavior is controlled by the sequence of transitions e, b, off, c, d.

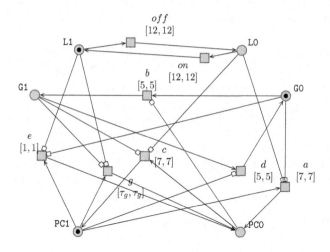

Fig. 2. Time Petri net model of circadian clock

3   Resilience of Biological Oscillatory Systems

The behavior of the mammalian circadian clock appears to be robust [11] with respect to the change of external stress conditions, namely the length of day and night. The authors consider the two possible scenarios (when the duration of night is 18 and 6 hours) and show that in both cases the oscillatory dynamics of the system does not suffer. Here, we expose the model to the various kinds of external stress and formalize the way of how to use model-checking procedure to reason about resilience properties. We address the properties that are related to resistance and functionality [31] and deal with the response of the system to external fluctuations together with learning the limits of such fluctuations strength. We also provide the methodology that allows to generalize given properties so that they can be constructed for other models and verified using *PTPN-TCTL*.

3.1   Property Specification

For each property we give the natural language formulation first and then the formalization in *PTPN-TCTL*. The example of application to the circadian clock model is given prior to more general specification that can be applied to other P-TPN models. The authors of [5] introduced the set of properties to characterize oscillatory behavior in biochemical systems that are modeled using the stochastic approach. They serve as an initial inspiration for the properties we introduce here.

Permanent oscillation. Let us first consider the permanent oscillation property on the example of circadian clock for both protein PC and gene G.

Property A. *State of the protein PC always oscillates.*

$$\phi_A = \big(M(p_{PC0}) = 1 \rightsquigarrow_{[0,\tau_{0,1}]} M(p_{PC1}) = 1\big)$$
$$\wedge \big(M(p_{PC1}) = 1 \rightsquigarrow_{[0,\tau_{1,0}]} M(p_{PC0}) = 1\big)$$

The two parametric intervals $[0, \tau_{0,1}]$ and $[0, \tau_{1,0}]$ determine how long it takes to change the state of PC in each case. According to the semantics of $\rightsquigarrow$ operator, the values of $\tau_{0,1}$ and $\tau_{1,0}$ refer to the longest time period over all possible model executions. For the model $\mathbb{N}_{CC}$ this property is satisfied with $\tau_{0,1} \geq 18$ and $\tau_{1,0} \geq 6$, which guarantees that the state of PC changes from 1 to 0 after 18 time units (however it may change in less amount of time).

Property B. *State of the gene G always oscillates.*

$$\phi_B = \big(M(p_{G0}) = 1 \rightsquigarrow_{[0,\tau_{0,1}]} M(p_{G1}) = 1\big) \wedge \big(M(p_{G1}) = 1 \rightsquigarrow_{[0,\tau_{1,0}]} M(p_{G0}) = 1\big)$$

For the model $\mathbb{N}_{CC}$ this property is satisfied with $\tau_{0,1} \geq 6$ and $\tau_{1,0} \geq 18$.

For a given entity $g \in \mathbf{G}_\mathcal{N}$ we can check the permanent oscillation between the two qualitative levels k and l by

$$\big(M(p_{g,k}) = 1 \rightsquigarrow_{[0,\tau_{k,l}]} M(p_{g,l}) = 1\big) \wedge \big(M(p_{g,l}) = 1 \rightsquigarrow_{[0,\tau_{l,k}]} M(p_{g,k}) = 1\big),$$

where $(\tau_{k,l} + \tau_{l,k})$ corresponds to the longest period of the oscillation (there may exist execution traces with shorter periods), and the value $|k - l|$ refers to the amplitude of oscillation (i.e. to the difference between the qualitative levels).

Entrainment behavior of circadian clock. One of the immutable properties of circadian clocks mechanism is the ability to be entrained by the external stress.

Property C. *State of the gene G always changes from 0 to 1 when the protein PC is not expressed.*

$$\phi_C = \big(M(p_{G0}) = 1 \wedge M(p_{PC0}) = 1 \rightsquigarrow_{[0,\tau_{0,1}]} M(p_{G1}) = 1\big)$$

Obviously, this property is satisfied by the model $\mathbb{N}_{CC}$ with $\tau_{0,1} \geq 5$ which corresponds to the transitions t_b and t_f.

Property D. *State of the protein PC always changes from 0 to 1 when there is no light and it always changes from 1 to 0 when there is light.*

$$\phi_D = \big(M(p_{L0}) = 1 \wedge M(p_{PC0}) = 1 \rightsquigarrow_{[0,\tau_{0,1}]} M(p_{PC1}) = 1\big)$$
$$\wedge \big(M(p_{L1}) = 1 \wedge M(p_{PC1}) = 1 \rightsquigarrow_{[0,\tau_{1,0}]} M(p_{PC0}) = 1\big)$$

This property is satisfied by the model $\mathbb{N}_{CC}$ with $\tau_{0,1} \geq 7$ and $\tau_{1,0} \geq 1$ that corresponds to the transitions t_c and t_e.

For given subsets of entities $G_c, G_e \subseteq \mathbf{G}_\mathcal{N}$, where G_c corresponds to the controller entities and G_e corresponds to the controllable ones, we construct the entrainment property as

$$\left(\phi \leadsto_{[0,\tau]} \psi\right),$$

where $\phi = \bigwedge_{g \in G_c} (M(p_g) = 1)$ describes the control condition and $\psi = \bigwedge_{g \in G_e} (M(p_g) = 1)$ refers to the expected response caused by the entrainment.

3.2   Properties with Observers

All properties introduced so far did not require to add additional elements to the Petri nets. However, there are limits for the expressivity of *PTPN-TCTL*, especially because nested properties are excluded from this logics. The addition of observers to the model itself can thus help to mitigate these restrictions (properties shall be expressed without using the nested temporal operators). For example, we can check the consistency of transitions in the model (Property E).

Property E. *All transitions are eventually fired at least once.*

$$\phi_E = \left(\bigwedge_{t \in \mathbf{T}} EF_{[0,\infty]} M(p_{O,t}) > 0\right)$$

The set of observers $\mathbf{O}$ is added to the model, where each observer O_t is associated with a transition $t \in \mathbf{T}$ such that $O_t = \{p_{O,t}, t_{O,t}\}$, $M(p_{O_t}) = 0$, ${}^\bullet t_{O,t} = P_{O,t}$, $t^\bullet_{O,t} = \emptyset$ and $J_s(t_{O,t}) = [0,0]$. We also add the place $p_{O,t}$ to the set $t^\bullet$. This property is not satisfied by the model $\mathbb{N}_{CC}$ since transitions t_a and t_g are never fired.

Property F. *Each 24 time units the system visits the state where $M_{L1} = 1$, $M_{G0} = 1$ and $M_{PC1} = 1$ (the initial state of $\mathbb{N}_{CC}$).*

$$\phi_F = (M(p_O) = 1) \leadsto_{[0,0]} (M(p_{L1}) = 1 \wedge M(p_{G0}) = 1 \wedge M(p_{PC1}) = 1)$$

The observer O (shown in Fig. 3) is keeping track of 24 time units intervals therefore we can judge about properties in global time.

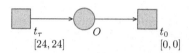

Fig. 3. Periodic observer with 24 time units period.

For a given expected response $\psi = \bigwedge_{g \in G_e} (M(p_g) = 1)$ and an observation interval time τ, we construct the periodic observation property as

$$M(p_O) = 1 \leadsto_{[0,0]} \psi$$

where the observer consists of the observer place p_O, $M(p_O) = 0$, the timer transition t_τ, ${}^\bullet t_\tau = \emptyset$, $t_\tau^\bullet = p_O$, $J_s(t_\tau) = [\tau, \tau]$ and the cleanup transition t_0, ${}^\bullet t_\tau = p_O$, $t_\tau^\bullet = \emptyset$, $J_s(t_0) = [0,0]$. This property can be extended such that the system is observed not only at the exact moment (each τ time units) but for δ time units by $J_s(t_0) = [0, \delta]$, $\delta < \tau$.

Another extension includes the initial delay α by adding the observer $'O = \{p'_{O,0}, t_\alpha, p'_{O,1}\}$, where ${}^\bullet t_\alpha = p'_{O,0}$, $t_\alpha^\bullet = p'_{O,1}$, $J_s(t_\alpha) = [\alpha, \alpha]$ and modifying an observer O such that ${}^\Box t_\tau = p'_{O,1}$. The example is shown in Fig. 4.

Fig. 4. Initial delay and interval time observers.

Properties A-F describe a certain set of behaviors that is normally exposed by the circadian clock model $\mathbb{N}_{CC}$. However, we can study the applicability of the model using the parameters in the transition firing interval function. The main external stress in the framework of mammalian circadian clock is light (sunlight or artificial light). The distortion of the normal day-night cycle affects the nominal behavior which causes negative effects like jet-lag. We address the corresponding properties and the model elicitation issues in the Supplementary Information [3].

4   Contribution and Future Work

In this paper we consider the model of mammalian circadian clock given in [11]. It serves as an initial inspiration for the translation of gene regulatory networks to parametric Petri net models.

We have proposed a methodology to assess the resilience properties of the gene regulatory networks. The corresponding properties are formalized in the TCTL logic and applied to the oscillatory system of mammalian circadian clock. They allow to conduct the model elicitation and gain new insights about the standard behavior of circadian clock as well as the limitations of its applicability under the perturbed environmental conditions. The latter also addresses the effect of artificial jet-lag and gene knock-out.

The properties introduced in the paper are formalized using observers that are easy to be extended and applied to other gene regulatory networks represented as parametric time Petri nets.

Future developments of this work include the consideration of more flexible formalism that allows for any delay in the given interval, as well as the comparison to the stochastic modeling formalism. We aim at verifying similar properties in the scope of stochastic Petri nets with exponentially distributed delays and extended generalized stochastic Petri nets that allow any valid probability distribution for the delay. Finally, the resilience properties shall be checked against the larger and more realistic model of circadian clock with the support of wet-lab experimental data.

References

1. Ahmad, J., Bernot, G., Comet, J.P., Lime, D., Roux, O.: Hybrid modelling and dynamical analysis of gene regulatory networks with delays. ComPlexUs **3**(4), 231–251 (2007)
2. Alur, R., Courcoubetis, C., Dill, D.: Model-checking for real-time systems. In: 1990 Proceedings of the Fifth Annual IEEE Symposium on Logic in Computer Science, LICS 1990, pp. 414–425. IEEE (1990)
3. Andreychenko, A., Magnin, M., Inoue, K.: Modeling of resilience properties in oscillatory biological systems using parametric time petri nets, supplementary information (2015). arXiv preprint arXiv:1506.06299 [cs.LO]
4. Ballarini, P.: Analysing oscillatory trends of discrete-state stochastic processes through hasl statistical model checking (2014). arXiv preprint arXiv:1410.4027
5. Ballarini, P., Mardare, R., Mura, I.: Analysing biochemical oscillation through probabilistic model checking. Electron. Notes Theoret. Comput. Sci. **229**(1), 3–19 (2009)
6. Bérard, B., Cassez, F., Haddad, S., Lime, D., Roux, O.H.: The expressive power of time Petri nets. Theoret. Comput. Sci. **474**, 1–20 (2013)
7. Boucheneb, H., Gardey, G., Roux, O.H.: TCTL model checking of time Petri nets. J. Logic Comput. **19**(6), 1509–1540 (2009)
8. Chaouiya, C., Remy, É., Thieffry, D.: Qualitative Petri net modelling of genetic networks. In: Priami, C., Plotkin, G. (eds.) Transactions on Computational Systems Biology VI. LNCS (LNBI), vol. 4220, pp. 95–112. Springer, Heidelberg (2006)
9. Chaouiya, C., Remy, E., Thieffry, D.: Petri net modelling of biological regulatory networks. J. Discrete Algorithms **6**(2), 165–177 (2008)
10. Clarke, E.M., Emerson, E.A., Sistla, A.P.: Automatic verification of finite-state concurrent systems using temporal logic specifications. ACM Trans. Program. Lang. Syst. **8**(2), 244–263 (1986)
11. Comet, J.P., Bernot, G., Das, A., Diener, F., Massot, C., Cessieux, A.: Simplified models for the mammalian circadian clock. Procedia Comput. Sci. **11**, 127–138 (2012)
12. Comet, J.-P., Klaudel, H., Liauzu, S.: Modeling multi-valued genetic regulatory networks using high-level Petri nets. In: Ciardo, G., Darondeau, P. (eds.) ICATPN 2005. LNCS, vol. 3536, pp. 208–227. Springer, Heidelberg (2005)
13. Edery, I.: Circadian rhythms in a nutshell. Physiol. Genomics **3**(2), 59–74 (2000)
14. Gilbert, D., Heiner, M., Lehrack, S.: A unifying framework for modelling and analysing biochemical pathways using Petri nets. In: Calder, M., Gilmore, S. (eds.) CMSB 2007. LNCS (LNBI), vol. 4695, pp. 200–216. Springer, Heidelberg (2007)
15. Golombek, D.A., Rosenstein, R.E.: Physiology of circadian entrainment. Physiol. Rev. **90**(3), 1063–1102 (2010)
16. Grimm, V., Calabrese, J.M.: What is resilience? A short introduction. In: Deffuant, D., Gilbert, N. (eds.) Viability and Resilience of Complex Systems, pp. 3–13. Springer, Heidelberg (2011)
17. Heiner, M., Gilbert, D., Donaldson, R.: Petri nets for systems and synthetic biology. In: Bernardo, M., Degano, P., Zavattaro, G. (eds.) SFM 2008. LNCS, vol. 5016, pp. 215–264. Springer, Heidelberg (2008)
18. Holling, C.S.: Resilience and stability of ecological systems. Annu. Rev. Ecol. Syst. **4**, 1–23 (1973)
19. Jovanović, A., Lime, D., Roux, O.H.: Integer parameter synthesis for timed automata. In: Piterman, N., Smolka, S.A. (eds.) TACAS 2013 (ETAPS 2013). LNCS, vol. 7795, pp. 401–415. Springer, Heidelberg (2013)

20. Koch, I., Heiner, M.: Petri Nets. In: Junker, B.H., Schreiber, F. (eds.) Biological Network Analysis, 7. Wiley Book Series on Bioinformatik, pp. 139–179. Wiley, New York (2008)

21. Larsen, K.G., Pettersson, P., Yi, W.: Model-checking for real-time systems. In: Reichel, H. (ed.) FCT 1995. LNCS, vol. 965, pp. 62–88. Springer, Heidelberg (1995)

22. Leloup, J.C., Goldbeter, A.: Toward a detailed computational model for the mammalian circadian clock. Proc. Natl. Acad. Sci. **100**(12), 7051–7056 (2003)

23. Leveson, N., Dulac, N., Zipkin, D., Cutcher-Gershenfeld, J., Carroll, J., Barrett, B.: Engineering resilience into safety-critical systems

24. Lime, D., Roux, O.H., Seidner, C., Traonouez, L.-M.: Romeo: a parametric model-checker for Petri nets with stopwatches. In: Kowalewski, S., Philippou, A. (eds.) TACAS 2009. LNCS, vol. 5505, pp. 54–57. Springer, Heidelberg (2009)

25. Maruyama, H., Legaspi, R., Minami, K., Yamagata, Y.: General resilience: taxonomy and strategies. In: 2014 International Conference and Utility Exhibition on Green Energy for Sustainable Development (ICUE), pp. 1–8 (2014)

26. Matsuno, H., Inouye, S.I.T., Okitsu, Y., Fujii, Y., Miyano, S.: A new regulatory interaction suggested by simulations for circadian genetic control mechanism in mammals. J. Bioinform. Comput. Biol. **4**(01), 139–153 (2006)

27. Merlin, P.M., Farber, D.J.: Recoverability of communication protocols-implications of a theoretical study. IEEE Trans. Commun. **24**(9), 1036–1043 (1976)

28. Oster, H., Yasui, A., van der Horst, G.T.J., Albrecht, U.: Disruption of mCry2 restores circadian rhythmicity in mPer2 mutant mice. Genes Dev. **16**(20), 2633–2638 (2002)

29. Pnueli, A.: The temporal logic of programs. In: Proceedings of the 18th Annual Symposium on Foundations of Computer Science, SFCS 1977, pp. 46–57. IEEE Computer Society, Washington, DC, USA (1977)

30. Rudell, R.L., Sangiovanni-Vincentelli, A.: Multiple-valued minimization for PLA optimization. IEEE Trans. Comput. Aided Des. Integr. Circuits Syst. **6**(5), 727–750 (1987)

31. Schwind, N., Okimoto, T., Inoue, K., Chan, H., Ribeiro, T., Minami, K., Maruyama, H.: Systems resilience: a challenge problem for dynamic constraint-based agent systems. In: Proceedings of the 2013 International Conference on Autonomous Agents and Multi-agent Systems. pp. 785–788. International Foundation for Autonomous Agents and Multiagent Systems (2013)

32. Spieler, D.: Characterizing oscillatory and noisy periodic behavior in Markov population models. In: Joshi, K., Siegle, M., Stoelinga, M., D'Argenio, P.R. (eds.) QEST 2013. LNCS, vol. 8054, pp. 106–122. Springer, Heidelberg (2013)

33. Tavana, M., Busch, T.E., Davis, E.L.: Modeling operational robustness and resiliency with high-level Petri nets. Technical report, DTIC Document (2012)

34. Toussaint, J., Simonot-Lion, F., Thomesse, J.P.: Time constraint verifications methods based time Petri nets. In: 6th Workshop on Future Trends in Distributed Computing Systems (FTDCS 1997), pp. 262–267, Tunis, Tunisia (1997)

35. Traonouez, L.M., Lime, D., Roux, O.H.: Parametric model-checking of stopwatch Petri nets. J. Univers. Comput. Sci. **15**(17), 3273–3304 (2009)

Parameter Synthesis by Parallel Coloured CTL Model Checking

Luboš Brim, Milan Češka(⊠), Martin Demko,
Samuel Pastva, and David Šafránek

Systems Biology Laboratory, Faculty of Informatics, Masaryk University,
Botanická 68a, 602 00 Brno, Czech Republic
{brim,xceska,xdemko,xpastva,xsafran1}@fi.muni.cz

Abstract. We propose a new distributed-memory parallel algorithm for parameter synthesis from CTL hypotheses. The algorithm colours the state space transitions by different parameterisations and extends CTL model checking to identify the maximal set of parameters that guarantee the satisfaction of the given CTL property. We experimentally confirm good scalability of our approach and demonstrate its applicability in the case study of a genetic switch controlling decisions in the cell cycle.

1 Introduction

Constructing computational models that describe dynamics of biochemical processes is a key step towards understanding of existing and even yet undiscovered behavioural and physiological phenotypes occurring in biology. Model-based prediction and analysis make cornerstones of systems biology. While the structure of dynamical models of some biochemical processes is already available at the qualitative level represented by known entities and interactions, most of the quantitative aspects of the systems dynamics, such as reaction rates or initial concentration values, cannot be easily determined. Such quantitative attributes are usually reflected in the model as *parameters*. In order to obtain reliable models, parameters need to be specified exactly. For a typical model, a fraction of the parameter values can be determined from the literature or experimental data, leaving many parameter values uncertain or completely unknown. The reason is, that many parameters are hard to measure *in vitro/in vivo*.

The *algorithmic discovery* of unknown parameter values (also referred to as *parameter estimation, parameter identification*, the *inverse problem*, or *model calibration*) remains thus one of the main challenges in computational systems biology. Besides the traditional approaches to tackle the inverse problem (e.g., [15–17,24]), there have recently appeared alternative techniques grounded in formal verification [2,4,21]. These methods typically focus on identifying reliable subsets of parameter space instead of finding singular parameter values.

This work has been partially supported by the Czech Science Foundation grant GA15-110895S and the Czech Ministry of Education, Youth, and Sport project No. CZ.1.07/2.3.00/30.0009 (Milan Češka).

O. Roux and J. Bourdon (Eds.): CMSB 2015, LNBI 9308, pp. 251–263, 2015.
DOI: 10.1007/978-3-319-23401-4_21

Hypotheses mined from biological literature as well as time-series experiments from wet-labs can be considered as dynamical constraints restricting the admissible set of model parameter values. Apart from a concrete kind of dynamical models, these constraints can be sufficiently captured in terms of temporal logic formulae (for review of approaches see, e.g., [7]). A common computational method that decides the question whether for a given parametrisation the model meets the temporal constraints is *model checking*. The inverse problem is then generalised to *parameter synthesis* [2,12] – to find the maximal subset of parameter values such that they meet the stated dynamical constraints.

The general advantage of temporal specification for parameter synthesis is its ability to focus on certain qualitative aspects of observed behaviour [23] (e.g., temporal ordering of events qualitatively characterising important moments in the systems dynamics). In particular, temporal properties can be viewed as *global properties* independent of particular setting of initial conditions (initial values of the state variables). The global view provides biologists a tool which, for a given model and a given property, computes the maximal set of parameter values and initial conditions for which the model entirely fulfils the property. Such an approach is complementary to traditional approaches based on monitoring a numerical simulation [11,25] or local sensitivity analysis [13].

To capture biologically-relevant temporal hypotheses both branching-time operators and linear-time operators are needed [6]. In this paper we focus on *branching logic* CTL. The reason is that many relevant questions in systems biology need branching operators to express them properly. For instance, switching mechanisms and multi-stability are present in genetic regulatory networks and drive many key biological phenotypes such as, e.g., irreversible decisions in cell division, cell differentiation or programmed cell death. However, it is difficult (or often impossible) to express relevant properties in linear temporal logics. Other reason for usage of CTL is related to the particular procedure for model checking. This procedure allows to effectively identify all system states where the given property is satisfied. Thus CTL procedure leads inherently to global analysis of systems dynamics as opposed to LTL procedure, which requires a single initial state (or iterates over a given set of initial states).

Contribution of the Paper. Several methods for parameter synthesis based on model checking have been proposed recently, targeting different kinds of models and different temporal logics (e.g., [2,5,11,12,19]). In [2] we proposed a parameter synthesis method for LTL hypotheses established on our automata-based *coloured* LTL *model checking* algorithm.

In this paper we extend that work in several directions. First, we consider CTL hypotheses. Second, we propose a distributed-memory parallel coloured CTL *model checking* algorithm, keeping thus both the advantage of having an explicit representation and the effectiveness of parallel solution in distributed-memory. Third, we propose a novel *heuristics for partitioning the state space* that effectively uses specifics of rectangularly abstracted ODE models (the abstraction is described in [10]). We have experimentally confirmed good scalability of our approach and demonstrated its applicability in a case study of a genetic switch employing rectangular abstraction [4,10] of an already existing ODE model [26].

2   Parallel Parameter Synthesis Algorithm

In this paper we propose a formal framework for parameter synthesis of biochemical models from branching time temporal logic formulae. Here, the term parameter refers to initial conditions of the model and to dynamical parameters. The method presumes a finite state space. For discrete models such as boolean networks, this can be ensured directly by the definition. For continuous models, like ODE models, a finite discrete abstraction of the state space is necessary. The existing abstractions typically lead to over- or under-approximation (or a mixture of both) of the dynamics of the original system [10]. This has generally some consequences regarding the interpretation of computed results. We will discuss this issue later. The method also presumes a finite parameter space. In the case of continuous parameter spaces an appropriate finite abstraction, like an interval abstraction in the case of ODE models, must be used.

It is important to note that there are two levels of complexity that significantly affect the tractability of parameter synthesis for biological models. First, the procedure requires consideration of all possible settings of parameters – points in the parameter space. The size of the parameter space grows exponentially with the number of unknown parameters. However, in reality the number of parameters to be considered should be small. A model with too many parameters is hard to falsify - it can fit almost any data. Second, the state space of the model, which has to be explored by the parameter synthesis algorithm, grows exponentially with the number of state variables (state space explosion).

Given the complexity of the problem and the need for comprehensive large-scale models, there is a natural call for development of techniques prepared to perform efficiently on high-performance computing platforms [1,7]. The complexity caused by the state space size can be reduced by either symbolic or enumerative parallel techniques. The achieved efficiency is again highly dependent on the modelling approach, character of models, and the properties considered. In the case of biological models, symbolic techniques were successfully employed for abstract logical (qualitative) models [5,14] whereas enumerative parallel techniques have proved to be fruitful for quantitative models [1,3].

Coloured CTL Model Checking

We start by introducing the notion of a parametrised Kripke structure that encapsulates a family of Kripke structures built over the same model but with different valuations of individual parameters.

Let AP be a set of atomic propositions. A *parametrised Kripke structure* (over AP) is a tuple $\mathcal{K} = (\mathcal{P}, S, I, \rightarrow, L)$, where $\mathcal{P}$ denotes the finite *set of parameter values (parameterisations)*, i.e., all the possible valuations of the parameters,

S is the finite set of states, $I \subseteq S$ is the set of initial states, $L : S \to 2^{AP}$ is a labelling of states by atomic propositions, $\to \subseteq S \times \mathcal{P} \times S$ is a transition relation labelled by parameter valuations (not required to be total). We write $s \xrightarrow{p} s'$ instead of $(s, p, s') \in \to$. Fixing a parametrisation $p \in \mathcal{P}$ reduces the parametrised Kripke structure $\mathcal{K}$ to the standard (non-parametrised) Kripke structure $\mathcal{K}(p) = (S, I, \xrightarrow{p}, L)$.

To express properties (hypotheses) about the dynamics of systems, we consider formulae of CTL defined by the following abstract syntax:

$$\varphi ::= Q \mid \neg\varphi \mid \varphi_1 \wedge \varphi_2 \mid \mathbf{AX}\varphi \mid \mathbf{EX}\varphi \mid \mathbf{A}(\varphi_1 \, \mathbf{U} \, \varphi_2) \mid \mathbf{E}(\varphi_1 \, \mathbf{U} \, \varphi_2)$$

where Q ranges over *atomic propositions* taken from a set AP. Let φ be a CTL formula. We denote by $cl(\varphi)$ the set of all subformulae of φ and by $tcl(\varphi)$ the set of all (temporal) subformulae of φ of the form $\mathbf{EX}\varphi$, $\mathbf{E}(\varphi_1 \, \mathbf{U} \, \varphi_2)$, $\mathbf{AX}\varphi$ or $\mathbf{A}(\varphi_1 \, \mathbf{U} \, \varphi_2)$. We use the standard abbreviations like $\mathbf{EF}\varphi$ which stands for $\mathbf{E}(true \, \mathbf{U} \, \varphi)$ or $\mathbf{AG}\varphi$ which stands for $\neg\mathbf{EF}\neg\varphi$. Examples of some typical CTL formulae are [14]:

- $\mathbf{EF} \, \varphi$ expresses a reachability of a state where the condition φ holds,
- $\mathbf{AG} \, \varphi$ expresses a stabilisation with φ being continually true,
- $\mathbf{EFAG}\varphi_1 \wedge \mathbf{EFAG}\varphi_2$ expresses a bistable switch (two different stable situations φ_1, φ_2 can be reached).

Most frequent types of temporal properties investigated for biochemical models have been collected in [23]. There are two important fragments of CTL relevant for biological models. A formula is said to be *positive* if it does not contain any negations. We say that a formula is *existential* (or in ECTL) if it is positive and only contains existential temporal operators. We say that a formula is *universal* (or in ACTL) if it is positive and only contains universal temporal operators.

It is important to note, that model abstraction based on over-approximation preserves truth of universally-quantified CTL properties (ACTL), i.e. if an ACTL property holds in the abstract model, it is guaranteed to hold in the concrete one. Dually, under-approximation preserves falsity of ACTL. The situation is reversed for existentially-quantified CTL properties (ECTL): over-approximation preserves falsity while under-approximation preserves truth.

The parameter synthesis problem is defined in the following way. Suppose we are given a parametrised Kripke structure $\mathcal{K}$ and a CTL formula Ψ. For each state $s \in S$ let $P_s = \{p \in \mathcal{P} \mid s \models_{\mathcal{K}(p)} \Psi\}$, where $s \models_{\mathcal{K}(p)} \Psi$ denotes that Ψ is satisfied in the state s of $\mathcal{K}(p)$. The *parameter synthesis problem* requires to compute the function $\mathcal{F}_\Psi^{\mathcal{K}} : S \to 2^P$ such that $\mathcal{F}_\Psi^{\mathcal{K}}(s) = P_s$. Often we are especially interested in computing the set $\cap_{s \in I} \mathcal{F}_\Psi^{\mathcal{K}}(s)$.

The algorithm for computing $\mathcal{F}_\Psi^{\mathcal{K}}$ is a modification of the (explicit) labelling CTL model checking algorithm [9]. It labels states with "coloured" subformulae of Ψ that are satisfied in the state of the Kripke structure $\mathcal{K}(p)$ for the "colour" $p \in \mathcal{P}$. Typically the structures $\mathcal{K}(p)$ have similar transition relations, thus leading to a significant acceleration of the parameter synthesis. The reason is

---

Algorithm 1. Compute parameters

Require: parametrised KS $\mathcal{K}$ and CTL formula Ψ
Ensure: $\mathcal{F}_\Psi^\mathcal{K}$
    for all $i \le |\Psi|$ **do**   ▷ compute the sets $ColSat(\Phi) = \{(p,s) \in \mathcal{P} \times S \mid s \models_{\mathcal{K}(p)} \Phi\}$
        for all $\Phi \in cl(\Psi)$ *with* $|\Phi| = i$ **do**
            compute $ColSat(\Phi)$ from $ColSat(\Phi')$ ▷ for maximal genuine $\Phi' \in cl(\Phi)$
    return $\{(p,s) \in \mathcal{P} \times S \mid (p,s) \in ColSat(\Psi)\}$

---

that a small change in a value of a single parameter causes only a local change in the transition relation.

The algorithm operates recursively on the structure of Ψ starting from atomic propositions. Its basic idea is described by the Algorithm 1. The recursive computation of the satisfaction sets $ColSat(\Psi) = \{(p,s) \in \mathcal{P} \times S \mid s \models_{\mathcal{K}(p)} \Psi\}$ follows the parse tree of the formula Ψ.

Kripke Fragments

Our aim is to perform the parameter synthesis algorithm as a distributed-memory algorithm on a cluster of n nodes (workstations) in order to enlarge the available memory to accommodate larger models. To this end we use a *partition function* $f : S \rightarrow \{1, \ldots, n\}$ to partition the state space among n nodes. After partitioning, each node owns a part of the original state space. Concrete techniques for the state space partitioning are discussed in the next subsection.

We adapt the assumption based distributed CTL model checking paradigm [8] as the basis of our work. We represent the state space owned by one node using a parametrised Kripke structure with *border states* (also called a *fragment*). Intuitively, border states, that are added to the states assigned by f, are states that in fact belong to other station and represent the missing parts of the state space (placed in the memory of other nodes and not directly accessible). For structure $\mathcal{K}$, the set of its border states is defined as $border(\mathcal{K}) = \{s \in S \mid \neg\exists(p, s').s \xrightarrow{p} s'\}$. A *fragment* $\mathcal{K}_i$ of $\mathcal{K}$ is a substructure of $\mathcal{K}$ satisfying the property that every state in $\mathcal{K}_i$ has either no successor in $\mathcal{K}_i$ or it has exactly the same successors as in $\mathcal{K}$. Partitioning the given structure $\mathcal{K}$ results in a finite set $\mathcal{K}_1, \ldots, \mathcal{K}_n$ of fragments each handled by one node. A border state is thus stored several times: as original one on the node that owns it and as duplicates on nodes they own its predecessors.

To define the semantics of CTL formulae over fragments we need to adapt the standard semantic definition. We define the notion of the truth under assumptions associated with border states. An *assumption function* for a parametrised Kripke structure $\mathcal{K}$ and a CTL formula ψ is defined as a partial function of type $\mathcal{A} : \mathcal{P} \times S \times cl(\psi) \rightarrow Bool$. The values $\mathcal{A}(p, s, \varphi)$ are called *assumptions*. We use the notation $\mathcal{A}(p, s, \varphi) = \perp$ to say that the value of $\mathcal{A}(p, s, \varphi)$ is undefined. By $\mathcal{A}_\perp$ we denote the assumption function which is undefined for all inputs. Intuitively, $\mathcal{A}(p, s, \varphi) = \mathtt{tt}$ if we can assume that φ holds in the state s under parametrisation p, $\mathcal{A}(p, s, \varphi) = \mathtt{ff}$ if we can assume that φ does not hold in the state s under parametrisation p, and $\mathcal{A}(p, s, \varphi) = \perp$ if we cannot assume anything. Let us denote by $AS_{\mathcal{K}}^{\psi}$ the set of all assumption functions for a formula ψ and a parametrised Kripke structure $\mathcal{K}$

We consider a new semantic function $\mathcal{C}_{\mathcal{K}}^{\psi} : AS_{\mathcal{K}}^{\psi} \to AS_{\mathcal{K}}^{\psi}$ that takes an input assumption function $\mathcal{A}_{in}$ and returns a new assumption function $\mathcal{A}$. If $s \in border(\mathcal{K})$ and $\varphi \in |tcl|(\psi)$ then $\mathcal{A}(p, s, \varphi) = \mathcal{A}_{in}(p, s, \varphi)$. If $s \notin border(\mathcal{K})$ and $\varphi \in |tcl|(\psi)$ then $\mathcal{A}(p, s, \varphi)$ is defined recursively. We provide here only the definition for the most complicated case of $\mathbf{A}(\varphi_1 \mathbf{U} \varphi_2)$. $\mathcal{A}(p, s, \mathbf{A}(\varphi_1 \mathbf{U} \varphi_2)) =$

$$\begin{cases} \text{tt if for all p-paths } \pi = s_0 s_1 s_2 \ldots \text{ with } s = s_0 \text{ there exists an index} \\ \quad x < |\pi| \text{ such that: either } \mathcal{A}(p, s_x, \varphi_2) = \text{tt or } [s_x \in border(\mathcal{K}) \text{ and} \\ \quad \mathcal{A}(p, s_x, \mathbf{A}(\varphi_1 \mathbf{U} \varphi_2)) = \text{tt})], \text{ and } \forall y : 0 \le y < x : \mathcal{A}(p, s_y, \varphi_1) = \text{tt} \\ \text{ff if there exist a p-path } \pi = s_0 s_1 s_2 \ldots \text{ with } s = s_0 \text{ and an index} \\ \quad x < |\pi| \text{ such that: } [\mathcal{A}(p, s_x, \varphi_1) = \text{ff and } \forall y \le x : \mathcal{A}(p, s_y, \varphi_2) = \text{ff}] \\ \quad \text{or } \forall x < |\pi| : [\mathcal{A}(p, s_x, \varphi_2) = \text{ff and } (|\pi| = \infty \text{ or } (s_{|\pi|-1} \in border(\mathcal{K}) \\ \quad \text{and } \mathcal{A}(p, s_{|\pi|-1}, \mathbf{A}(\varphi_1 \mathbf{U} \varphi_2)) = \text{ff}))] \\ \perp \text{ otherwise} \end{cases}$$

Here a p-path π from a state s_0 is a sequence $\pi = s_0 s_1 \ldots$ such that $\forall i \ge 0 : s_i \in S$ and $s_i \xrightarrow{p} s_{i+1}$. The truth of a formula is relative to given assumptions $\mathcal{A}_{in}$ and it is defined as $\mathcal{C}_{\mathcal{K}}^{\psi}(\mathcal{A}_{in})(p, s, \psi)$. The value of an assumption function $\mathcal{A}_{in}(p, s, \varphi)$ for a state $s \notin border(\mathcal{K})$ does not influence the value $\mathcal{C}_{\mathcal{K}}^{\psi}(\mathcal{A}_{in})$. Hence, for any *total* parametrised Kripke structure $\mathcal{K}$ (i.e. $border(\mathcal{K}) = \emptyset$), CTL formula ψ and an arbitrary assumption function $\mathcal{A} \in AS_{\mathcal{K}}^{\psi}$, we have that $s \models_{\mathcal{K}(p)} \psi \Leftrightarrow \mathcal{C}_{\mathcal{K}}^{\psi}(\mathcal{A})(p, s, \psi) = \text{tt}$. In particular, $ColSat(\psi) = \{(p, s) \in \mathcal{P} \times S \mid \mathcal{C}_{\mathcal{K}}^{\psi}(\mathcal{A})(p, s, \psi) = \text{tt}\}$ and thus we can solve the parameter synthesis problem by computing the assumption function $\mathcal{C}_{\mathcal{K}}(\mathcal{A}_{\perp})$.

Distributed Algorithm

We are now ready to describe the algorithm for distributed parameter synthesis. In order to compute $\mathcal{C}_{\mathcal{K}}(\mathcal{A}_{\perp})$ in a distributed environment, we iteratively compute assumption functions that are defined on fragments of the system $\mathcal{K}$.

The algorithm starts by partitioning the given state space of $\mathcal{K}$ among the nodes using a partition function f. Each node performs Algorithm 1 modified in such way, that it is also able to cope with "undefined values". Moreover, it computes both the positive and negative results. This means that if a state s has a successor for which φ is true for parametrisation p, it can be concluded both that s satisfies $\mathbf{EX}\varphi$ and that s does not satisfy $\mathbf{AX}\neg\varphi$ under p, even when the validity of φ in other successors of s is undefined (unknown) yet.

The main idea of the entire distributed computation, summarised in Algorithm 2, is the following. Each fragment $\mathcal{K}_i$ is managed by a separate process (node) P_i. These processes are running in parallel (simultaneously on each node). Each process P_i initialises the assumption function $\mathcal{A}_i$ to the undefined assumption function $\mathcal{A}_{\perp}$. After initialisation, it computes the semantic function $\mathcal{C}_{\mathcal{K}_i}(\mathcal{A}_i)$

Algorithm 2. Main Idea of the Distributed Algorithm

Require: parametrised KS $\mathcal{K}$, CTL formula Ψ, function f
Ensure: $\mathcal{F}_{\Psi}^{\mathcal{K}}$

    Partition $\mathcal{K}$ into $\mathcal{K}_1, \ldots, \mathcal{K}_n$
    for all $\mathcal{K}_i$ where $i \in \{1, \ldots, n\}$ **do in parallel**
        Take the initial assumption function
        repeat
            Compute the semantic function using the node algorithm;
            Exchange relevant information with other nodes;
            Modify assumption function;
        until all processes reach fixpoint

---

using the node algorithm. If new assumptions have been computed for some border states, this result is sent directly to appropriate processes. Similarly, if such information is received from another process, the assumption function is modified to reflect these new results. This procedure is repeated until all running processes are "deadlocked", i.e. until no new information (value of an assumption function) can be computed using the node algorithm or by exchanging assumptions among processes. We say that the *fixpoint has been reached* ("global" stabilisation has occurred). In our experimental implementation, the deadlock is detected by additional communication among processes (the code has been skipped for clarity).

After stabilisation (reaching the fixpoint) there may still remain a state s and a formula φ, for which $\mathcal{A}_i(s, \varphi) = \perp$. This can happen in the case of the **U** operator. However, if the results for all subformulas of φ have already been computed in all states on all nodes and the fixpoint has been reached then we can conclude that φ does not hold in s.

State Space Partitioning

The key ingredient of distributed model checking algorithms is a suitable state space partitioning that minimises the communication overhead and equally distributes the workload. In particular, the partitioning should provide (1) a regular *load-balancing* ensuring that each node is responsible for a proportional part of the state space and (2) a good *locality* minimising the number of cross transitions where the source and target states are assigned to two different nodes.

The computation of the optimal partitioning for the given state space typically brings a significant overhead and thus various heuristics are considered. For computer and engineering systems, a hash-based partitioning is usually used, since it does not require any prior knowledge about the structure of the state

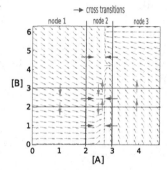

Fig. 1. State space partitioning.

space. It constructs a hash function mapping each state to a node. This approach usually provides very good load-balancing following from an uniformity of the hash function. However, these heuristics are not able to control the locality and thus they introduce a considerable communication overhead.

In our approach we utilise the regular structure of the state space for biochemical models [20]. We use structural properties of the rectangular abstraction of the given parametric piece-wise multi-affine ODE model [4,10]. The approximation is formed by an n-dimensional hyper-rectangular state space defined by m state variables and by a set of thresholds for each variable. The partitioning decomposes the state space into n hyper-rectangular subspaces (n is the number of nodes) such that each subspace has similar volume. Figure 1 depicts such partitioning for $m = 2$ and $n = 3$ where the volume for each subspace is 3. Our heuristic usually provides a good load balancing, since the volume reflects the number of states. The construction of the discretised state space further ensures there are only transitions between the adjacent states with respect to the hyper-rectangular structure. Therefore, our partitioning naturally provides almost the minimal number of cross transitions, since only cross transitions between the border states are introduced as illustrated in Fig. 1. Comparing to the hash-based partitioning we significantly decrease the communication overhead. Note that, the final load balancing can be negatively affected by the backward connectivity of the state space. However, our experiments demonstrate the connectivity is significantly increased due to the fact that we have to consider all parameterisations of the model. Additional heuristics are used to improve the load balancing by reflecting the atomic propositions in the CTL formula.

3   Experimental Evaluation

We first consider a suitable model that enables us to thoroughly evaluate the scalability of the proposed distributed algorithm. Afterwards, we apply our approach to a relevant and interesting model describing the regulation in a cell cycle transition.

Scalability

The scalability of the algorithm is evaluated on a catalytic reaction model. The model allows to scale the number of intermediate products/variables (N), discretisation thresholds (T) and unknown parameters. For each variable we assume a same number of thresholds and thus the total number of states is $(T-1)^N$. We employ the state space partitioning that reflects the model structure and thus it provides a good load-balancing and locality.

We use homogeneous cluster with 12 nodes each equipped with 16 GB of RAM and a quad-core Intel Xeon 2 GHz processor. In order to provide a fair evaluation we utilise only a single core on each node (although our implementation can effectively utilise multi-core nodes). The reported runtime has been obtained as the arithmetic mean from several experiments.

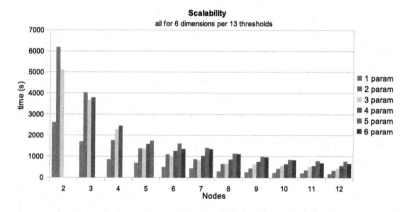

Fig. 2. The scalability with respect to the number of unknown parameters.

Figure 2 illustrates the results for $N = 6$ and $T = 13$ (i.e. almost 3 millions states) and different number of unknown parameters. The figure demonstrates a significant acceleration of the parameter synthesis when more nodes are used. Note that the missing columns indicate that the corresponding experiment run out of memory. The number of unknown parameters changes the structure of the state space and its partitioning. Therefore, in some cases, a higher number of parameters can decrease the runtime.

Case Study: Regulation of G_1/S Cell Cycle Transition

To demonstrate applicability of our framework, we investigate a well-known ODE model [26] representing a two-gene regulatory network that describes the interaction of the tumour suppressor protein pRB and the central transcription factor $E2F1$ (see Fig. 3 (left)). This network represents the crucial mechanism governing the transition from G_1 to S phase in the mammalian cell cycle. In the G_1-phase the cell makes an important decision. In high concentration levels, $E2F1$ activates the G_1/S transition mechanism. In low concentration of $E2F1$, committing to S-phase is refused and that way the cell avoids DNA replication.

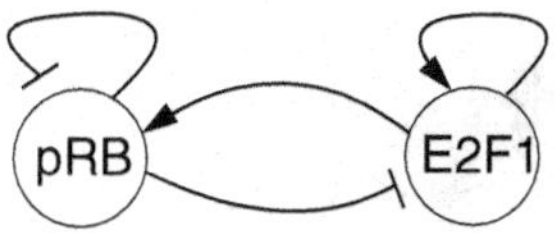

$$\frac{d[pRB]}{dt} = k_1 \frac{[E2F1]}{K_{m1}+[E2F1]} \frac{J_{11}}{J_{11}+[pRB]} - \phi_{pRB}[pRB]$$

$$\frac{d[E2F1]}{dt} = k_p + k_2 \frac{a^2+[E2F1]^2}{K_{m2}^2+[E2F1]^2} \frac{J_{12}}{J_{12}+[pRB]} - \phi_{E2F1}[E2F1]$$

$a = 0.04$, $k_1 = 1$, $k_2 = 1.6$, $k_p = 0.05$, $\phi_{E2F1} = 0.1$
$J_{11} = 0.5$, $J_{12} = 5$, $K_{m1} = 0.5$, $K_{m2} = 4$

Fig. 3. G_1/S transition regulatory network and its ODE model taken from [26].

The mechanism is an example of a *bistable switch*, an irreversible decision to finally reach some of the two different stable states. In particular, we are interested in the existence of two different stable equilibria on $E2F1$. Activity of pRB is rapidly modulated by phosphorylation/dephosphorylation turn-over controlled by growth factor signals transferred to cyclin-dependent kinases each acting on a specific subset of pRB phosphorylation sites [22]. This control is captured in the model by means of the degradation rate parameter ϕ_{pRB}.

In [26] the authors have provided bifurcartion analysis investigating $E2F1$ equilibria depending on ϕ_{pRB}. As shown in Fig. 4(left), by non-trivial elaboration with numerical analysis methods expecting the previous knowledge of the equilibria they constructed equilibrium point curve for $E2F1$ in proportion to ϕ_{pRB} and discovered two saddle-node bifurcation points. For ϕ_{pRB} smaller then 0.007 the system converges to a single low-concentration stable equilibrium whereas for values higher than 0.027 it converges to a single high-concentration equilibrium. In between the two bifurcation points the system is bistable provided that there always exists an unstable equilibrium for which there is an ϵ-ball that makes a basin of attraction for both stable equilibria.

To employ our framework for this non-linear model, we have first created the piece-wise multi-affine approximation (PMA) of the ODE model [18]. We approximate each non-linear function in the right-hand side of ODEs with an optimal sequence of piece-wise affine ramp functions (in our case we have set the precision to 70 affine segments per each non-linear function). For the resulting PMA we have employed rectangular abstraction [4] to obtain a finite (rectangular) automaton over-approximating the PMA (the intuition is shown in Fig. 1). Finally, we have run the parallel coloured CTL model checking algorithm for the formula $\varphi \equiv EFAG$ high $\wedge$ $EFAG$ low and the initial parameter space $\phi_{pRB} \in [0.001, 0.025]$. The atomic propositions low and high characterise the location of expected regions of $E2F1$ stability. Based on the results reported in [26] we define the stable regions as high $\equiv (E2F1 > 4 \wedge E2F1 < 7.5)$ and low $\equiv (E2F1 > 0.5 \wedge E2F1 < 2.5)$ that determine the expected regions of the two stable attractors including (a subset of) their surrounding attracted points.

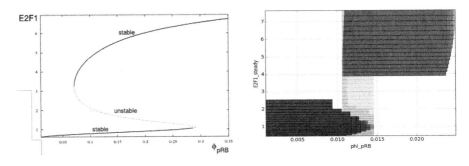

Fig. 4. (left) Equilibrium point curve taken from [26] (we believe there is a typo in the scale of ϕ_{pRB} in the original figure, the range of ϕ_{pRB} should read 0.005-0.035). (right) Model checking results. Red and blue correspond to the high and low stable regions, respectively. Yellow are the states where the *bistable switch* formula φ holds (Color figure online).

Results of the analysis are depicted in Fig. 4(right) in comparison with the equilibrium curve, Fig. 4(left), provided in [26]. The blue region is the place where AG low is satisfied, in particular, it says the $E2F1$ low concentration is *guaranteed* to stabilise for the corresponding values of ϕ_{pRB} in the PMA. The guarantee comes from the fact that the abstraction employed is over-approximation [10]. In particular, for each trajectory in the PMA there must exist a corresponding path in the rectangular automaton. For example, the model checking result says that for a fixed parameter value 0.005 there is no path in the rectangular automaton that would exit the concentration bounds $0.5 \leq E2F1 \leq 2.5$ and hence there is no such trajectory in the PMA. However, although there is no red region identified at $\phi_{pRB} = 0.005$ we are not sure this holds also in the PMA since it might be the property introduced by the abstraction. For a given ACTL formula, the abstraction thus causes the parameter space synthesised by model checking to be under-approximated [4]. For example, with ϕ_{pRB} getting closer to the bistable region the guarantee of low stabilisation becomes limited to a smaller subset of the low region until it disappears at $\phi_{pRB} > 0.0145$. The analogous explanation fits the red region obtained for AG high, note that in that case the effect of parameter value under-approximation is negligible when compared with equilibrium point curve. For $\phi_{pRB} \in [0.012, 0.0145]$, the system is bistable (there exist two stable regions, i.e., AG low $\wedge AG$ high is guaranteed).

The yellow region covers points where φ holds. Since an EF-formula might be satisfied within a spurious behaviour introduced by the abstraction, this result does not provide any guarantees but rather estimates parameter values and initial conditions under which both stable regions might be reached. The diagram projects pRB values by means of fill opacity. Grey region reflects the fact there are values of pRB from which the red or the blue region is not reachable. This information is again guaranteed.

4   Conclusions

We have developed a fully automatic method for synthesising parameters that guarantee the satisfaction of a given CTL hypothesis. The method uses a novel distributed-memory parallel algorithm that extends the CTL model-checking algorithm by colouring the transitions in the underlying state space. We have demonstrated a very good scalability of the algorithm as well as the usefulness of the method on a biological problem of bistable switch. This is an example of a wide range of possible applications. The case study can be compared to numerical bifurcation analysis methods that require good initial estimate of the equilibria and do not scale up with the number of unknown parameters. Our method does not require so detailed initial knowledge about the system and scales well with the number of unknown parameters.

References

1. Ballarini, P., Guido, R., Mazza, T., Prandi, D.: Taming the complexity of biological pathways through parallel computing. Brief. Bioinform **10**(3), 278–288 (2009)
2. Barnat, J., Brim, L., Krejci, A., Streck, A., Safranek, D., Vejnar, M., Vejpustek, T.: On parameter synthesis by parallel model checking. IEEE/ACM Trans. Comput. Bio. Bioinform. **9**(3), 693–705 (2012)
3. Barnat, J., Brim, L., Safránek, D.: High-performance analysis of biological systems dynamics with the divine model checker. Brief. Bioinform. **11**(3), 301–312 (2010)
4. Batt, G., Belta, C., Weiss, R.: Model checking liveness properties of genetic regulatory networks. In: Grumberg, O., Huth, M. (eds.) TACAS 2007. LNCS, vol. 4424, pp. 323–338. Springer, Heidelberg (2007)
5. Batt, G., Page, M., Cantone, I., Gössler, G., Monteiro, P., de Jong, H.: Efficient parameter search for qualitative models of regulatory networks using symbolic model checking. Bioinformatics **26**(18), 603–610 (2010)
6. Batt, G., Ropers, D., Jong, H.D., Geiselmann, J., Mateescu, R., Schneider, D.: Validation of qualitative models of genetic regulatory networks by model checking: analysis of the nutritional stress response in escherichia coli. Bioinformatics **21**, 19–28 (2005)
7. Brim, L., Češka, M., Šafránek, D.: Model checking of biological systems. In: Bernardo, M., de Vink, E., Di Pierro, A., Wiklicky, H. (eds.) SFM 2013. LNCS, vol. 7938, pp. 63–112. Springer, Heidelberg (2013)
8. Brim, L., Yorav, K., Zidkova, J.: Assumption-based distribution of CTL model checking. STTT **7**(1), 61–73 (2005)
9. Clarke, E.M., Emerson, E.A., Sistla, A.P.: Automatic verification of finite-state concurrent systems using temporal logic specifications. ACM Trans. Program. Lang. Syst. **8**, 244–263 (1986)
10. Collins, P., Habets, L.C., van Schuppen, J.H., Černá, I., Fabriková, J., Šafránek, D.: Abstraction of biochemical reaction systems on polytopes. In: IFAC World Congress, pp. 14869–14875. IFAC (2011)
11. Donaldson, R., Gilbert, D.: A model checking approach to the parameter estimation of biochemical pathways. In: Heiner, M., Uhrmacher, A.M. (eds.) CMSB 2008. LNCS (LNBI), vol. 5307, pp. 269–287. Springer, Heidelberg (2008)
12. Donzé, A., Clermont, G., Langmead, C.J.: Parameter synthesis in nonlinear dynamical systems: application to systems biology. J. Comput. Biol. **17**(3), 325–336 (2010)
13. Donzé, A., Fanchon, E., Gattepaille, L.M., Maler, O., Tracqui, P.: Robustness analysis and behavior discrimination in enzymatic reaction networks. PLoS ONE **6**(9), e24246 (2011)
14. Fages, F., Soliman, S.: Formal cell biology in biocham. In: Bernardo, M., Degano, P., Zavattaro, G. (eds.) SFM 2008. LNCS, vol. 5016, pp. 54–80. Springer, Heidelberg (2008)
15. Fröhlich, F., Theis, F.J., Hasenauer, J.: Uncertainty analysis for non-identifiable dynamical systems: profile likelihoods, bootstrapping and more. In: Mendes, P., Dada, J.O., Smallbone, K. (eds.) CMSB 2014. LNCS, vol. 8859, pp. 61–72. Springer, Heidelberg (2014)
16. Gábor, A., Banga, J.R.: Improved parameter estimation in kinetic models: selection and tuning of regularization methods. In: Mendes, P., Dada, J.O., Smallbone, K. (eds.) CMSB 2014. LNCS, vol. 8859, pp. 45–60. Springer, Heidelberg (2014)

17. Gilbert, D., Breitling, R., Heiner, M., Donaldson, R.: An introduction to biomodel engineering, illustrated for signal transduction pathways. In: Corne, D.W., Frisco, P., Păun, G., Rozenberg, G., Salomaa, A. (eds.) WMC 2008. LNCS, vol. 5391, pp. 13–28. Springer, Heidelberg (2009)

18. Grosu, R., Batt, G., Fenton, F.H., Glimm, J., Le Guernic, C., Smolka, S.A., Bartocci, E.: From cardiac cells to genetic regulatory networks. In: Gopalakrishnan, G., Qadeer, S. (eds.) CAV 2011. LNCS, vol. 6806, pp. 396–411. Springer, Heidelberg (2011)

19. Jha, S.K., Langmead, C.J.: Synthesis and infeasibility analysis for stochastic models of biochemical systems using statistical model checking and abstraction refinement. Theor. Comput. Sci. **412**(21), 2162–2187 (2011)

20. Jha, S., Shyamasundar, R.K.: Adapting biochemical kripke structures for distributed model checking. In: Priami, C., Ingólfsdóttir, A., Mishra, B., Riis Nielson, H. (eds.) Transactions on Computational Systems Biology VII. LNCS (LNBI), vol. 4230, pp. 107–122. Springer, Heidelberg (2006)

21. Liu, B., Kong, S., Gao, S., Zuliani, P., Clarke, E.M.: Parameter synthesis for cardiac cell hybrid models using δ-decisions. In: Mendes, P., Dada, J.O., Smallbone, K. (eds.) CMSB 2014. LNCS, vol. 8859, pp. 99–113. Springer, Heidelberg (2014)

22. Mittnacht, S.: Control of prb phosphorylation. Curr. Opin. Genet. Dev. **8**(1), 21–27 (1998)

23. Monteiro, P.T., Ropers, D., Mateescu, R., Freitas, A.T., de Jong, H.: Temporal logic patterns for querying qualitative models of genetic regulatory networks. In: ECAI. FAIA, vol. 178, pp. 229–233. IOS Press (2008)

24. Raue, A., Karlsson, J., Saccomani, M.P., Jirstrand, M., Timmer, J.: Comparison of approaches for parameter identifiability analysis of biological systems. Bioinformatics **30**, 1440–1448 (2014)

25. Rizk, A., Batt, G., Fages, F., Soliman, S.: A general computational method for robustness analysis with applications to synthetic gene networks. Bioinformatics **25**(12), 169–178 (2009)

26. Swat, M., Kel, A., Herzel, H.: Bifurcation analysis of the regulatory modules of the mammalian G1/S transition. Bioinformatics **20**(10), 1506–1511 (2004)

Analysing Cell Line Specific EGFR Signalling via Optimized Automata Based Model Checking

Adam Streck[(✉)], Kirsten Thobe, and Heike Siebert

Freie Universität Berlin, Berlin, Germany
adam.streck@fu-berlin.de

Abstract. Building models with a high degree of specificity, e.g. for particular cell lines, is becoming an important tool in the advancement towards personalised medicine. Constraint-based modelling approaches allow for utilizing general system knowledge to generate a set of possible models that can be further filtered with more specific data. Here, we exploit such an approach in a Boolean modelling framework to investigate EGFR signalling for different cancer cell lines, motivated by a study from Klinger et al. [8]. To optimize performance of the underlying model checking procedure, we present a number of constraint encodings tailored to describing common data types and experimental set-ups. This results in a significant increase in the performance of the approach.

Keywords: Boolean Networks · Model checking · EGFR · Cancer

1 Introduction

Mathematical modelling in systems medicine and biology has long since proved its worth in gaining a deeper understanding of the functionalities of complex biological systems. However, modellers are often confronted with data uncertainty, e.g., originating in a big span in the quality of the available data but also simply in lack of information on specific components or interactions. Often, to combat lack of specific knowledge, cellular pathway or network models utilize a collection of information integrating data derived from different experimental settings, tissues or even organisms. This can be problematic, in particular when analysis is focused on questions pertaining to networks in very specific settings. A prime example for such a situation is the evaluation of drug target effectiveness that necessarily aims at particular cancer types or cell lines [9].

In this paper, we tackle this problem in the context of the epidermal growth factor receptor (EGFR) signalling pathway motivated by a study by Klinger et al. [8]. This receptor drives cell proliferation and cell growth, but is also involved in the regulation of cell death and is found to carry prominent mutations in cancer cells (BRAF, PIK3CA). However, the exact topology of this regulatory system is not completely clear, not least since mutations can cause major changes in the inner regulations. Klinger et al. presented a combined experimental and theoretical approach to identify the cell line specific topology of the network starting

© Springer International Publishing Switzerland 2015
O. Roux and J. Bourdon (Eds.): CMSB 2015, LNBI 9308, pp. 264–276, 2015.
DOI: 10.1007/978-3-319-23401-4_22

from a literature model aggregating information from various sources. To this end, human colorectal cancer cell lines were treated with stimuli and inhibitors to produce a rich data set that was evaluated using a semi-quantitative modelling approach. However, this approach necessitates a steady state assumption for the data points, which can be seen as problematic due to the interplay of various feedback effects [9]. In addition, while quite comprehensive, the method still relies on parameter estimation steps and statistical cut-offs.

In a fully qualitative modelling formalism, as e.g. the Boolean Networks [6], recently developed methods allow to consider all models consistent with given constraints pertaining the underlying network topology. Additional data can be used to further narrow down this pool of models. Once all available information has been integrated, features shared between all remaining models as well as distinguishing characteristics of interest for experimental design can be extracted [3,4]. Here, we use such an approach to analyse the EGFR signalling network for different cell lines, which was focus of qualititative studies before, however, only steady-state and not transient behavior was examined [4,10]. Utilizing the rich data set of [8], we aim at a comparison of the results delivered by the two methods, both in the case of adding steady state assumption for the data points and without it. Going beyond the study by Klinger et al., we use comparative analysis of the model pools of different cell lines to evaluate the differences in not only network topology but also regulatory mechanisms generated by different genotypes.

As the system is expected to exhibit non-linear behaviour, we rely on the strongly expressive model checking method [1] that, together with the problem of parameter uncertainty, places high demands on the computational power. To streamline application and improve performance we developed several convenient constraint encodings tailored to data types and experimental set-ups often encountered when modelling biological regulatory systems. Here we present all the methods we employed to improve performance of our custom model-checker[1]. Without these improvements, the procedure would be barely possible, as we illustrate by a comparison of performance with a state of the art model checker.

2 Background

The topology of a biological system is encoded as a directed graph $G = (V, E)$ where V is a set of named *components* and $E \subseteq V \times V$ is a set of *regulations*. Each component can occur in one of two qualitatively distinct states (0 and 1) representing e.g. being or not being phosphorylated, resulting in a Boolean Network (BN). The set of all possible configurations of a system, called *state space*, is denoted and defined $S^G = \mathbb{B}^{|V|}$ with $\mathbb{B} = \{0, 1\}$. The behaviour of a BN in its state space is then described via a *parametrization function* $K : S^G \to S^G$ with coordinate functions $K_v : S^G \to \mathbb{B}$, $v \in V$, as described below. The pair (G, K) then constitutes a unique *model*.

---

[1] The tool used here, called TREMPPI, is available in a development version at `github.com/xstreck1/TREMPPI` and is expected to be fully released in 2015.

We consider an *asynchronous* update rule for a BN, meaning that at any state of the network we can change at most one component value per time step, even if multiple value changes are indicated by the behavioural rules. Formally we use a transition relation $\to^G \subseteq S^G \times S^G$ s.t. for a parametrization K we have that $s \to^G s'$ if and only if one of the following holds:

$$\forall v \in V : s'_v = s_v \wedge K_v(s) = s_v,$$
$$\exists u \in V, \forall v \in V \setminus \{u\} : s'_v = s_v \wedge s_u \neq s'_u = K_u(s).$$

The pair $(S^G, \to^G)$ is called a *transition system* (TS) of a model.

To illustrate the technical notions of this and the following section, we use an example of a trivial BN with a single self-regulating node, i.e. $V = \{v\}, E = \{(v, v)\}$. A parametrization for our trivial example and its respective TS has e.g. the following form (brackets not used for 1-dimensional vectors):

$$S^G = \{0, 1\}, K_v(0) = 1, K_v(1) = 0, \to^G = \{(0, 1), (1, 0)\} \tag{1}$$

At this point, there is not yet a strict relation between the given network structure and the parametrization function. To make this link, a BN is equipped with an *edge labelling*, which are the predicates $+ : E \to \mathbb{B}$ and $- : E \to \mathbb{B}$ s.t.:

$$+(u, v) \iff \exists s \in S, s_u = 0 : K_v(s) < K_v(s + e_u),$$
$$-(u, v) \iff \exists s \in S, s_u = 0 : K_v(s) > K_v(s + e_u),$$

where $e_u \in \mathcal{B}^{|V|}$ is the u-th unit vector. The predicate $+$ on (u, v) can be interpreted as u being an *activator* of v and $-$ as u being an *inhibitor*. Additionally we say that an edge (u, v) is *functional* iff $+(u, v)$ or $-(u, v)$. Using these predicates we can encode respective biological knowledge as a predicate formula l over the domain E called *labelling*. The model (G, K) is then valid for the labelling l *iff* $(G, K) \models l$ where $\models$ is the standard logical validity [5]. The set of all parametrization of a BN G where l is valid is then denoted $\mathcal{K}^{G,l}$. For our example (1) we have that $+(v, v) = false$, $-(v, v) = true$ and therefore labelling $l = \neg + (v, v) \wedge -(v, v)$ is valid in (G, K).

Clearly, two models (G, K) and (G, K') for $K \neq K'$ will in general exhibit behavioural differences. During modelling one is only interested in those models whose behaviour fits the experimental observations. We use the term *property* for such behavioural observation of the system, e.g. time series data, and use the *Büchi Automata* [1] (BA) based model checking to decide whether a property holds in a model. A BA is a four-tuple $A = (S^A, \xrightarrow{\mathcal{L}(G)}, I^A, F^A)$, where:

- S^A is a set of states,
- $\xrightarrow{\mathcal{L}(G)}$ is a transition relation with propositions s.t.:
  $\mathcal{L}(G) = \mathcal{P}(\{v * n \mid G = (V, E, \rho), v \in V, * \in \{\leq, \geq, <, >, =\}, n \in [0, \rho(v)]\})$,
- $I^A \subseteq S^A, F^A \subseteq S^A$ are a set of initial and final states, respectively.

The intuition here is that the automaton controls validity of certain statements. Once a statement becomes valid, the automaton changes its state as a form of memory. Examples are given in Sect. 3. A property is resolved on a synchronous product of an automaton and a transition system $P = A \times (S^G, \to^G) = (S, \to, I, F)$, which is obtained as follows:

- $S = S^A \times S^G, I = I^A \times S^G, F = F^A \times S^G$,
- $(s^A, s^G) \to (r^A, r^G) \iff (s^G \to r^G) \wedge (s^A \xrightarrow{\phi} r^A) \wedge (s^G \models \phi)$.

In the general case the property encoded by A is satisfied *iff* there is a path $(i, \ldots, f, \ldots, f)$ in P s.t. $i \in I, f \in F$. The requirement for a cycle on f stems from expectation that some properties repeat infinitely, e.g. stable behaviour in an attractor. For a class of so-called reachability properties, e.g. a time series, a path $(i, \ldots, f)$ is sufficient. Then A is called a *terminal Büchi Automaton* (TBA).

Having a property ϕ, we are interested in the set $\mathcal{K}^{G,\phi}$ of parametrizations that satisfy the property ϕ. For our running example it holds $|\mathcal{K}^G| = 4$ and $|\mathcal{K}^{G,l}| = 1$ where l is as described above. Additionally, multiple experiments are usually considered. Have ϕ, ψ properties, then $\mathcal{K}^{G,\phi} \cap \mathcal{K}^{G,\psi} = \mathcal{K}^{G,\phi \wedge \psi}$ can be used to obtain models that satisfy both ϕ and ψ. We use this simple observation later to obtain sets of models that fit the data of all considered experiments.

3 Methods

In this section we show the reduction methods we used in the encoding of the data, which, albeit quite technical, represents intuitive biological notions.

As the knowledge about a system is usually obtained by measuring concentration or activity of a component, we use measurements as a basic unit of our property system. A *measurement* M in a TS $(S^G, \to^G)$ is a predicate over S^G, i.e. $M : S^G \to \{true, false\}$. Interpreted as a set, we also intuitively have that $M \subseteq S^G$, meaning that a measurement is the set of states that match the data.

A sequence of measurements $M = (M_1, \ldots, M_k)$ can be encoded via a TBA that loops in its current state until its respective measurement is matched, then it proceeds to a next state. To implement this for the last measurement, an arbitrary state is added after the last measurement. Formally we use a TBA $A = (S^A, \xrightarrow{\mathcal{L}(G)}, I^A, F^A)$ where $S^A = \{s_1^A, \ldots, s_{k+1}^A\}, I^A = \{s_1^A\}, F^A = \{s_{k+1}^A\}$, $\forall i \in [1, k] : s_i^A \xrightarrow{M_i} s_{i+1}^A \wedge s_i^A \xrightarrow{\neg M_i} s_i^A$. Consider our trivial example and the measurements $M_1(s) \iff s_v^G = 0, M_2(s) \iff s_v^G = 1, M = (M_1, M_2)$. Then the TBA A that controls whether the TS (1) is capable of reproducing M is:

$$(\{s_1^A, s_2^A, s_3^A\}, \{s_1^A \xrightarrow{(1)} s_1^A, s_1^A \xrightarrow{(0)} s_2^A, s_2^A \xrightarrow{(0)} s_2^A, s_2^A \xrightarrow{(1)} s_3^A\}, \{s_1^A\}, \{s_3^A\}). \quad (2)$$

Clearly the size of an automaton is linear w.r.t. to the number of measurements. This is advantageous as for an arbitrary property the resulting automaton can be exponential in the worst case [1]. We can however further reduce the size by encoding the initial and accepting states directly in the transitions system. This primarily gives us the advantage that we can start the search from the states that are relevant for the property directly and not from all the states of the TS as in standard model checking. When building a product of some TS T with the above TBA A we put $S = (S^A \setminus \{s_1^A, s_{k+1}^A\}) \times S^G; I = \{s_2^A\} \times M_1; F = \{s_k^A\} \times M_k$. E.g. the reduced product of (1) with (2) in our example would be:

$$(s_2^A, 0) \rightarrow (s_2^A, 1), (s_2^A, 1) \rightarrow (s_2^A, 0), I = \{(s_2^A, 0)\}, F = \{(s_2^A, 1)\}, S = I \cup F \quad (3)$$

Such encoding is sufficient if we want to pass through a set of measurements. Sometimes it is however expected that the last measurement represents a stable state of the system, as e.g. in many *perturbation experiments*. This can be simply achieved by further reducing the set of final states s.t. $F = \{s_k^A\} \times \{s^G | s^G \in M_k \wedge s^G \rightarrow s^G\}$. Note that in (3) this means $F = \emptyset$, as the system never stabilizes.

Until now we were focusing on passing through measurement points, without any specifications on the behaviour between them. There are multiple related biologically relevant constraints that can be implemented by simplifying the product structure. In particular, we may want to require a component not to change in between two measurements, or to change only once prohibiting unobserved oscillations. We define an additional constraint related to a measurement called *component delta*, $\delta : V \rightarrow \{up, down, stay, none\}$. This constraint is resolved on the transition system as follows:

$$\begin{aligned}(s^G, r^G) \models \delta \iff \forall v \in V : \ &(\delta(v) = stay \wedge (s^G)_v = (r^G)_v) \vee \\ &(\delta(v) = up \wedge (s^G)_v \leq (r^G)_v) \vee \\ &(\delta(v) = down \wedge (s^G)_v \geq (r^G)_v) \vee \\ &(\delta(v) = none)\end{aligned}$$

and the product is extended s.t.:

$$(s^A, s^G) \rightarrow (r^A, r^G) \iff (s^G \rightarrow r^G) \wedge (s^A \xrightarrow{\phi, \delta} r^A) \wedge (s^G \models \phi) \wedge ((s^G, r^G) \models \delta).$$

We then apply this constraint to the encoding of measurements. In particular, consider a measurement vector M and a δ that must be satisfied when transitioning from M_i to M_{i+1} for some $i \in [1, |M|)$. Then the automaton A encoding M is extended so that $s_i^A \xrightarrow{M_i, \delta} s_{i+1}^A$ and $s_{i+1}^A \xrightarrow{\neg M_{i+1}, \delta} s_{i+1}^A$. This slightly involved encoding follows from the fact that s_i^A is left only after M_i was satisfied, therefore we already require δ when leaving it and the requirement is kept until M_{i+1} is satisfied.

Note that the monotonicity is only one-sided, i.e. we can either require for a component that it is monotonously increasing (*up*) or monotonously decreasing (*down*). The general monotonicity, is more complicated and we do not discuss it in the article, for reference see [7].

Lastly we focus on a configuration of an experiment. Experimentally, measurements are conducted under specific conditions, e.g. presence of certain nutrients in the medium or addition of known inhibitors. Such conditions are usually expected to stay constant for the duration of the experiment. If they are explicitly modelled, e.g. with a component representing an inhibitor, we can enforce them by removing the states that do not match the corresponding component value from the TS, together with the respective transitions. For a property ϕ we denote $Exp_\phi : V \rightarrow \{\{0\}, \{1\}, \{0, 1\}\}$ the function that provides the range of values a component may attain in the current experiment. We then restrict the state space s.t. $S^G = \prod_{v \in V} Exp_\phi(v)$.

4 EGFR Signalling Pathway Study

Exploiting the efficient property encodings just presented, we now conduct a thorough analysis of the EGFR signalling pathway with particular emphasis on the comparison of network structure and regulatory mechanisms in different cell lines. As discussed in the introduction, we utilize a comprehensive data set provided by Klinger et al. in [8]. There, human colorectal cancer cell lines were treated with stimuli and inhibitors in order to elucidate the underlying network structure of the pathway using a semi-quantitative modelling approach. Here, we generate and analyse comprehensive model pools for the different cell lines[2] and compare our results with those by Klinger et al. In addition, we discuss further results from our analysis unrelated to the original study.

4.1 Model Building

Based on the model of [8] we constructed a BN, depicted in Fig. 1. We kept the original components and regulations, with a few exceptions. As the IGF1 stimulus is the only regulator of IGFIR we know that IGFIR copies its value and therefore we modelled the stimulation directly on IGFIR, removing IGF1 completely. Additionally, p70S6K is depicted as activator of IRS1, however based on [11] we modelled it as an inhibition. The same for AKT which is known to repress IRS1 indirectly through mTorC1 [11]. Note that these changes are to regulations of IRS1 only, which is an output component and therefore can not affect the upstream feedback loops. Any resulting inconsistencies with [8] should therefore be localised to IRS1. Since the data originates from cancer cells, we accounted for possible disruptions in the network due to mutations by not requiring regulations to be functional, i.e. activations are labelled as $\neg-$ and inhibitions as $\neg+$. However, stimuli and inhibitions as well as components with a single regulator (MEK, AKT) were set as always functional. In the data there are two stimuli, $TGFa$ and $IGF1$, and two effective inhibitors, MEK inhibitor AZD6244 and the PI3K inhibitor LY294002. There are two more inhibitors in the original data set on GSK3 and IKK, which were found to be non-effective and therefore neglected here. In our model, we set the stimuli as $Exp_\phi(TGFa) = \{1\}$ if $TGFa$ is stimulated in ϕ and $Exp_\phi(TGFa) = \{0\}$ otherwise, and the same for IGFIR. The inhibitors do not remove the targets from the system, only prohibit their effect on the down-stream components. We therefore added them as extra components LY and AZD, and modelled them analogously to stimuli.

---

[2] For spatial reasons, only samples from the results are provided in the article. All the data are listed in the supplementary archive or at `dibimath.github.io/CMSB_2015`.

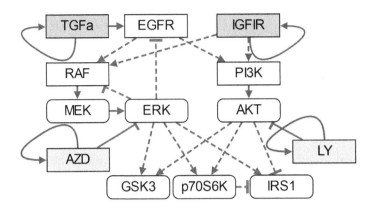

Fig. 1. EGFR Boolean Network. The full green edges are labelled $+ \wedge \neg-$, the dashed $\neg-$, the full red $- \wedge \neg+$, the dashed $\neg+$. Stimuli are in green, inhibitors in yellow. The measured nodes are semi-oval (Color figure online).

Additionally we set the regulatory functions $K_{ERK}(s) \iff s_{MEK} = 1 \wedge s_{AZD} = 0$ and $K_{AKT}(s) \iff s_{PI3K} = 1 \wedge s_{LY} = 0$ to enforce the correct inhibition semantics. After having resolved all the edge constraints, we obtained a model pool $\mathcal{K}^l$ with 259200 models. Note that the inhibitors and stimuli are fixed components, they do not contribute to the size of the state space, which then only has $2^9 = 512$ states instead of 2^{13}.

In their experiments, Klinger et al. used a high-throughput immunoblotting method, called Luminex assay, which measures intensities of labelled antibodies that bind the phosphorylated components, showing their activity (for a detailed description see [8]). Here, we used a reduced data set containing experiments on 5 human colorectal cancer cell lines. Each of the cell lines was treated with each pairwise combination of one stimuli ($TGFa$, $IRS1$, no stimulus) and one inhibitor (AZD, LY, no inhibitor), which were then compared to the measurements before treatment. Since the configuration without stimulus and inhibitor is not expected to change, we did not include it.

Prior to their usage, the data needed to be discretized to fit the Boolean formalism. Additionally, for some experiments, multiple measurements were available. In such a case we took the mean of those. For the discretization one usually uses a software which creates a threshold value that separates the range of measured values for each component. In our data set some of the values however almost do not change between measurements being e.g. at a plateau and therefore should not be assigned with different states. To aviod this separation, we focused on a fold change, which shows the measured activity of the treated sample relative to the measurement before the treatment. Here, we rely on an assumption that a fold change of two or more is significant, which is to the best of our knowledge a common practice and in our case seems to produce a good separation. Since the focus of this study is on evaluating the effect of regulatory influences, we assigned Boolean values to the component measurements consistent with the nature of the fold changes found in the data. If we observed an increase by a factor of at least two, we assigned the value 0 to the measurement before and 1 after the treatment. Analogously, we encoded a decrease by a factor of at least two. If the change factor is less than two, we did not specify the value, but required the component to be stable, as explained in Sect. 3. In this approach, interpretation of the qualitative dynamics heavily focuses on the component changes indicating actively regulated behaviour, as is our intention. Note that it therefore differs from the often employed interpretation of the Boolean component values as an abstraction for ranges of quantitative values. In our approach the same quantitative value might be assigned different Boolean counterparts depending on the observed component behaviour in the respective experiments. In our opinion, this does not pose a problem, since we are focusing on the qualitative dynamics and thus the values 0 and 1 can be viewed as labels of qualitative change, rather than ranges of quantitative values. Presumably, if a component can undergo both a significant increase and a decrease in its concentration, such mechanics should be allowed by the network without contradicting the effects of the regulations.

As we considered 8 treatments for 5 cell lines, we obtained altogether 40 measurement pairs. In [8] the authors argue that at the time of the measurements the system is expected to reach a stable plateau. However, Figure S1 therein shows that the kinetics of some components have an unstable behaviour after the time point of measurement. To investigate the impact of the steady state assumption, we created a *stable* and *transient* (i.e. not required to be stable) version of each time series, as explained in Sect. 3. Additionally, we were interested in effects of monotonicity constraints on the results. We therefore also considered for each property a version where all the components that are measured and not stable are required to be monotonous in their behaviour. By combining the treatments, cell lines and constraints we obtained 160 properties. The properties are listed in the supplementary files. Note that as each of these properties is a two-step experiment, we can reduce the encoding TBA just to 1 state, as explained in Sect. 3, keeping the size of the product at the 512 states.

4.2   Results

Initially we found that each of the cell lines shows inconsistencies in at least one measurement pair. In each of these, the experimental set-up, listed in Table 1, requires that a component whose activator was inhibited undergoes itself an activation, which is logically inconsistent. For example cell line SW403 shows with IGF1 stimulus an over 4-fold increase in concentration of AKT under inhibition of PI3K, its only activator. This is still comparably lower than the about 12-fold increase without the inhibition, showing that the inhibitor is working, but the dose is not sufficient to lower the activity of AKT to the threshold of being inactive after discretizing. Since dose-dependent processes are not considered in this formalism, we removed the respective experiments from the testing set. After the removal we have sets of 7 measurement pairs for each cell line except LIM1215 where there are only 6. We therefore further used only 34 measurement

Table 1. Experimental set up causing logical inconsistencies after discretization.

Cell line	TGFa	IGFIR	AZD	LY
LIM1215	1	0	1	0
LIM1215	0	1	0	1
HCT116	0	1	0	1
SW403	0	1	0	1
SW480	0	1	0	1
HT29	1	0	1	0

Table 2. Sizes of a parametrization sets matching the data from all the consistent experiments for each cell. Monotone property sets are not listed as monotonicity did not cause any reduction.

Cell line	transient	partially stable	stable
LIM1215	180000	6100	40
HCT116	129600	5580	2
SW403	180000	111000	840
SW480	136800	74670	36
HT29	163800	101010	216

pairs, yielding 136 properties when combined with different path constraints. In Table 2 column *transient*, which represents the weakest assumption concerning the stability of the system, shows how many members of $\mathcal{K}^l$ fit all the measurements for each respective time series. Note that each set remains more than one half in size compared to the set of models consistent with the constraints derived from the network structure, suggesting that the topology itself already strongly determines the dynamics.

In [8] the modular response analysis (MRA) method was used to identify non-functional connections in the network for the different cell lines. Here, we aimed to compare the topologies of their resulting networks with the topologies that occur in our model pools. To improve comparability, we used a stability requirement for the measurements in each cell line to account for the steady-state assumption necessary for the MRA approach. The sizes of the parametrization sets are listed in Table 2-*stable*. Note that there is a much stronger reduction than in the transient case, suggesting that the stability requirement is indeed very strong for this network, presumably due to the negative feedback mediated by ERK. However, each of the resulting parametrization sets is non-empty, therefore we can compare which edges are required/allowed to be functional. The results for two examples, SW480 and HT29, are shown in Table 3, where in **A** the functions in the pool fit well to the results of Klinger et al. However, all other cell lines such as SW480 in **B** our results match [8] only in part. This is likely to be in part due to negative feedback from ERK which is a source of instability in the Boolean framework, but in the real system may lead to damped oscillation and consequently to a quasi-stability. Additionally, the effect of ERK on IRS1 creates an incoherent feed-forward motif, which was not captured in [8] as there the semantics of the regulations of IRS1 are consistent.

As our method allows for testing transient states, and the time series measurement in Figure S1 of [8] illustrates that AKT and ERK may not be in steady state at the time point of measurement, we also created a *partially stable* selection. Here, those components which are not stimulated are assumed to be in steady-state. Stimulated samples are allowed to be in a transient state, since their last treatment was shortly before sampling. In our opinion, this scenario accounts for the most biologically realistic assumptions and we used it as the basis for the subsequent analysis (see Table 2-*partially stable*).

Table 3. Presence of regulators in the individual cell lines. The Must column contains the set of edges that are functional in all parametrizations fitting the data. The Klinger et al. column contains the ones reported in [8]. The May column contains the edges that are functional in at least one parametrization fitting the data. If May includes edges of [8], Match is set to yes.

A: HT29				
Target	Must	Klinger et al.	May	Match
EGFR	TGFa	TGFa, ERK	TGFa, ERK	yes
RAF	∅	EGFR, IGFIR	EGFR, IGFIR	yes
PI3K	∅	EGFR, IGFIR	EGFR, IGFIR	yes
GSK3	∅	AKT	ERK, AKT	yes
p70S6K	∅	ERK, AKT	ERK, AKT	yes
IRS1	ERK	ERK	ERK, AKT, p70S6K	yes

B: SW480				
Target	Must	Klinger et al.	May	Match
EGFR	TGFa	TGFa, ERK	TGFa	no
RAF	∅	EGFR, IGFIR, ERK	EGFR, IGFIR	no
PI3K	EGFR, IGFIR	EGFR, IGFIR	EGFR, IGFIR	total
GSK3	∅	AKT	ERK, AKT	yes
p70S6K	ERK, AKT	ERK, AKT	ERK, AKT	total
IRS1	ERK	p70S6K	ERK, AKT, p70S6K	no

Table 4. Comparison of occurrence of different regulatory functions between the partially stable pools. For each pair the difference of the first member when compared with the second member is described. The notation $y = 1$ is a shorthand for $K_v(s) = 1$ for any $s \in S$ where $v \in V$ is the Target. For most of the cases, the same set of functions was present, but the frequency of their occurrence in the set differed.

Target	A: LIM1215-HCT116	B: HCT116-SW480	C: SW403-HT29
EGFR	differences in frequency	almost the same	no difference
RAF	LIM allows for 15 (out of 20) functions, HCT only for $y = 1$	HCT allows only for $y = 1$, SW for 15 functions	no difference
PI3K	strong increase in $y = 1$	differences in frequency	no difference
GSK3	strong increase in $y = 1$	no difference	almost the same
p70S6K	$y = 1$ appears	almost the same	almost the same
IRS1	no difference	almost the same	almost the same

Focusing on a comparison of the cell lines carrying different genotypes we expected to find topological and functional differences between the pools causing the observed variations in the measurements. The pools of all 5 cell lines were compared, resulting in 10 different tables. Due to limitation of space, only three comparisons are presented. Table 4 **A** shows the comparison of the pools corresponding to LIM1215 and HCT116, where major differences can be seen. The most striking variation is the function for RAF, which is always active in all the models for HCT116, meaning that the component is completely independent from the receptors and their stimulation. This observation can be explained by considering the genotype of HCT116 listed in Table 1 in [8] where mutations in KRAS, RTK and PI3K are noted. KRAS is a kinase of RAF and PI3K, and is regulated by RTK. This mutation may lead to constant activation of RAF in this cell line. Similarly, PI3K is constantly 1 in more than 70 % of models of LIM1215, which again can be attributed to a mutation in KRAS present in this cell line. Note that the KRAS mutation differs between these cell lines and therefore could cause different effects. HCT116 though does not show a specific tendency in the regulation of PI3K, although it carries a mutation in this component.

Not all the comparisons are showing such clear differences between the pools. Table 4**B** and **C** compare the pools of HCT116 with SW480 as well as SW403 with HT29 without resulting in any clear variations. For cell lines HCT116 and SW480 this could be explained by looking again at the genotypes, which show many commonly shared mutations (see Table 1 in [8]). SW403 and HT29 in **C** have the most similar pool of all 10 comparisons, without sharing any mutation concerning components in our model. However, they do share an identical mutation in p53, which is a prominent oncogene and might govern the behaviour in these cell lines [2].

Aside from the biologically motivated analysis, we also used the case study to evaluate performance of our new constraint encodings. We have executed the validation in batches for each set of 40 properties with different path constraints. The execution time for the transient properties was 7600 s, for the stable 8184 s, for the monotone 8778 s, and for the stable monotone 10156 s. The program did not use more than 7 MB of memory at any time. The program was executed as a single-threaded instance on a Debian 3.2.65 workstation with a processor i5-2400S, 2.5 GHz, and 4 GB RAM. We have also tested execution with a script [7] that called the NuSMV model checker using a respective LTL formula for one of the 160 properties and the computation took roughly 5 days, illustrating that customization was necessary for the problem to be solvable in a reasonable time.

5   Conclusion

Generating and analysing model pools using constraints encoding the available knowledge for a given system allows to evaluate data uncertainty and guides the step from generic to more specific models. Here, we utilized this approach to investigate cell line specific properties of the EGFR signalling pathway. Motivated by a study of Klinger et al. [8], we first aimed at a comparison of our fully qualitative approach with the semi-quantitative method employed in the original study. While obtaining good agreement of the results in some cases, others did not match very well. We expect that this emerges mainly from the semantics of the edge labels required by us. Since in the results in [8], some edges exert both positive and negative response, we feel that the points of difference are good candidates for further investigation.

Going beyond the results of the original study we dropped the stability requirement for several components based on the available experimental data. By comparing the resulting model pools we tried to find differences between cell lines and to examine whether variations in the measurements can be connected with the topology or even the genotype. Interesting insights can be derived for the cell lines LIM1215 and HCT116, where the valid regulatory functions of PI3K in LIM1215 and of RAF in HCT116 could indicate an activating mutation in that component or upstream, in these cases probably KRAS. Such results give suggestions about dominant players, like KRAS here, of great interest for the development of therapeutic strategies. Other comparisons, e.g. of cell lines SW403 and HT29, show only slight differences, although they do not share a mutation in components of the pathway. However, a shared mutation can be detected in the oncogene p53. This kinase is not directly linked to the EGFR pathway, but nevertheless might govern the behaviour of these cells. Thus, a model expansion by adding p53 and new p53 measurement data could help to clarify this result.

To conduct such an analysis in a reasonable time frame, efficient encodings are crucial. To this end, we have implemented a number of reductions on the level of the model checking procedure that proved very effective. These could be further extended to account for properties like oscillation or to broader formalisms, like multi-valued instead of Boolean Networks. Also, while they were sufficiently effective for our study, additional reductions and subsequently performance improvement may be possible.

References

1. Baier, C., Katoen, J.-P.: Principles of Model Checking. The MIT Press, Cambridge (2008)
2. Feng, Z., Levine, A.J.: The regulation of energy metabolism and the igf-1/mtor pathways by the p53 protein. Trends Cell Biol. **20**(7), 427–434 (2010)
3. Gallet, E., Manceny, M., Le Gall, P., Ballarini, P.: An LTL model checking approach for biological parameter inference. In: Merz, S., Pang, J. (eds.) ICFEM 2014. LNCS, vol. 8829, pp. 155–170. Springer, Heidelberg (2014)
4. Guziolowski, C., Videla, S., Eduati, F., Thiele, S., Cokelaer, T., Siegel, A., Saez-Rodriguez, J.: Exhaustively characterizing feasible logic models of a signaling network using answer set programming. Bioinformatics **29**, 2320–2326 (2013)
5. Huth, M., Ryan, M.: Logic in Computer Science: Modelling and reasoning about systems. Cambridge University Press, Cambridge (2004)
6. Kauffman, S.: Metabolic stability and epigenesis in randomly constructed genetic nets. J. Theor. Biol. **22**(3), 437–467 (1969)
7. Klarner, H.: Contributions to the Analysis of Qualitative Models of Regulatory Networks. Ph.D. thesis, Freie Universität Berlin, Germany (2015)

8. Klinger, B., Sieber, A., Fritsche-Guenther, R., Witzel, F., Berry, L., Schumacher, D., Yan, Y., Durek, P., Merchant, M., Schäfer, R., et al.: Network quantification of EGFR signaling unveils potential for targeted combination therapy. Mol. Syst. Biol. **9**(1), 673 (2013)

9. Rozengurt, E., Soares, H.P., Sinnet-Smith, J.: Suppression of feedback loops mediated by pi3k/mtor induces multiple overactivation of compensatory pathways: An unintended consequence leading to drug resistance. Mol. Cancer Ther. **13**(11), 2477–2488 (2014)

10. Samaga, R., Saez-Rodriguez, J., Alexopoulos, L.G., Sorger, P.K., Klamt, S.: The logic of EGFR/ErbB signaling: theoretical properties and analysis of high-throughput data. PLoS Comput. Biol. **5**(8), e1000438 (2009)

11. Tanti, J.-F., Jager, J.: Cellular mechanisms of insulin resistance: role of stress-regulated serine kinases and insulin receptor substrates (irs) serine phosphorylation. Curr. Opin. Pharmacol. **9**(6), 753–762 (2009)

Short Papers

OPINION PAPER Evolutionary Constraint-Based Formulation Requires New Bi-level Solving Techniques

Marko Budinich[✉], Jérémie Bourdon, Abdelhalim Larhlimi,
and Damien Eveillard

LINA, UMR 6241 CNRS, EMN, Université de Nantes,
2 rue de la Houssinière, Nantes, France
marko.budinich@univ-nantes.fr

Abstract. Constraint Based Methods had been successfully used to simulate genome-scale metabolic behaviors over a range of experimental conditions. In most applications, environmental constraints are parameterized, and the use of metabolic reactions and corresponding genes is the direct consequence of the tuning of these parameters.

However, in evolutionary studies, the problem is different: one knows the relative importance of reactions and one seeks environmental conditions that could explain such a biological fitness.

This study details this modeling paradigm change and discuss a putative formalization of such a biological problem in the form of a Mixed Integer Bi-level Linear Problem (MIBLP). Unfortunately, solving a MIBLP is difficult, paving the way for the need of further constraint based method developments for understanding evolutionary processes.

Constraint Based Methods (CBMs) are considered as efficient approaches to predict phenotypic responses and explore the structure of genome-scale networks of a variety of organisms [1,2]. For instance, they tackle effects of genetic mutations (resp. gene deletions [3,4] and gene insertion [5]) on metabolic behaviors, whereas complementary analysis focused on gene transfers [6], gene dispensability [7] or nutrient adaptation [8]. Similarly, high-throughput sequencing allows today to compare lineages and biological studies to infer evolutionary patterns [9], paving the way to bridge evolutionary studies and CBMs.

From an evolutionary viewpoint, environment exerts or relaxes pressure in biological systems. Thus, in front of detrimental or beneficial environments, organisms adapt themselves by gaining or loosing functions [10,11]. Those knowledge being available nowadays, it is of great interest to decipher the environmental conditions that maximize lineage evolution, pointing conditions that could lead to metabolic reaction losses [12].

When CBM is applied in evolutionary contexts, environment usually is first parameterized and its effect is then studied and interpreted *via* a range of simulations [6,13]. Herein, instead of standard approaches, we propose to focus on selecting environmental conditions that make most reactions unable to carry

© Springer International Publishing Switzerland 2015
O. Roux and J. Bourdon (Eds.): CMSB 2015, LNBI 9308, pp. 279–281, 2015.
DOI: 10.1007/978-3-319-23401-4_23

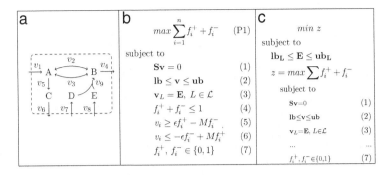

Fig. 1. Evolutionary problem formulation. Considering a putative metabolic network (a), we assume the production of metabolite B as a fitness proxy. If A is the only substrate in a particular environment, we expect that genes coding for v_7, v_8 and v_9 disappear upon evolution. b) The inner Problem (P1) identify blocked reactions, i.e., those that can not carry a non-zero flux under steady-state conditions. A variation of (P1) is used in [14,15]. c) A mixed integer bi-level linear problem seeking for an environmental setting (i.e., defined values for environmental variables in $\mathcal{L}$, see text) **E** that maximizes the number of blocked reactions.

fluxes (see Fig. 1a). Indeed, recent evolutionary studies hypothesize that such blocked reactions are likely to be lost as functions due to evolution [12].

Formalization of the previous statements leads to an optimization problem as shown in Fig. 1b. Constraints in (1) and (2) are mass balance and boundary conditions. Equations in (3) represent environmental variables as a subset of reaction fluxes indexed by $\mathcal{L}$.

To identify blocked reactions, we introduce for each reaction i two binary variables f_i^+ and f_i^- (resp. forward and reverse flux) in (7). Constraints in (4), (5) and (6) guarantee that a reaction i is blocked if and only if $f_i^+ = f_i^- = 0$. By M (resp. ϵ), we denote a large (resp. small) number. Given an environmental setting **E**, maximizing $\sum f_i^+ + f_i^-$ identifying all blocked reactions.

As a next step in our study, we propose to use the Mixed Integer Bi-level Linear Problem (MIBLP) shown in Fig. 1c in order to select an environmental setting **E** that maximizes the number of blocked reactions. The main difference with other bi-level approaches is the focus on controlling metabolic networks using only environmental variables and not genetic manipulations [16].

Unfortunately, despite several tentatives [17,18], no general solution is available for this type of problem [19], emphasizing the need for an *ad-hoc* algorithm implementation to solve this new evolutionary problem. Furthermore, for the sake of generalization, any method that handle this type of bi-level program, will lead to theoretical and practical advances in system biology.

From an evolutionary viewpoint, we expect that solving this problem will pinpoint the environmental conditions that are responsible for the specification of lineages or microbial strains. This question is particularly vivid considering drastic environmental condition changes that are expected in a near future.

References

1. Bordbar, A., Monk, J.M., King, Z.A., Palsson, B.O.: Constraint-based models predict metabolic and associated cellular functions. Nat. Rev. Genet. **15**, 107–120 (2014)
2. Lewis, N.E., Nagarajan, H., Palsson, B.O.: Constraining the metabolic genotype-phenotype relationship using a phylogeny of in silico methods. Nat. Rev. Microbiol. **10**, 291–305 (2012)
3. Burgard, A.P., Pharkya, P., Maranas, C.D.: Optknock: a bilevel programming framework for identifying gene knockout strategies for microbial strain optimization. Biotechnol. Bioeng. **84**, 647–657 (2003)
4. Tepper, N., Shlomi, T.: Predicting metabolic engineering knockout strategies for chemical production: accounting for competing pathways. Bioinformatics **26**, 536–543 (2010)
5. Larhlimi, A., Basler, G., Grimbs, S., Selbig, J., Nikoloski, Z.: Stoichiometric capacitance reveals the theoretical capabilities of metabolic networks. Bioinformatics **28**, i502–i508 (2012)
6. Pál, C., Papp, B., Lercher, M.J.: Adaptive evolution of bacterial metabolic networks by horizontal gene transfer. Nat. Genet. **37**, 1372–1375 (2005)
7. Papp, B., Pál, C., Hurst, L.D.: Metabolic network analysis of the causes and evolution of enzyme dispensability in yeast. Nature **429**, 661–664 (2004)
8. Ibarra, R.U., Edwards, J.S., Palsson, B.O.: Escherichia coli K-12 undergoes adaptive evolution to achieve in silico predicted optimal growth. Nature **420**, 186–189 (2002)
9. Koonin, E.V.: The Logic of Chance: The Nature and Origin of Biological Evolution. FT Press, New Jersey (2011)
10. Van Valen, L.: A new evolutionary law. Evol. Theory **1**, 1–30 (1973)
11. Van Valen, L.: Molecular evolution as predicted by natural selection. J. Mol. Evol. **3**, 89–101 (1974)
12. Morris, J.J., Lenski, R.E., Zinser, E.R.: The black queen hypothesis: evolution of dependencies through adaptive gene loss. MBio **3**(2), e00036-12 (2012)
13. Yang, H., Roth, C.M., Ierapetritou, M.G.: A rational design approach for amino acid supplementation in hepatocyte culture. Biotechnol. Bioeng. **103**, 1176–1191 (2009)
14. de Figueiredo, L.F., Podhorski, A., Rubio, A., Kaleta, C., Beasley, J.E., Schuster, S., Planes, F.J.: Computing the shortest elementary flux modes in genome-scale metabolic networks. Bioinformatics **25**, 3158–3165 (2009)
15. Goldstein, Y.A.B., Bockmayr, A.: A lattice-theoretic framework for metabolic pathway analysis. In: Gupta, A., Henzinger, T.A. (eds.) CMSB 2013. LNCS, vol. 8130, pp. 178–191. Springer, Heidelberg (2013)
16. Chowdhury, A., Zomorrodi, A.R., Maranas, C.D.: Bilevel optimization techniques in computational strain design. Comput. Chem. Eng. **72**, 363–372 (2015)
17. Saharidis, G.K., Ierapetritou, M.G.: Resolution method for mixed integer bi-level linear problems based on decomposition technique. J. Glob. Optim. **44**, 29–51 (2008)
18. Xu, P., Wang, L.: An exact algorithm for the bilevel mixed integer linear programming problem under three simplifying assumptions. Comput. Oper. Res. **41**, 309–318 (2014)
19. Saharidis, G.K.D., Conejo, A.J., Kozanidis, G.: Exact solution methodologies for linear and (mixed) integer bilevel programming. In: Talbi, E.-G. (ed.) Metaheuristics for Bi-level Optimization. SCi, vol. 482, pp. 221–245. Springer, Heidelberg (2013)

SBMLDock: Docker Driven Systems Biology Tool Development and Usage

Etienne Z. Gnimpieba[1(✉)], Mathialakan Thavappiragasam[1],
Abalo Chango[2], Bill Conn[1], and Carol M. Lushbough[1]

[1] Computer Science Department, University of South Dakota,
Vermillion, SD, USA
{Etienne.Gnimpieba,Mathialakan.Thavappi,Bill.Conn,
Carol.Lushbough}@usd.edu
[2] UPSP EGEAL, Institut Polytechnique LaSalle Beauvais, Beauvais, France
abalo.chango@lasalle-beauvais.fr

Abstract. A glut of Systems Biology tools and their lack of accessibility has significantly delayed bioscience advances that depend on the analysis of large scale systems with big datasets and High Performance Computing (HPC) resources. This work presents SBMLDock, the first Systems Biology Docker image that aims to advance scalability, usability and reproducibility in Systems Biology by making tools much more immediately available to the biological domain scientist, student, and educator, without requiring special training for use, and without losing the reproducibility aspect of their research. SBMLDock consists of one Docker image containing basic tools developed for Systems Biology Model manipulation (parallel model similarity analyzer, model checker, model splitter, model annotation, model extractor). The user can then pull up the Docker image, customize it and/or run each tool as service. Stored on the Docker hub, the image version is managed to assure research reproducibility. SBMLDock is available as a Docker file under CC licence at github https://github.com/USDBioinformatics/SBMLDock and the Docker image can be found in Docker hub at https://registry.hub.docker.com/u/usdbioinformatics/sbmldock/ with supplementary documents.

Keywords: SBMLDocker · Docker image · Systems biology · Reproducible research

1 Introduction

Emerging developments in Big Data, Systems Biology, and Integrative Biology introduce an increasing number of challenges in life science research. The primary objective of Software as a Service (SaaS) and platform as a service (PaaS) initiatives such as Workflow Management Systems (WMS) or Docker is to simplify researchers' ability to access, apply, and share analytic tools, workflows and data [1]. Executing an analytic tool can be very difficult if the researchers are not well prepared. Additionally, it is not always optimal to use systems biology tools due to deployment times that degrade the tool usability [1].

© Springer International Publishing Switzerland 2015
O. Roux and J. Bourdon (Eds.): CMSB 2015, LNBI 9308, pp. 282–285, 2015.
DOI: 10.1007/978-3-319-23401-4_24

The development of a container system (Docker) allows bioscience tool developers to hide complexity from researchers by providing a distributed container to embed any development module (service, tool, workflow, data storage) (https://www.docker.com/whatisdocker/). This method has been adopted in bioinformatics areas including the Galaxy infrastructure [2].

System Biology Markup Language(SMBL) is a machine-readable XML format for representing computational models of biological processes [3]. Software tools that support SBML as a format for reading and writing biological systems models facilitate their cooperative sharing, evaluation, and development. The XML-based SBML is the de facto standard file format for the storage and exchange of quantitative computational models in systems biology, supported by more than 220 software packages to date (March 2014) [3]. This includes several biological systems modeling tools (e.g. Systems Biology toolbox for Matlab, COPASI, EPISIM, Virtual Cell) and several databases for the representation and knowledge sharing (e.g. BioModels, BRENDA, KEGG).

2   SBMLDock

SBMLDock is the first systems biology Docker container for researchers, educators, and developers. We developed the first set of tools for SBML file manipulation including SBMLSplit, SBMLModeler, SBMLAnnotate. In order to complete our toolkit, we integrated recently published tools in the same series, such as ParaABioS [4], SBMLMerge, SBMLChecker [5], SBMLCompare [6]. Each tool has been integrated into a Docker image with a test dataset. The researcher can use this test data set to test each *tool*.

ParaABioS is an implementation of a parallel algorithm for bioscience elements similarity estimation [4]. This parallelization is critical when you involve the synonyms of bioscience terms because the curse of dimensionality becomes worse and requires HPC resources. ParaABioS uses heuristic techniques to measure similarity parameter values (distance and ratio) of the elements. The algorithm was implemented using SIMD data parallelization techniques in java.

ParaABioS requires four parameters to run, and provides the similarity results in a text file. Running in Docker, the syntax is `ParaABioS <inputfile1> <input-file2> <distance> <ratio>` Where `<inputfile1>`, `<inputfile2>` are two bioscience element lists (metabolite, compound, protein, gene, etc.), `<distance>` and `<ratio>` are threshold values for edit distance and the ratio respectively.

E.g. `docker run -v /home/wjconn/SBMLDock/mount:/tmp -w /tmp usdbioinformatics/sbmldock ParaABioS file1.txt file2.txt 6 0.7`

SBMLChecker is a Systems Biology Markup Language model checker. SBML-Checker improves the online SBML validator by integrating meaning using semantic (ontology and database) checking [5, 7]. It uses the annotated URL ids of each element to measure the semantic strength of the reliability score. In order to execute SBML-Checker in Docker use the following command `SBMLChecker <sbmlinputfile>`. This will return a checking report printed in the system out or in a report output files store on your mounted directory.

E.g. `docker run -v /home/wjconn/SBMLDock/mount:/tmp -w /tmp usdbioinformatics/sbmldock SBMLChecker one.xml`

SBMLCompare is an implementation of ParABioS algorithm specific for SBML model comparison. In addition to naming similarity techniques used in ParABioS, SBMLCompare use biological annotated meanings to ensure the semantic similarity between models. SBMLCompare on the Docker can be use as follow `SBMLCompare <inputfile1> <inputfile2>` . This will provide a comparison report in 3 formats (text, excel or xml) in files named *sbml_compare_report*.

E.g. `docker run -v /home/wjconn/SBMLDock/mount:/tmp -w /tmp usdbioinformatics/sbmldock SBMLCompare one.xml two.xml`

SBMLMerge is an automatic merging tool for SBML models. Other existing merging tools for SBML models require human interaction. Using a heuristic algorithm, SBMLmerge provides a consistent merged model. This tool helps biologists combine sub-model from different sub-biosystems into a targeted biosystem. To execute SBMLMerge on Docker use the following syntax `SBMLMerge <edit distance int[0-10]> <similarity ratio float[0-1]> <inputfile1> <inputfile2> <optional input files up to 6>`. This will provide a merged SBML model mergedmodel.xml file in your mounted directory.

E.g. `docker run -v /home/wjconn/SBMLDock/mount:/tmp -w /tmp usdbioinformatics/sbmldock SBMLMerge 6 0.7 /opt/SBMLMerge/one.xml /opt/SBMLMerge/two.xml`

SBMLSplit is an SBML model extractor. A researcher can extract a sub-model based on reaction or compound (metabolite, species) list. SBMLSplit can be run on the Docker as `SBMLSplit <flag> <inputfile>` where your `<flag>` is C or R to split on Compound or Reaction respectively, and the `<inputfile>` is the SBML file you want to split. This provide 2 split SBML files (e.g. *S0.xml* and *S1.xml*), that are stored in your mounted folder.

E.g. `docker run -v /home/wjconn/SBMLDock/mount:/tmp -w /tmp usdbioinformatics/sbmldock SBMLSplit C one.xml`

SBMLModeler is an implementation of a data mining workflow for SBML model design from multiple data repositories (e.g. KEGG, SABIO-RK, BRENDA, …), using a top down approach with the pathway name as the entry. The current version of SBMLModeler focuses on a short pathway list for accuracy purposes. The list named Pathwayslist.txt can be found in the directory `/opt/SBMLModeler/` in the SBMLDock image. Once you have your pathway picked out you can run SBMLModeler using the following command `SBMLModeler <Path to store file> <Pathway name>`.

E.g. `docker run -v/home/wjconn/SBMLDock/mount:/tmp -w/tmp usdbioinformatics/sbmldock SBMLModeler. "folate biosynthesis"`

SBMLAnnotate is an automatic annotation tool for SBML models. SBMLAnnotate evaluates the existing annotation degree of your SBML model (i.e. number of element annotated with ontologies or common databases such as SBO, KEGG) and proposes a reliable annotation to improve the model. To execute SBMLAnnotate use: `SBMLAnnotate <inputfile> <outputfile>`. This will save an `out.xml` file in your mounted directory as output.

```
E.g. docker run -v/home/wjconn/SBMLDock/mount:/tmp -w/tmp
usdbioinformatics/sbmldock SBMLAnnotate one.xml out.xml
```

3 Conclusion

Systems integration in life science research has become a complex challenge as data sets have grown. The ability to minimize the tools usage can be a tremendous asset for bioscience scientist. SBMLDock provides systems biology tools that allow developers and users to work together in minimizing the complexity of tool deployment and version management. This also greatly contributes toward the development of reproducible research.

Acknowledgement. This work has been partially supported by the National Science Foundation/EPSCoR Award No. IIA-1355423 and by the state of South Dakota, through BioSNTR.

References

1. Beasley, J.M., Coronado, G.D., Livaudais, J., Angeles-Llerenas, A., Ortega-Olvera, C., Romieu, I., Lazcano-Ponce, E., Torres-Mejía, G.: Alcohol and risk of breast cancer in Mexican women. Cancer Causes Control **21**, 863–870 (2010)
2. Cock, P.J.A., Grüning, B.A., Paszkiewicz, K., Pritchard, L.: Galaxy tools and workflows for sequence analysis with applications in molecular plant pathology. Peer J. **1**, e167 (2013)
3. Hucka, M.: Systems biology markup language (SBML). In: Dubitzky, W., Wolkenhauer, O., Cho, K.-H., Yokota, H. (eds.) Encyclopedia of Systems Biology SE – 1091, pp. 2057–2063. Springer, New York (2013)
4. Thavappiragasam, M., Lushbough, C.M., Gnimpieba, E.Z.: Heuristic parallelizable algorithm for similarity based biosystems comparison. In: Proceedings of the 5th ACM Conference on Bioinformatics, Computational Biology, and Health Informatics - BCB 2014, pp. 782–789 (2014)
5. Thavappiragasam, M., Lushbough, C., Gnimpieba, E.: SBMLChecker, a Semantic approach for SBML model reliability evaluation, 2–5 (2014). worldcomp-proceedings.com
6. Thavappiragasam, M., Lushbough, C.M., Gnimpieba, E.Z.: Automatic biosystems comparison using semantic and name similarity. In: Proceedings of the 5th ACM Conference on Bioinformatics, Computational Biology, and Health Informatics - BCB 2014, pp. 790–796 (2014)
7. Dräger, A., Rodriguez, N., Dumousseau, M., Dörr, A., Wrzodek, C., Le Novère, N., Zell, A., Hucka, M.: JSBML: a flexible Java library for working with SBML. Bioinformatics **27**, 2167–2168 (2011)

Author Index

Printed in the United States
By Bookmasters